D1072161

Thermodynamics for the Practicing Engineer

Thermodynamics for the Practicing Engineer

Louis Theodore
Francesco Ricci
Timothy Van Vliet

WILEY

A JOHN WILEY & SONS, INC., PUBLICATION

Published by John Wiley & Sons, Inc., Hoboken, New Jersey
Published simultaneously in Canada

For general information on our other products and services or for technical support, please contact our Customer Care Department within the United States at (800) 762-2974, outside the United States at (317) 572-3993 or fax (317) 572-4002.

Wiley also publishes its books in a variety of electronic formats. Some content that appears in print may not be available in electronic formats. For more information about Wiley products, visit our web site at www.wiley.com.

Library of Congress Cataloging-in-Publication Data:

Theodore, Louis.
    Thermodynamics for the practicing engineer / Louis Theodore.
        p. cm.
    Includes bibliographical references and index.
    ISBN 978-0-470-44468-9 (cloth)
 1. Thermodynamics. 2. Energy conversion. I. Title.
    TJ265.T455 2009
    621.402′1—dc22

                                                        2009016146

Printed in the United States of America

10 9 8 7 6 5 4 3 2 1

Thermodynamics for the Practicing Engineer

A. Edward Newton [1863–1940]

I wish that some one would give a course in how to live. It can't be taught in the colleges: that's perfectly obvious, for college professors don't know any better than the rest of us.

—This Book-Collecting Game

Louis Theodore
Francesco Ricci
Timothy Van Vliet

Agnes Repplier [1858–1950]

That little band of authors who, unknown to the wide careless world, remain from generation to generation the friends of a few fortunate readers.

—*Preface to James Howell*

André Gide [1869–1951]

A unanimous chorus of praise is not an assurance of survival; authors who please everyone at once are quickly exhausted. I would prefer to think that a hundred years hence people will say we did not properly understand him.

—*Pretexts*

To my family and friends for their love and support, and to the Manhattan College Chemical Engineering Department for its commitment to greatness—without either of which, my dreams could never be realized (FR)

To George Scott, my high school technology teacher, for introducing me to this wonderful profession (TVV)

To Cecil K. Walkins, a friend who has contributed mightily to basketball and the youth of America (LT)

Plato [427–347 B.C.]

The beginning is the most important part of the work.

—The Republic, Book II

Contents

Ezra Pound [1885–1972]

Men do not understand books until they have had a certain amount of life, or at any rate no man understands a deep book, until he has seen and lived at least part of its contents.

—A, B, C of Reading. Page 88

Part IV   OTHER TOPICS 317

15. Economic Considerations 319

16. Open-Ended Problems 333

17. Other ABET Topics 343

Preface

Sir Walter Scott [1771–1832]
Good wine needs neither bush nor preface to make it welcome.

—Peveril of the Peak

This project was a rather unique undertaking. Rather than prepare a textbook on thermodynamics in the usual and traditional format, the authors considered writing a book that highlighted applications rather then theory. The book would hopefully serve as a training tool for those individuals in academia and industry involved directly, or indirectly, with this topic. Despite the significant reduction in theoretical matter, it addresses both technical and pragmatic problems in this field. While this book can be viewed as a text in thermodynamics, it also stands alone as a self-teaching aid.

The book is divided into four parts:

I. Introduction

II. Enthalpy Effects

III. Equilibrium Thermodynamics

IV. Other Topics

The first part of the book serves as an introduction to the subject of thermodynamics and reviews such topics as units and dimensions, the conservation laws, gas laws, and the second law of thermodynamics. The second part of the book is concerned with enthalpy effects and reviews such topics as sensible, latent, mixing, and chemical enthalpy effects. The third part of the book examines equilibrium thermodynamics. Topics here include both phase and chemical reaction equilibrium. The fourth section of the book addresses the general all purpose title of other topics. Subjects reviewed here include economics, open-ended problems, environmental concerns, health and safety management, numerical methods, ethics, and exergy analysis.

The authors cannot claim sole authorship to all the problems and material in this book. The present text has evolved from a host of sources, including: notes, homework problems and exam problems prepared by L. Theodore for a required one-semester, three-credit "Chemical Engineering Thermodynamics" undergraduate course offered at Manhattan College; *Introduction to Hazardous Waste Incineration*, 2nd Edition, J. Santoleri, J. Reynolds, and L. Theodore, John Wiley & Sons; *Chemical Reaction Kinetics*, L. Theodore, a Theodore Tutorial; and, *Introduction to Chemical Engineering Thermodynamics*, 3rd Edition, J.M. Smith and H.C. Van Ness, McGraw-Hill. Although the bulk of the problems are original and/or taken from the sources that the authors have been directly involved with, every effort has been made to acknowledge material drawn from other sources.

The policy of most technical societies and publications is to use SI (metric) units or to list both the common British engineering unit and its SI equivalent. However, British units are primarily used in this book for the convenience of the majority of the reading audience. Readers who are more familiar and at ease with SI units are advised to refer to the Appendix of this book.

It is hoped that this writing will place in the hands of academic and industrial individuals a book covering the principles and applications of thermodynamics in a thorough and clear manner. Upon completion of the text, the reader should have acquired not only a working knowledge of the principles of thermodynamics but also experience in their application; and, the reader should find himself/herself approaching advanced texts, engineering literature, and industrial applications (even unique ones) with more confidence.

Sincere thanks are extended to Shannon O'Brien at Manhattan College for her invaluable help in solving some of the problems in the text, preparing part of the initial draft of the solutions manual, and proofing the manuscript. Special thanks are due Eric Huang and Pat Abulencia for their technical assistance in preparing parts of the manuscript.

L. THEODORE
F. RICCI
T. VAN VLIET

February 2009

Part I

Introduction

Nicolò Machiavelli [1469–1527]

There is nothing more difficult to take in hand, more perilous to conduct, or
more uncertain in its success, than to take the lead in the introduction of a new
order of things.

—*The Prince. Chap. 6*

Part I serves as the introductory section to this book. It reviews engineering and
science fundamentals that are an integral part of the field of thermodynamics. It
consists of six chapters, as noted below:

1 Basic Calculations

2 Process Variables

3 Gas Laws

4 Conservation Laws

5 Stoichiometry

6 The Second Law of Thermodynamics

Those individuals with a strong background in the above area(s) may choose to bypass
this Part.

Thermodynamics for the Practicing Engineer. By L. Theodore, F. Ricci, and T. Van Vliet
Copyright © 2009 John Wiley & Sons, Inc.

Chapter 1

Basic Calculations

Johann Wolfgang Von Goethe [1749–1832]

The sum which two married people owe to one another defies calculation. It is an infinite debt, which can only be discharged through all eternity.

—Elective Affinities [1808]. Book I, Chap. 9

INTRODUCTION

This first chapter provides a review of basic calculations and the fundamentals of measurement. Four topics receive treatment:

1 Units and Dimensions

2 Conversion of Units

3 The Gravitational Constant, g_c

4 Significant Figures and Scientific Notation

The reader is directed to the literature in the reference section of this chapter if additional information on these four topics is deemed necessary.[1–3]

UNITS AND DIMENSIONS

The units used in this text are consistent with those adopted by the engineering profession in the United States. For engineering work, SI (*Système International*) and *English* units are most often employed; in the United States, the *English engineering* units are generally used, although efforts are still underway to obtain universal adoption of SI units for all engineering and science applications. The SI units have the advantage of being based on the decimal system, which allows for more convenient conversion of units within the system.

There are other systems of units. Some of the more common of these are shown in Table 1.1; however, English engineering units are primarily used in this text. Tables 1.2 and 1.3 present units for both the English and SI systems, respectively.

Thermodynamics for the Practicing Engineer. By L. Theodore, F. Ricci, and T. Van Vliet
Copyright © 2009 John Wiley & Sons, Inc.

Table 1.1 Common Systems of Units

System	Length	Time	Mass	Force	Energy	Temperature
SI	meter	second	kilogram	Newton	Joule	Kelvin, degree Celsius
egs	centimeter	second	gram	dyne	erg, Joule, or calorie	Kelvin, degree Celsius
fps	foot	second	pound	poundal	foot poundal	degree Rankine, degree Fahrenheit
American engineering	foot	second	pound	pound (force)	British thermal unit, horsepower · hour	degree Rankine, degree Fahrenheit
British engineering	foot	second	slug	pound (force)	British thermal unit, foot pound (force)	degree Rankine, degree Fahrenheit

Table 1.2 English Engineering Units

Physical quantity	Name of unit	Symbol for unit
Length	foot	ft
Time	second	s
Mass	pound (mass)	lb
Temperature	degree Rankine	°R
Temperature (alternative)	degree Fahrenheit	°F
Moles	pound · mole	lbmol
Energy	British thermal unit	Btu
Energy (alternative)	horsepower · hour	hp · h
Force	pound (force)	lb_f
Acceleration	foot per second square	ft/s^2
Velocity	foot per second	ft/s
Volume	cubic foot	ft^3
Area	square foot	ft^2
Frequency	cycles per second, hertz	cycles/s, Hz
Power	horsepower, Btu per second	hp, Btu/s
Heat capacity	British thermal unit per (pound mass · degree Rankine)	Btu/lb · °R
Density	pound (mass) per cubic foot	lb/ft^3
Pressure	pound (force) per square inch	psi
	pound (force) per square foot	psf
	atmospheres	atm
	bar	bar

Table 1.3   SI Units

Physical unit	Name of unit	Symbol for unit
Length	meter	m
Mass	kilogram, gram	kg, g
Time	second	s
Temperature	Kelvin	K
Temperature (alternative)	Celsius	°C
Moles	gram · mole	gmol
Energy	Joule	J, $kg \cdot m^2/s^2$
Force	Newton	N, $kg \cdot m/s^2$, J/m
Acceleration	meters per square second	m/s^2
Pressure	Pascal, Newton per square meter	Pa, N/m^2
Pressure (alternative)	bar	bar
Velocity	meters per second	m/s
Volume	cubic meter, liters	m^3, L
Area	square meter	m^2
Frequency	Hertz	Hz, cycles/s
Power	Watt	W, $kg \cdot m^2 \cdot s^3$, J/s
Heat capacity	Joule per kilogram · Kelvin	$J/kg \cdot K$
Density	kilogram per cubic meter	kg/m^3
Angular velocity	radians per second	rad/s

Table 1.4   Prefixes for SI Units

Multiplication factors	Prefix	Symbol
$1{,}000{,}000{,}000{,}000{,}000{,}000 = 10^{18}$	exa	E
$1{,}000{,}000{,}000{,}000{,}000 = 10^{15}$	peta	P
$1{,}000{,}000{,}000{,}000 = 10^{12}$	tera	T
$1{,}000{,}000{,}000 = 10^{9}$	giga	G
$1{,}000{,}000 = 10^{6}$	mega	M
$1{,}000 = 10^{3}$	kilo	k
$100 = 10^{2}$	hecto	h
$10 = 10^{1}$	deka	da
$0.1 = 10^{-1}$	deci	d
$0.01 = 10^{-2}$	centi	c
$0.001 = 10^{-3}$	milli	m
$0.000\ 001 = 10^{-6}$	micro	μ
$0.000\ 000\ 001 = 10^{-9}$	nano	n
$0.000\ 000\ 000\ 001 = 10^{-12}$	pico	p
$0.000\ 000\ 000\ 000\ 001 = 10^{-15}$	femto	f
$0.000\ 000\ 000\ 000\ 000\ 001 = 10^{-18}$	atto	a

Table 1.5 Decimal Equivalents

Inch in fractions	Decimal equivalent	Millimeter equivalent
	A. 4ths and 8ths	
1/8	0.125	3.175
1/4	0.250	6.350
3/8	0.375	9.525
1/2	0.500	12.700
5/8	0.625	15.875
3/4	0.750	19.050
7/8	0.875	22.225
	B. 16ths	
1/16	0.0625	1.588
3/16	0.1875	4.763
5/16	0.3125	7.938
7/16	0.4375	11.113
9/16	0.5625	14.288
11/16	0.6875	17.463
13/16	0.8125	20.638
15/16	0.9375	23.813
	C. 32nds	
1/32	0.03125	0.794
3/32	0.09375	2.381
5/32	0.15625	3.969
7/32	0.21875	5.556
9/32	0.28125	7.144
11/32	0.34375	8.731
13/32	0.40625	10.319
15/32	0.46875	11.906
17/32	0.53125	13.494
19/32	0.59375	15.081
21/32	0.65625	16.669
23/32	0.71875	18.256
25/32	0.78125	19.844
27/32	0.84375	21.431
29/32	0.90625	23.019
31/32	0.96875	24.606

Some of the more common prefixes for SI units are given in Table 1.4, and decimal equivalents are provided in Table 1.5. Conversion factors between SI and English units and additional details on the SI system are provided in the Appendix III.

Two units that appear in dated literature are the *poundal* and *slug*. By definition, one poundal force will give a one pound mass an acceleration of one ft/s^2. Alternatively, one slug can be defined as the mass that will accelerate one ft/s^2 when acted upon by a one pound force; thus, a slug is equal to 32.2 pounds mass.

CONVERSION OF UNITS

Converting a measurement from one unit to another can be conveniently accomplished by using *unit conversion factors*; these factors are obtained from simple equations that relate the two units numerically. For example, from

$$1 \text{ foot (ft)} = 12 \text{ inches (in)} \qquad (1.1)$$

the following conversion factor can be obtained:

$$12 \text{ in}/1 \text{ ft} = 1 \qquad (1.2)$$

Since this factor is equal to unity, multiplying some quantity (e.g., 18 ft) by this factor cannot alter its value. Hence

$$18 \text{ ft } (12 \text{ in}/1 \text{ ft}) = 216 \text{ in} \qquad (1.3)$$

Note that in Equation (1.3), the old units of *feet* on the left-hand side cancel out leaving only the desired units of *inches*.

Physical equations must be dimensionally consistent. For the equality to hold, each term in the equation must have the same dimensions. This condition can be and should be checked when solving engineering problems. Throughout the text, great care is exercised in maintaining the dimensional formulas of all terms and the dimensional homogeneity of each equation. Equations will generally be developed in terms of specific units rather than general dimensions (e.g., feet rather than length). This approach should help the reader to more easily attach physical significance to the equations presented in these chapters.

ILLUSTRATIVE EXAMPLE 1.1

Convert units of acceleration in cm/s^2 to $miles/yr^2$.

SOLUTION: The procedure outlined above is applied to the units of cm/s^2:

$$\left(\frac{1 \text{ cm}}{s^2}\right)\left(\frac{3600^2 \text{ s}^2}{1 \text{ h}^2}\right)\left(\frac{24^2 \text{ h}^2}{1 \text{ day}^2}\right)\left(\frac{365^2 \text{ day}^2}{1 \text{ yr}^2}\right)\left(\frac{1 \text{ in}}{2.54 \text{ cm}}\right)\left(\frac{1 \text{ ft}}{12 \text{ in}}\right)\left(\frac{1 \text{ mile}}{5280 \text{ ft}}\right)$$

$$= 6.18 \times 10^9 \text{ miles}/yr^2$$

Thus, 1.0 cm/s^2 is equal to 6.18×10^9 $miles/yr^2$.   ∎

THE GRAVITATIONAL CONSTANT, g_C

The momentum of a system is defined as the product of the mass and velocity of the system:

$$\text{Momentum} = (\text{mass})(\text{velocity}) \tag{1.4}$$

One set of units for momentum are therefore $\text{lb} \cdot \text{ft}/\text{s}$. The units of the time rate of change of momentum (hereafter referred to as rate of momentum) are simply the units of momentum divided by time, i.e.,

$$\text{Rate of momentum} = \frac{\text{lb} \cdot \text{ft}}{\text{s}^2}$$

The above units can be converted to lb_f if multiplied by an appropriate constant. As noted earlier, a conversion constant is a term that is used to obtain units in a more convenient form; all conversion constants have magnitude and units in the term, but can also be shown to be equal to 1.0 (unity) with *no* units.

A defining equation is

$$1 \, \text{lb}_f = 32.2 \left(\frac{\text{lb} \cdot \text{ft}}{\text{s}^2} \right) \tag{1.5}$$

If this equation is divided by lb_f, one obtains

$$1.0 = 32.2 \left(\frac{\text{lb} \cdot \text{ft}}{\text{lb}_f \cdot \text{s}^2} \right) \tag{1.6}$$

This serves to define the conversion constant g_c. If the rate of momentum is divided by g_c as $32.2 \, \text{lb} \cdot \text{ft}/\text{lb}_f \cdot \text{s}^2$—this operation being equivalent to dividing by 1.0—the following units result:

$$\text{Rate of momentum} \equiv \left(\frac{\text{lb} \cdot \text{ft}}{\text{s}^2} \right) \left(\frac{\text{lb}_f \cdot \text{s}^2}{\text{lb} \cdot \text{ft}} \right)$$

$$\equiv \text{lb}_f \tag{1.7}$$

It can be concluded from the above dimensional analysis that a force is equivalent to a rate of momentum.

SIGNIFICANT FIGURES AND SCIENTIFIC NOTATION[3]

Significant figures provide an indication of the precision with which a quantity is measured or known. The last digit represents, in a qualitative sense, some degree of doubt. For example, a measurement of 8.32 inches implies that the actual quantity is somewhere between 8.315 and 8.325 inches. This applies to calculated and measured quantities; quantities that are known exactly (e.g., pure integers) have an infinite number of significant figures.

The significant digits of a number are the digits from the first nonzero digit on the left to either (a) the last digit (whether it is nonzero or zero) on the right if there is a decimal point, or (b) the last nonzero digit of the number if there is no decimal point. For example:

370	has 2 significant figures
370.	has 3 significant figures
370.0	has 4 significant figures
28,070	has 4 significant figures
0.037	has 2 significant figures
0.0370	has 3 significant figures
0.02807	has 4 significant figures

Whenever quantities are combined by multiplication and/or division, the number of significant figures in the result should equal the lowest number of significant figures of any of the quantities. In long calculations, the final result should be rounded off to the correct number of significant figures. When quantities are combined by addition and/or subtraction, the final result cannot be more precise than any of the quantities added or subtracted. Therefore, the position (relative to the decimal point) of the last significant digit in the number that has the lowest degree of precision is the position of the last permissible significant digit in the result. For example, the sum of 3702., 370, 0.037, 4, and 37. should be reported as 4110 (without a decimal). The least precise of the five numbers is 370, which has its last significant digit in the *tens* position. The answer should also have its last significant digit in the *tens* position.

Unfortunately, engineers and scientists rarely concern themselves with significant figures in their calculations. However, it is recommended that—at least for this chapter—the reader attempt to follow the calculational procedure set forth in this subsection.

In the process of performing engineering calculations, very large and very small numbers are often encountered. A convenient way to represent these numbers is to use *scientific notation*. Generally, a number represented in scientific notation is the product of a number (<10 but $>$ or $=1$) and 10 raised to an integer power. For example,

$$28{,}070{,}000{,}000 = 2.807 \times 10^{10}$$

$$0.000\,002\,807 = 2.807 \times 10^{-6}$$

A positive feature of using scientific notation is that only the significant figures need appear in the number.

REFERENCES

1. R. Perry and D. Green (editors), "*Perry's Chemical Engineers' Handbook*," 8th edition, McGraw-Hill, New York, 2008.

2. J. REYNOLDS, J. JERIS, and L. THEODORE, "*Handbook of Chemical and Environmental Engineering Calculations,*" John Wiley & Sons, Hoboken, NJ, 2004.

3. J. SANTOLERI, J. REYNOLDS, and L. THEODORE, "*Introduction to Hazardous Waste Incineration,*" 2nd edition, John Wiley & Sons, Hoboken, NJ, 2000.

NOTE: Additional problems for each chapter are available for all readers at www. These problems may be used for additional review or homework purposes.

Chapter **2**

Process Variables

Seneca [8 B.C.–A.D. 65]
The best ideas are common property.

—Epistles. 12, 11

INTRODUCTION

The authors originally considered the title "State, Physical, and Chemical Properties" for this chapter. However, since these three properties have been used interchangeably and have come to mean different things to different people, it was decided to simply employ the title "Process Variables." The three aforementioned properties were therefore integrated into this all-purpose title and eliminated the need for differentiating between the three.

This second chapter provides a review of some basic concepts from physics, chemistry, and engineering in preparation for material that is covered in later chapters. All of these topics are vital to thermodynamics and thermodynamic applications. Because many of these topics are unrelated to each other, this chapter admittedly lacks the cohesiveness that chapters covering a single topic might have. This is usually the case when basic material from such widely differing areas of knowledge as physics, chemistry, and engineering is surveyed. Though these topics are widely divergent and covered with varying degrees of thoroughness, all of them will find later use in this text. If additional information on these review topics is needed, the reader is directed to the literature in the reference section of this chapter.

ILLUSTRATIVE EXAMPLE 2.1

Discuss the traditional difference (in definition) between chemical and physical properties.

SOLUTION: Every compound has a unique set of *properties* that allows one to recognize and distinguish it from other compounds. These properties can be grouped into two main

Thermodynamics for the Practicing Engineer. By L. Theodore, F. Ricci, and T. Van Vliet
Copyright © 2009 by John Wiley & Sons, Inc.

categories: physical and chemical. *Physical properties* are defined as those that can be measured without changing the identity and composition of the substance. Key properties include viscosity, density, surface tension, melting point, boiling point, etc. *Chemical properties* are defined as those that may be altered via reaction to form other compounds or substances. Key chemical properties include upper and lower flammability limits, enthalpy of reaction, autoignition temperature, etc.

These properties may be further divided into two categories—intensive and extensive. *Intensive properties* are not a function of the quantity of the substance, while *extensive properties* depend on the quantity of the substance.   ∎

TEMPERATURE

Whether in the gaseous, liquid, or solid state, all molecules possess some degree of kinetic energy, i.e., they are in constant motion—vibrating, rotating, or translating. The kinetic energies of individual molecules cannot be measured, but the combined effect of these energies in a very large number of molecules can. This measurable quantity is known as *temperature*; it is a macroscopic concept only and as such does not exist on the molecular level.

Temperature can be measured in many ways; the most common method makes use of the expansion of mercury (usually encased inside a glass capillary tube) with increasing temperature. (In thermal applications, however, thermocouples or thermistors are more commonly employed.) The two most commonly used temperature scales are the Celsius (or Centigrade) and Fahrenheit scales. The Celsius is based on the boiling and freezing points of water at 1-atm pressure; to the former, a value of 100°C is assigned, and to the latter, a value of 0°C. On the older Fahrenheit scale, these temperatures correspond to 212°F and 32°F, respectively. Equations (2.1) and (2.2) show the conversion from one scale to the other:

$$°F = 1.8(°C) + 32 \qquad (2.1)$$

$$°C = (°F - 32)/1.8 \qquad (2.2)$$

where   °F = a temperature on the Fahrenheit scale
        °C = a temperature on the Celsius scale

Experiments with gases at low-to-moderate pressures (up to a few atmospheres) have shown that, if the pressure is kept constant, the volume of a gas and its temperature are linearly related via Charles' law (see next chapter) and that a decrease of 0.3663% or (1/273) of the initial volume is experienced for every temperature drop of 1°C. These experiments were not extended to very low temperatures, but if the linear relationship were extrapolated, the volume of the gas would *theoretically* be zero at a temperature of approximately −273°C or −460°F. This temperature has become known as *absolute zero* and is the basis for the definition of two *absolute* temperature scales. (An *absolute* scale is one which does not allow negative quantities.) These absolute temperature scales are the Kelvin (K) and Rankine (°R) scales; the former is defined by shifting the Celsius scale by 273°C so that 0 K is equal to −273°C; Eq. (4.3.3) shows this relationship:

$$K = {}^\circ C + 273 \qquad (2.3)$$

The Rankine scale is defined by shifting the Fahrenheit scale 460°, so that

$$^\circ R = {}^\circ F + 460 \qquad (2.4)$$

The relationships among the various temperature scales are shown in Fig. 2.1.

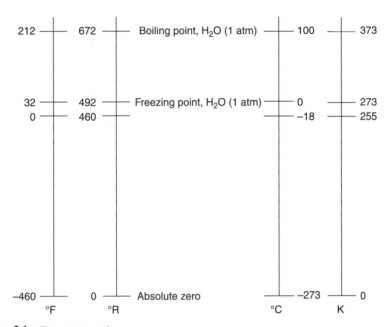

Figure 2.1   Temperature scales.

ILLUSTRATIVE EXAMPLE 2.2

Perform the following temperature conversions:

    1 Convert 55°F to (a) Rankine, (b) Celsius, and (c) Kelvin.

    2 Convert 55°C to (a) Fahrenheit, (b) Rankine, and (c) Kelvin.

SOLUTION

    1 **(a)** $^\circ R = {}^\circ F + 460 = 55 + 460 = 515$
        (b) $^\circ C = \frac{5}{9}({}^\circ F - 32) = \frac{5}{9}(55 - 32) = 12.8$
        (c) $K = \frac{5}{9}({}^\circ F + 460) = \frac{5}{9}(55 + 460) = 286$

2   **(a)** $°F = 1.8(°C) + 32 = 1.8(55) + 32 = 131$

   (b) $°R = 1.8(°C) + 492 = 1.8(55) + 492 = 591$

   (c) $K = °C + 273 = 55 + 273 = 328$ ∎

PRESSURE

Molecules in the gaseous state possess a high degree of translational kinetic energy, which means they are able to move quite freely throughout the body of the gas. If the gas is in a container of some type, the molecules are constantly bombarding the walls of the container. The macroscopic effect of this bombardment by a tremendous number of molecules—enough to make the effect measurable—is called *pressure*. The natural units of pressure are force per unit area. In the example of the gas in a container, the *unit area* is a portion of the inside solid surface of the container wall and the *force*, measured perpendicularly to the unit area, is the result of the molecules hitting the unit area and giving up momentum during the sudden change of direction.

There are a number of different methods used to express a pressure measurement. Some of them are natural units, i.e., based on a force per unit area, e.g., pound (force) per square inch (abbreviated lb_f/in^2 or psi) or dyne per square centimeter (dyn/cm^2). Others are based on a fluid height, such as inches of water (in H_2O) or millimeters of mercury (mm Hg); units such as these are convenient when the pressure is indicated by a difference between two levels of a liquid as in a *manometer* or *barometer*. *Barometric pressure* and *atmospheric pressure* are synonymous and measure the ambient air pressure. *Standard barometric pressure* is the average atmospheric pressure at sea level, 45° north latitude at 32°F. It is used to define another unit of pressure called the atmosphere (atm). Standard barometric pressure is 1 atm and is equivalent to 14.696 psi and 29.921 in Hg. As one might expect, barometric pressure varies with weather and altitude.

Measurements of pressure by most gauges indicate the difference in pressure either above or below that of the atmosphere surrounding the gauge. *Gauge pressure* is the pressure indicated by such a device. If the pressure in the system measured by the gauge is greater than the pressure prevailing in the atmosphere, the gauge pressure is expressed positively. If lower than atmospheric pressure, the gauge pressure is a negative quantity; the term *vacuum* designates a negative gauge pressure. Gauge pressures are often identified by the letter g after the pressure unit; for example, psig (pounds per square inch gauge) is a gauge pressure in psi units.

Since gauge pressure is the pressure relative to the prevailing atmospheric pressure, the sum of the two gives the *absolute pressure*, indicated by the letter a after the unit [e.g., psia (pounds per square inch absolute)]:

$$P = P_a + P_g \qquad (2.5)$$

where P = absolute pressure (psia)

   P_a = atmospheric pressure (psia)

   P_g = gauge pressure (psig)

The absolute pressure scale is absolute in the same sense that the absolute temperature scale is absolute, i.e., a pressure of zero psia is the lowest possible pressure theoretically achievable—a perfect vacuum.

ILLUSTRATIVE EXAMPLE 2.3

Consider the following pressure calculations.

1 A liquid weighing 100 lb held in a cylindrical column with a base area of 3 in^2 exerts how much pressure at the base in lb$_f$/ft^2?

2 If the pressure is 35 psig (pounds per square inch gauge), what is the absolute pressure?

SOLUTION

1 See Chapter 1.

$$F = mg/g_c = 100\,lb(1\,lb_f/lb)$$
$$= 100\,lb_f$$

Note: As discussed in Chapter 1, g_c is a conversion factor equal to 32.2 lb · ft/lb$_f$ · s^2; g is the gravitational acceleration, which is equal, or close to, 32.2 ft/s^2 on Earth's surface. Therefore,

$$P = F/\text{area} = 100\,lb_f/3\,in^2$$
$$= 33.33\,lb_f/in^2$$
$$= 4800\,lb_f/\,ft^2$$

2 $P = P_g + P_a = 35 + 14.7$
$\quad = 49.7\,\text{psia}$ ∎

MOLES AND MOLECULAR WEIGHTS

An atom consists of protons and neutrons in a nucleus surrounded by electrons. An electron has such a small mass relative to that of the proton and neutron that the weight of the atom (called the *atomic weight*) is approximately equal to the sum of the weights of the particles in its nucleus. Atomic weight may be expressed in *atomic mass units (amu) per atom* or *in grams per gram · atom*. One gram · atom contains 6.02×10^{23} atoms (Avogadro's number). The atomic weights of the elements are listed in Table 2.1.

The *molecular weight* (MW) of a compound is the sum of the atomic weights of the atoms that make up the molecule. Atomic mass units per molecule (amu/molecule)

Table 2.1 Atomic Weights of the Elements[a,b]

Element	Symbol	Atomic weight
Actinium	Ac	227.0278
Aluminum	Al	26.9815
Americium	Am	(243)
Antimony	Sb	121.75
Argon	Ar	39.948
Arsenic	As	74.9216
Astatine	At	(210)
Barium	Ba	137.34
Berkelium	Bk	(247)
Beryllium	Be	9.0122
Bismuth	Bi	208.980
Boron	B	10.811
Bromine	Br	79.904
Cadmium	Cd	112.40
Calcium	Ca	40.08
Californium	Cf	(251)
Carbon	C	12.01115
Cerium	Ce	140.12
Cesium	Cs	132.905
Chlorine	Cl	35.453
Chromium	Cr	51.996
Cobalt	Co	58.9332
Copper	Cu	63.546
Curium	Cm	(247)
Dysprosium	Dy	162.50
Einsteinium	Es	(252)
Erbium	Er	167.26
Europium	Eu	151.96
Fermium	Fm	(257)
Fluorine	F	18.9984
Francium	Fr	(223)
Gadolinium	Gd	157.25
Gallium	Ga	69.72
Germanium	Ge	72.59
Gold	Au	196.967
Hafnium	Hf	178.49

(*Continued*)

Table 2.1 *Continued*

Element	Symbol	Atomic weight
Helium	He	4.0026
Holmium	Ho	164.930
Hydrogen	H	1.00797
Indium	In	114.82
Iodine	I	126.9044
Iridium	Ir	192.2
Iron	Fe	55.847
Krypton	Kr	83.80
Lanthanum	La	138.91
Lawrencium	Lr	(260)
Lead	Pb	207.19
Lithium	Li	6.939
Lutetium	Lu	174.97
Magnesium	Mg	24.312
Manganese	Mn	54.9380
Mendelevium	Md	(258)
Mercury	Hg	200.59
Molybdenum	Mo	95.94
Neodymium	Nd	144.24
Neon	Nc	20.183
Neptunium	Np	237.0482
Nickel	Ni	58.71
Niobium	Nb	92.906
Nitrogen	N	14.0067
Nobelium	No	(259)
Osmium	Os	190.2
Oxygen	O	15.9994
Pallasium	Pd	106.4
Phosphorus	P	30.9738
Platinum	Pt	195.09
Plutonium	Pu	(244)
Polonium	Po	(209)
Potassium	K	39.102
Praseodymium	Pr	140.907
Promethium	Pm	(145)
Protactinium	Pa	231.0359

(*Continued*)

Table 2.1   *Continued*

Element	Symbol	Atomic weight
Radium	Ra	226.0254[c]
Radon	Rn	(222)
Rhenium	Re	186.2
Rhodium	Rh	102.905
Rubidium	Rb	84.57
Rutherium	Ru	101.07
Samarium	Sm	150.35
Scandium	Sc	44.956
Selenium	Se	78.96
Silicon	Si	28.086
Silver	Ag	107.868
Sodium	Na	22.9898
Strontium	Sr	87.62
Sulfur	S	32.064
Tantalum	Ta	180.948
Technetium	Tc	(98)
Tellurium	Te	127.60
Terbium	Tb	158.924
Thallium	Tl	204.37
Thorium	Th	232.038
Thulium	Tm	168.934
Tin	Sn	118.69
Titanium	Ti	47.90
Tungsten	W	183.85
Uranium	U	238.03
Vanadium	V	50.942
Xenon	Xe	131.30
Ytterbium	Yb	173.04
Yttrium	Y	88.905
Zinc	Zn	65.37
Zirconium	Zr	91.22

[a]Atomic weights apply to naturally occurring isotopic compositions and are based on an atomic mass of $^{12}C = 12$.

[b]A value given in parenthesis for radioactive elements is the atomic mass number of the isotope of longest known half-life.

[c]Geologically exceptional samples are known in which the element has an isotopic composition outside the limits for normal material.

or grams per gram-mole (g/gmol) are used for molecular weight. One gram · mole (gmol) contains an Avogadro number of molecules. For the English system, a pound · mole (lbmol) contains $454 \times 6.023 \times 10^{23}$ molecules.

Molal units are used extensively in thermodynamic calculations as they greatly simplify material balances where chemical (including combustion) reactions are occurring. For mixtures of substances (gases, liquids, or solids), it is also convenient to express compositions in mole fractions or mole percentages instead of mass fractions. The mole fraction is the ratio of the number of moles of one component to the total number of moles in the mixture. Equations (2.6)–(2.9) express these relationships:

$$\text{moles of A} = \frac{\text{mass A}}{\text{molecular weight of A}}$$

$$n_A = \frac{m_A}{(MW)_A} \qquad (2.6)$$

$$\text{mole/ fraction A} = \frac{\text{moles A}}{\text{total moles}}$$

$$y_A = \frac{n_A}{n} \qquad (2.7)$$

$$\text{mass fraction A} = \frac{\text{mass A}}{\text{total mass}}$$

$$w_A = \frac{m_A}{m} \qquad (2.8)$$

$$\text{volume fraction A} = \frac{\text{volume A}}{\text{total volume}}$$

$$v_A = \frac{V_A}{V} \qquad (2.9)$$

The reader should note that, in general, mass fraction (or percent) is *NOT* equal to mole fraction (or percent).

ILLUSTRATIVE EXAMPLE 2.4

If a 55-gal tank contains 20.0 lb of water, (1) how many pound · moles of water does it contain? (2) how many gram · moles does it contain? and (3) how many molecules does it contain?

SOLUTION: The molecular weight of the water (H_2O) is

$$MW = (2)(1.008) + (15.999) = 18.015 \, \text{g/gmol}$$
$$= 18.015 \, \text{lb/lbmol}$$

Therefore,

$$\textbf{1}\ (20.0\,\text{lb})\left(\frac{\text{lbmol}}{18.015\,\text{lb}}\right) = 1.11\ \text{lbmol water}$$

$$\textbf{2}\ (20.0\,\text{lb})\left(\frac{454\,\text{g}}{1\,\text{lb}}\right)\left(\frac{\text{gmol}}{18.015\,\text{g}}\right) = 503.6\ \text{gmol water}$$

$$\textbf{3}\ (503.6\,\text{gmol})\left(\frac{6.023 \times 10^{23}\ \text{molecules}}{1\,\text{gmol}}\right) = 3.033 \times 10^{26}\ \text{molecules} \qquad\blacksquare$$

MASS AND VOLUME

The *density* (ρ) of a substance is the ratio of its mass to its volume and may be expressed in units of pounds per cubic foot (lb/ft^3), kilograms per cubic meter (kg/m^3), etc. For solids, density can be easily determined by placing a known mass of the substance in a liquid and determining the displaced volume. The density of a liquid can be measured by weighing a known volume of the liquid in a volumetric flask. For gases, the ideal gas law, to be discussed in Chapter 3, can be used to calculate the density from the pressure, temperature, and molecular weight of the gas.

Densities of pure solids and liquids are relatively independent of temperature and pressure and can be found in standard reference books.[1,2] The *specific volume* (v) of a substance is its volume per unit mass (ft^3/lb, m^3/kg, etc.) and is, therefore, the inverse of its density.

The *specific gravity* (SG) is the ratio of the density of a substance to the density of a reference substance at a specific condition:

$$SG = \rho/\rho_{\text{ref}} \qquad (2.10)$$

The reference most commonly used for *solids* and *liquids* is water at its maximum density, which occurs at $4°C$; this reference density is $1.000\ \text{g}/\text{cm}^3$, $1000\ \text{kg}/\text{m}^3$, or 62.43 lb/ft^3. Note that, since the specific gravity is a ratio of two densities, it is dimensionless. Therefore, any set of units may be employed for the two densities as long as they are consistent. The specific gravity of *gases* is used only rarely; when it is, air at the same conditions of temperature and pressure as the gas is usually employed as the reference substance.

Another dimensionless quantity related to density is the API (American Petroleum Institute) gravity, which is often used to indicate densities of fuel oils. The relationship between the API scale and specific gravity is

$$\text{degrees API} = \frac{141.5}{\text{SG}(60/60°\text{F})} - 131.5 \qquad (2.11)$$

where $\text{SG}(60/60°\text{F})$ = specific gravity of the liquid at $60°F$ using water at $60°F$ as the reference.

ILLUSTRATIVE EXAMPLE 2.5

The following information is given:

Specific gravity of liquid (methanol) = 0.92 (at 60°F)

Density of reference substance (water) = 62.4 lb/ft³ (at 60°F)

Determine the density of methanol in lb/ft³.

SOLUTION: Calculate the density of methanol in English units by multiplying the specific gravity by the density of water—see Equation (2.10)

$$\text{Density of methanol} = (\text{specific gravity})(\text{density of water})$$

$$= (0.92)(62.4)$$

$$= 57.4 \, \text{lb/ft}^3$$

The procedure is reversed in order to calculate specific gravity from density data. As noted above, the notation for density is usually, but not always, ρ. ■

VISCOSITY

Viscosity is a property associated with a fluid's resistance to flow; more precisely, this property accounts for energy losses which result from shear stresses that occur between different portions of the fluid that are moving at different velocities. The *absolute viscosity* (μ) has units of mass per length · time; the fundamental unit is the *poise* (P), which is defined as 1 g/cm · s. This unit is inconveniently large for many practical purposes and viscosities are frequently given in *centipoises* (0.01 poise), which is abbreviated cP. The viscosity of pure water at 68.6°F is 1.00 cP. In English units, absolute viscosity is expressed either as pounds (mass) per foot · second (lb/ft · s) or pounds per foot · hour (lb/ft · h). The absolute viscosity depends primarily on temperature and to a lesser degree on pressure. The *kinematic viscosity* (v) is the absolute viscosity divided by the density of the fluid and is useful in certain fluid flow problems; the units for this quantity are length squared per time, e.g., square foot per second (ft²/s) or square meters per hour (m²/h). A kinematic viscosity of 1 cm²/s is called a *stoke*, denoted as S. For pure water at 70°F, $v = 0.983$ cS (centistokes). Because fluid viscosity changes rapidly with temperature, a numerical value of viscosity has no significance unless the temperature is specified.

Liquid viscosity is usually measured by the amount of time it takes for a given volume of liquid to flow through an orifice. The *Saybolt universal viscometer* is the most widely used device in the United States for the determination of the viscosity of fuel oils and liquids. It should be stressed that Saybolt viscosities, which are expressed in *Saybolt seconds (SSU)*, are not even approximately proportional to absolute viscosities except in the range above 200 SSU; hence, converting units from Saybolt seconds to other units requires the use of special conversion tables. As the time of flow decreases, the deviation becomes more marked. In any event, viscosity is an important property because of potential flow problems that might arise with viscous liquids and/or fuel oil.

The viscosities of air at atmospheric pressure and water are presented in Tables 2.2 and 2.3, respectively, as functions of temperature. Viscosities of other substances are available in the literature.[1]

Table 2.2   Viscosity of Air at 1 Atmosphere[a]

T (°C)	Viscosity, Micropoise (μP)
0	170.8
18	182.7
40	190.4
54	195.8
74	210.2
229	263.8

[a]$1\,P = 100\,cP = 10^6\,\mu P;\ 1\,cP = 6.72 \times 10^{-4}\,lb/ft \cdot s.$

Table 2.3   Viscosity of Water

T (°C)	Viscosity, Centipoise (cP)
0	1.792
5	1.519
10	1.308
15	1.140
20	1.000
25	0.894
30	0.801
35	0.723
40	0.656
50	0.594
60	0.469
70	0.406
80	0.357
90	0.317
100	0.284

ILLUSTRATIVE EXAMPLE 2.6

What is the kinematic viscosity of a gas, the specific gravity and absolute viscosity of which are 0.8 and 0.02 cP, respectively?

SOLUTION

$$\left(\frac{0.02\,\text{cP}}{1}\right)\left(\frac{6.720 \times 10^{-4}\,\text{lb/ft} \cdot \text{s}}{1\,\text{cP}}\right) = 1.344 \times 10^{-5}\,\text{lb/ft} \cdot \text{s}$$

$$\rho = (SG)(\rho_{\text{ref}}) = (0.08)(62.43\,\text{lb/ft}^3) = 49.94\,\text{lb/ft}^3$$

$$\nu = \mu/\rho = (1.344 \times 10^{-5}\,\text{lb/ft} \cdot \text{s})/(49.94\,\text{lb/ft}^3)$$

$$= 2.691 \times 10^{-7}\,\text{ft}^2/\text{s}$$

■

HEAT CAPACITY

The *heat capacity* of a substance is defined as the quantity of heat required to raise the temperature of that substance by one degree on a unit mass (or mole) basis. The term *specific heat* is frequently used in place of *heat capacity*. This is not strictly correct, because specific heat has been defined traditionally as the ratio of the heat capacity of a substance to the heat capacity of water. However, since the heat capacity of water is approximately $1\,\text{cal/g} \cdot {}^\circ\text{C}$ or $1\,\text{Btu/lb} \cdot {}^\circ\text{F}$, the term *specific heat* has come to imply heat capacity.

For gases, the addition of heat to cause the 1° temperature rise may be accomplished either at constant pressure or at constant volume. Since the amounts of heat necessary are different for the two cases, subscripts are used to identify which heat capacity is being used—c_P for constant pressure and c_V for constant volume. For liquids and solids, this distinction does not have to be made since there is little difference between the two. Values of heat capacity are available in the literature[1] and are also provided in Chapter 7.

Heat capacities are often used on a *molar* basis instead of a *mass* basis, in which case the units become $\text{cal/gmol} \cdot {}^\circ\text{C}$ or $\text{Btu/lbmol} \cdot {}^\circ\text{F}$. To distinguish between the two bases, uppercase letters (C_P, C_V) will be used in this text to represent the molar-based heat capacities, and lowercase letters (c_P, c_V) will be used for the mass-based heat capacities or specific heats.

Heat capacities are functions of both the temperature and pressure, although the effect of pressure is generally small and is neglected in almost all engineering calculations. The effect of temperature on C_P can be described by

$$C_P = \alpha + \beta T + \gamma T^2 \tag{2.12}$$

or

$$C_P = a + bT + cT^{-2} \tag{2.13}$$

Values for α, β, γ, and a, b, c, as well as average heat capacity information are provided in tabular form in Chapter 7. *Average* or *mean* heat capacity data over specific temperature ranges are also available. These will find extensive application in Chapters 7–10, as well as some of the later chapters. Properties such as enthalpy (heat) of vaporization are also discussed in Chapter 8.

ILLUSTRATIVE EXAMPLE 2.7

The following is given:

$$\text{Heat capacity of methanol} = 0.61 \, \text{cal/g} \cdot \text{°C (at 60°F)}$$

Convert the heat capacities to English units.

SOLUTION: Note that 1.0 Btu/lb · °F is equivalent to 1.0 cal/g · °C. This also applies on a mole basis, i.e.,

$$1 \, \text{Btu/lbmol} \cdot \text{°F} = 1 \, \text{cal/gmol} \cdot \text{°C}$$

The heat capacity can be converted from units of cal/g · °C to Btu/lb · °F using appropriate conversion factors.

$$\left(\frac{0.61 \, \text{cal}}{\text{g} \cdot \text{°C}}\right)\left(\frac{454 \, \text{g}}{\text{lb}}\right)\left(\frac{\text{Btu}}{252 \, \text{cal}}\right)\left(\frac{\text{°C}}{1.8 \text{°F}}\right) = 0.61 \, \text{Btu/lb} \cdot \text{°F} \qquad \blacksquare$$

THERMAL CONDUCTIVITY

Experience has shown that when a temperature difference exists across a solid body, energy in the form of heat will transfer from the high-temperature region to the low-temperature region until thermal equilibrium (same temperature) is reached. This mode of heat transfer where vibrating molecules pass along kinetic energy through the solid is called *conduction*. Liquids and gases may also transport heat in this fashion. The property of *thermal conductivity* provides a measure of how fast (or how easily) heat flows through a substance. It is defined as the amount of heat that flows in unit time through a unit surface area of unit thickness as a result of a unit difference in temperature. Typical units for conductivity are $\text{Btu} \cdot \text{ft/h} \cdot \text{ft}^2 \cdot \text{°F}$ or $\text{Btu/h} \cdot \text{ft} \cdot \text{°F}$.

With regard to thermodynamic applications, this particular property finds application in designing heat exchangers (see Chapter 6).

ILLUSTRATIVE EXAMPLE 2.8

The following data is given:

$$\text{Thermal conductivity of methanol} = 0.0512 \, \text{cal/m} \cdot \text{s} \cdot \text{°C (at 60°F)}$$

Convert the thermal conductivity to English units.

SOLUTION: The factor for converting $\text{cal}/\text{m}\cdot\text{s}\cdot{}^{\circ}\text{C}$ to $\text{Btu}/\text{ft}\cdot\text{h}\cdot{}^{\circ}\text{F}$ can be shown to be 2.419.

The thermal conductivity of methanol can be converted to $\text{Btu}/\text{ft}\cdot\text{h}\cdot{}^{\circ}\text{F}$ from $\text{cal}/\text{m}\cdot\text{s}\cdot{}^{\circ}\text{C}$ as follows:

$$k = \left(\frac{0.0512\,\text{cal}}{\text{m}\cdot\text{s}\cdot{}^{\circ}\text{C}}\right)\left(\frac{\text{Btu}}{252\,\text{cal}}\right)\left(\frac{0.3048\,\text{m}}{\text{ft}}\right)\left(\frac{3600\,\text{s}}{\text{h}}\right)\left(\frac{{}^{\circ}\text{C}}{1.8{}^{\circ}\text{F}}\right)$$

$$= (0.0512)(2.419)$$

$$= 0.124\,\text{Btu}/\text{ft}\cdot\text{h}\cdot{}^{\circ}\text{F}$$

Note that the usual engineering notation for thermal conductivity is k.　　　　　■

REYNOLDS NUMBER

The Reynolds number, Re, is a dimensionless number that indicates whether a moving fluid is flowing in the laminar or turbulent mode. *Laminar* flow is characteristic of fluids flowing slowly enough so that there are no eddies (whirlpools) or macroscopic mixing of different portions of the fluid. (*Note:* In any fluid, there is always *molecular* mixing due to the thermal activity of the molecules; this is distinct from *macroscopic* mixing due to the swirling motion of different portions of the fluid.) In laminar flow, a fluid can be imagined to flow like a deck of cards, with adjacent layers sliding past one another. *Turbulent* flow is characterized by eddies and macroscopic currents. In practice, moving gases are generally in the turbulent region. For flow in a pipe, a Reynolds number above 2100 is an indication of turbulent flow.

The Reynolds number is dependent on the fluid velocity, density, viscosity, and some *length* characteristic of the system or conduit; for pipes, this characteristic length is the inside diameter:

$$\text{Re} = Dv\rho/\mu = Dv/\nu \tag{2.14}$$

where   Re = Reynolds number
　　　　D = inside diameter of the pipe (ft)
　　　　v = fluid velocity (ft/s)
　　　　ρ = fluid density (lb/ft³)
　　　　μ = fluid viscosity (lb/ft·s)
　　　　ν = fluid kinematic viscosity (ft²/s)

Any consistent set of units may be used with Equation (2.14).

ILLUSTRATIVE EXAMPLE 2.9

Calculate the Reynolds number for a fluid flowing through a 5-inch diameter pipe at 10 fps (feet per second) with a density of 50 lb/ft³ and a viscosity of 0.65 cP? Is the flow turbulent or laminar?

SOLUTION: By definition

$$Re = Dv\rho/\mu \qquad (2.14)$$

Substitution yields

$$Re = \left(\frac{50\,lb}{ft^3}\right)\left(\frac{10\,ft}{s}\right)\left(\frac{5\,in}{1}\right)\left(\frac{1\,ft}{12\,in}\right)\left(\frac{1}{0.65\,cP}\right)\left(\frac{1\,cP}{6.720 \times 10^{-4}\,lb/ft \cdot s}\right)$$

$$= (50\,lb/ft^3)(10\,ft/s)[(5/12)ft]/(0.65 \times 6.72 \times 10^{-4}\,lb/ft \cdot s)$$

$$= 477{,}000$$

The Reynolds number is >2100; therefore, the flow is turbulent.   ∎

pH

An important chemical property of an aqueous solution is its pH. The pH measures the acidity or basicity of the solution. In a neutral solution, such as pure water, the hydrogen (H^+) and hydroxyl (OH^-) ion concentrations are equal. At ordinary temperatures, this concentration is

$$C_{H^+} = C_{OH^-} = 10^{-7}\,g \cdot ion/L \qquad (2.15)$$

where C_{H^+} = hydrogen ion concentration
  C_{OH^-} = hydroxyl ion concentration

The unit $g \cdot ion$ stands for gram $\cdot$ ion, which represents an Avogadro number of ions. In all aqueous solutions, whether neutral, basic, or acidic, a chemical equilibrium or balance is established between these two concentrations, so that

$$K_{eq} = C_{H^+} C_{OH^-} = 10^{-14} \qquad (2.16)$$

where K_{eq} = equilibrium constant.
    The numerical value for K_{eq} given in Equation (2.16) holds for room temperature and only when the concentrations are expressed in gram $\cdot$ ion per liter ($g \cdot ion/L$). In acid solutions, C_{H^+} is $>C_{OH^-}$; in basic solutions, C_{OH^-} predominates.
    The pH is a direct measure of the hydrogen ion concentration and is defined by

$$pH = -\log C_{H^+} \qquad (2.17)$$

Thus, an acidic solution is characterized by a pH below 7 (the lower the pH, the higher the acidity), a basic solution by a pH above 7, and a neutral solution by a pH of 7. It should be pointed out that Equation (2.17) is not the exact definition of pH but is a close approximation to it. Strictly speaking, the *activity* of the hydrogen ion, a_{H^+}, and not the ion concentration belongs in Equation (2.17). The reader is directed to the literature[2,3] for a discussion of chemical activities.

ILLUSTRATIVE EXAMPLE 2.10

Calculate the hydrogen ion and the hydroxyl ion concentration of an aqueous solution if the pH of the solution is 1.0.

SOLUTION: Apply Equation (2.17). For a pH of 1.0

$$pH = -\log(C_{H^+})$$

$$C_{H^+} = 10^{-pH} = 10^{-1} = 0.1\,g\cdot ion/L$$

$$C_{H^+} \times C_{OH^-} = 10^{-14}$$

$$C_{OH^-} = \frac{10^{-14}}{C_{H^+}}$$

$$= 10^{-13}\,g\cdot ion/L$$

■

ILLUSTRATIVE EXAMPLE 2.11

Process considerations require pH control in a 50,000-gal storage tank used for incoming waste mixtures (including liquid plus solids) at a hazardous waste incinerator. Normally, the tank is kept at neutral pH. However, the operation can tolerate pH variations from 6 to 8. Waste arrives in 5000-gal shipments.

    Assume that the tank is completely mixed, contains 45,000 gal when the shipment arrives, the incoming acidic waste is fully dissociated, and that there is negligible buffering capacity in the tank. What is the pH of the most acidic waste shipment that can be handled without neutralization?

SOLUTION: The pH of the most acidic waste shipment that can be handled without neutralization is calculated as follows: 5000 gal of waste with a $[H^+] = X$ is diluted by 45,000 gal at pH = 7 or $[H^+] = 10^{-7}$. The minimum pH of 6 that can be tolerated is equivalent to an average $[H^+] = 10^{-6}$. From an ion balance:

$$[H^+] = 10^{-6} = (5000/50,000)X + (45,000/50,000)(10^{-7})$$

$$X = \left(\frac{50,000}{5000}\right)\left[10^{-6} - \frac{45,000(10^{-7})}{50,000}\right]$$

$$= 0.91 \times 10^{-6}$$

$$pH = 6.04$$

■

VAPOR PRESSURE

Vapor pressure is an important property of liquids, and, to a much lesser extent, of solids. If a liquid is allowed to evaporate in a confined space, the pressure in the vapor space increases as the amount of vapor increases. If there is sufficient liquid present, a point is eventually reached at which the pressure in the vapor space is exactly equal to the pressure exerted by the liquid at its own surface. At this point, a dynamic equilibrium exists in which vaporization and condensation take place at equal rates and the pressure in the vapor space remains constant.

The pressure exerted at equilibrium is defined as the vapor pressure of the liquid. The magnitude of this pressure for a given liquid depends on the temperature, but not on the amount of liquid present. Solids, like liquids, also exert a vapor pressure. Evaporation of solids (called *sublimation*) is noticeable only for those with appreciable vapor pressures. This topic is reviewed in more detail in Chapter 8.

PROPERTY ESTIMATION

This last section provides references detailing procedures that will allow the practitioner to estimate key physical and chemical properties of materials. Although the scientific community has traditionally resorted to experimental methods to accurately determine the aforementioned properties, that option may very well not be available when dealing with new materials.

Predictive methods, albeit traditional ones, may be the only option available to obtain a first estimate of these properties. It should be noted that significant errors may be involved since extrapolating (or extending) satisfactory estimation procedures at the macroscale level may not always be reasonable. Notwithstanding these concerns, procedures to estimate some of the key physical and chemical properties in thermodynamics given below are available in the literature.[3-6]

1 Vapor pressure

2 Latent enthalpy

3 Critical properties

4 Viscosity

5 Thermal conductivity

6 Heat capacity

References (3,4) are somewhat complementary, but each provides extensive information on this topic. This includes equations and procedures on several other properties not listed above. The interested reader should check these references for more details.

Is property estimation important? Absolutely. As indicated above, there are times and situations when experimental procedures cannot be implemented. For this scenario, one can turn to theoretical and semitheoretical methods and equations to obtain first estimates of important property information.

One could argue that the present procedures available to estimate the properties of materials are based on questionable approaches. Nonetheless, the traditional methods available for property estimation either may be applicable or may suggest alternative theoretical approaches.

Finally, it should be noted that the Periodic Law correlates properties of elements. Scientists and engineers came to realize that all matter is composed of a rather limited number of basic building blocks, and the desire to discover all the fundamental units/parts became apparent. As data on the properties of elements became available, a pattern in the physical and chemical properties grew discernible, and because the pattern repeated itself in a rather well-organized fashion, it became known as the Periodic Law. This law is one of the finest generalizations of science, and has proven extremely useful both in predicting and correlating physical and chemical properties. The law essentially states that the properties of the chemical elements are not arbitrary, but depend on the structure of the atom and vary systematically, i.e., periodically, with the atomic number. The main feature is that elements and compounds exhibit structural, physical, and chemical properties that are remarkably similar, and these properties establish the periodic relations among them.

REFERENCES

1. R. C. Weast (ed.), *CRC Handbook of Chemistry and Physics*, 80th edition, CRC Press, Boca Raton, FL, 1999.
2. S. MARON and C. PRUTTON, *Principles of Physical Chemistry*, 4th edition, MacMillan, New York, 1970.
3. R. Perry and D. Green (Eds.), *Perry's Chemical Engineers' Handbook*, 8th edition, McGraw-Hill, New York, 2008.
4. J. SMITH, H. VAN NESS, and M. ABBOTT, *Introduction to Chemical Engineering Thermodynamics*, 7th edition, McGraw-Hill, New York, 2005.
5. N. CHOPEY, *Handbook of Chemical Engineering Calculations*, 2nd edition, McGraw-Hill, New York, 1994.
6. R. BIRD, W. STEWART, and E. LIGHTFOOT, *Transport Phenomena*, 2nd edition, John Wiley and Sons, Hoboken, NJ, 2002.

NOTE: Additional problems for each chapter are available for all readers at www. These problems may be used for additional review or homework purposes.

Chapter **3**

Gas Laws

Epictetus (Circa A.D. 60)
The appearance of things to the mind is the standard of every action to man.

—That We Ought Not to Be Angry with Mankind. 27

INTRODUCTION

Observations based on physical experimentation often can be synthesized into simple mathematical equations called *laws*. These laws are never perfect and hence are only an approximate representation of reality. There is no universal all-purpose equation of state to describe the PVT (pressure–volume–temperature) behavior of a pure fluid. (An equation of state is an equation that relates pressure, volume, and temperature for any pure fluid.) Since most equations of state deal with gases, this chapter primarily highlights this phase. It should also be noted that liquids and solids exhibit little PVT variations; understandably, engineers and scientists are therefore rarely concerned with the PVT behavior of either of these two phases.

The *ideal gas law* (IGL) was derived from experiments in which the effects of pressure and temperature on gaseous volumes were measured over moderate temperature and pressure ranges. This law works well in the pressure and temperature ranges that were used in taking the data; extrapolations outside of the ranges have been found to work well in some cases and poorly in others. As a general rule, this law works best when the molecules of the gas are far apart, i.e., when the pressure is low and the temperature is high. Under these conditions, the gas is said to behave *ideally*, i.e., its behavior is a close approximation to the so-called *perfect* or *ideal gas*, a hypothetical entity that obeys the ideal gas law exactly. For engineering calculations, the ideal gas law is almost always assumed to be valid since it generally works well (usually within a few percent of the correct result) up to the highest pressures and down to the lowest temperatures used in most thermodynamic applications. Nonetheless, the chapter concludes with an introduction to non-ideal behavior.

Thermodynamics for the Practicing Engineer. By L. Theodore, F. Ricci, and T. Van Vliet
Copyright © 2009 John Wiley & Sons.

BOYLE'S AND CHARLES' LAWS

The two precursors of the ideal gas law were *Boyle's* and *Charles'* laws. Boyle found that the volume of a given mass of gas is inversely proportional to the *absolute* pressure if the temperature is kept constant:

$$P_1 V_1 = P_2 V_2 \tag{3.1}$$

where V_1 = volume of gas at absolute pressure P_1 and temperature T
           V_2 = volume of gas at absolute pressure P_2 and temperature T

Charles found that the volume of a given mass of gas varies directly with the *absolute* temperature at constant pressure:

$$\frac{V_1}{T_1} = \frac{V_2}{T_2} \tag{3.2}$$

where V_1 = volume of gas at pressure P and absolute temperature T_1
           V_2 = volume of gas at pressure P and absolute temperature T_2

Boyle's and Charles' laws may be combined into a single equation in which neither temperature nor pressure need be held constant:

$$\frac{P_1 V_1}{T_1} = \frac{P_2 V_2}{T_2} \tag{3.3}$$

For Equation (3.3) to hold, the mass of gas must be constant as the conditions change from (P_1, T_1) to (P_2, T_2). This equation indicates that for a given mass of a specific gas, PV/T has a constant value. Since, at the same temperature and pressure, volume and mass must be directly proportional, this statement may be extended to

$$\frac{PV}{mT} = C \tag{3.4}$$

where m = mass of a specific gas
           C = constant that depends on the gas
Note that volume terms may be replaced by volume rate (or volumetric flow rate), q.

ILLUSTRATIVE EXAMPLE 3.1

What is the final (f) volumetric flow rate of a gas that is heated at constant pressure from 100 to 300°F if its initial (i) flow is 3500 acfm.

SOLUTION: Apply Charles' law on a volume rate basis. See Equation (3.2) and be sure to employ absolute temperature units:

$$q_f = q_i(T_f/T_c)$$
$$= 3500[(300 + 460)/(100 + 460)]$$
$$= 4750 \, acfm$$

■

ILLUSTRATIVE EXAMPLE 3.2

What is the volumetric flow rate of the gas (100°F, 1 atm) in the previous example if it is compressed isothermally (constant temperature) to 3 atm?

SOLUTION: Apply Boyle's law. See Equation (3.1).

$$q_f = q_i(P_i/P_f)$$
$$= 3500(1.0/3.0)$$
$$= 1167 \, acfm$$

■

ILLUSTRATIVE EXAMPLE 3.3

What is the volumetric flow rate of the gas in the previous example if the final temperature is 300°F.

SOLUTION: Using the combined gas law,

$$q_f = q_i(P_i/P_f)(T_f/T_c)$$
$$= 3500(1/3)(760/560)$$
$$= 1583 \, acfm$$

■

THE IDEAL GAS LAW

Experiments with different gases showed that Equation (3.4) could be expressed in a far more generalized form. If the number of moles (n) is used in place of the mass (m), the constant is the same for all gases:

$$\frac{PV}{nT} = R \qquad (3.5)$$

where R = universal gas constant.

Equation (3.5) is called the ideal gas law. Numerically, the value of R depends on the units used for P, V, T, and n (see Table 3.1). In this text, gases are generally assumed to approximate ideal gas behavior. As is usually the case in engineering practice, unless otherwise stated the ideal gas law is assumed to be valid for all illustrative and assigned

Table 3.1   Values of R in Various Units

R	Temperature scale	Units of V	Units of n	Units of P	Unit of PV (energy)
10.73	°R	ft^3	lbmol	psia	—
0.7302	°R	ft^3	lbmol	atm	—
21.85	°R	ft^3	lbmol	in Hg	—
555.0	°R	ft^3	lbmol	mm Hg	—
297.0	°R	ft^3	lbmol	in H$_2$O	—
0.7398	°R	ft^3	lbmol	bar	—
1545.0	°R	ft^3	lbmol	psfa	—
24.75	°R	ft^3	lbmol	ft H$_2$O	—
1.9872	°R	—	lbmol	—	Btu
0.0007805	°R	—	lbmol	—	hp · h
0.0005819	°R	—	lbmol	—	kW · h
500.7	°R	—	lbmol	—	cal
1.314	K	ft^3	lbmol	atm	—
998.9	K	ft^3	lbmol	mm Hg	—
19.32	K	ft^3	lbmol	psia	—
62.361	K	L	gmol	mm Hg	—
0.08205	K	L	gmol	atm	—
0.08314	K	L	gmol	bar	—
8314	K	L	gmol	Pa	—
8.314	K	m^3	gmol	Pa	—
82.057	K	cm^3	gmol	atm	—
1.9872	K	—	gmol	—	cal
8.3144	K	—	gmol	—	J

problems in Parts I–III, unless otherwise indicated. If a case is encountered in practice where the gas behaves in a very nonideal fashion, e.g., a high-molecular-weight gas (such as a chlorinated organic) under high pressures, one of the many *real gas* correlations found later in this chapter as well as the approaches provided in Chapters 12 and 14 should be used.

Other useful forms of the ideal gas law are shown in Equations (3.6) and (3.7). Equation (3.6) applies to gas volume flow rate rather than to a gas volume confined in a container:[4]

$$Pq = \dot{n}RT \tag{3.6}$$

where q = gas volumetric flow rate (ft^3/h)
P = absolute pressure (psia)
$\dot{n}$ = molar flow rate (lbmol/h)
T = absolute temperature (°R)
R = 10.73 psia · ft^3/lbmol · °R

Equation (3.7) combines n and V from Equation (3.5) to express the law in terms of density:

$$P(MW) = \rho RT \tag{3.7}$$

where MW = molecular weight of gas (lb/lbmol)
    ρ = density of gas (lb/ft^3)

ILLUSTRATIVE EXAMPLE 3.4

1 What is the density of air at 75°F and 14.7 psia? The molecular weight of air is 29.

2 Calculate the volume (in ft^3) of 1.0 lbmol of any ideal gas at 60°F and 14.7 psia.

3 Calculate the density of a gas ($MW = 29$) in g/cm^3 at 20°C and 1.2 atm using the ideal gas law.

SOLUTION: This example is solved using the ideal gas law:

$$PV = nRT = \left(\frac{m}{MW}\right)RT \tag{3.7}$$

1 For the density,

$$\rho = \frac{P(MW)}{RT} = \frac{(14.7\,\text{psia})(29\,\text{lb/lbmol})}{(10.73\,\text{ft}^3 \cdot \text{psi/lbmol} \cdot °\text{R})(75 + 460)}$$

$$= 0.0743\,\text{lb/ft}^3$$

2 Solve the ideal gas law for V and calculate the volume:

$$V = \frac{nRT}{P} = \frac{(1)(10.73)(60 + 460)}{14.7}$$

$$= 379\,\text{ft}^3$$

This result is an important number to remember in many thermodynamic calculations—1 lbmol of any (ideal) gas at 60°F and 1 atm occupies 379 ft^3.

3 Calculate the density of the gas again using the ideal gas law:

$$PV = nRT = \left(\frac{m}{MW}\right)RT \tag{3.7}$$

$$\frac{m}{V} = \rho = \frac{P(MW)}{RT} = \frac{(1.2)(29)}{(82.06)(20 + 273)}$$

$$= 0.00145\,\text{g/cm}^3$$

The effects of pressure, temperature, and molecular weight on density can be obtained directly from the ideal gas law equation. Increasing the pressure and molecular weight increases the density; increasing the temperature decreases the density.   ∎

ILLUSTRATIVE EXAMPLE 3.5

A certain pure-component two-element ideal gas has a specific volume v of 10.58 ft^3/lb at 70°F and 14.7 psia. Determine the molecular weight of the gas and state its name. (Hint: The gas is acidic and soluble in water).

SOLUTION: Rewrite the ideal gas law equation in terms of MW,

$$MW = \left(\frac{m}{V}\right)\frac{RT}{P} = \frac{RT}{vP}$$

Then substitute

$$MW = \frac{(10.73)(460 + 70)}{(10.58)(14.7)}$$

$$= 36.56 \text{ lb/lbmol}$$

It appears that the gas is HCl, i.e., hydrogen chloride.   ∎

STANDARD CONDITIONS

Volumetric flow rates are often not given at the actual conditions of pressure and temperature but rather at arbitrarily chosen standard conditions (*STP, standard temperature and pressure*). To distinguish between flow rates based on the two conditions, the letters *a* and *s* are often used as part of the unit. The units acfm and scfm stand for actual cubic feet per minute and standard cubic feet per minute, respectively. The ideal gas law can be used to convert from *standard* to *actual* conditions, but since there are many standard conditions in use, the STP being used must be known or specified. Standard conditions most often used are shown in Table 3.2. The reader is cautioned on the incorrect use of acfm and/or scfm. The use of standard conditions is a convenience; actual conditions *must* be employed when predicting the performance of or designing equipment. Designs based on standard conditions can lead to disastrous results, with the unit usually underdesigned. For example, for a flue gas stream at

Table 3.2   Common Standard Conditions

System	Temperature	Pressure	Molar volume
SI	273K	101.3 kPa	22.4 m^3/kmol
Universal scientific	0°C	760 mm Hg	22.4 L/gmol
Natural gas industry	60°F	14.7 psia	379 ft^3/lbmol
American engineering	32°F	1 atm	359 ft^3/lbmol
Hazardous waste	60°F	1 atm	379 ft^3/lbmol
incinerator industry	70°F	1 atm	387 ft^3/lbmol

$2140°F$, the ratio of acfm to scfm (standard temperature $= 60°F$) for a thermal application is 5.0.

Equation (3.8), which is a form of Charles' law, can be used to correct flow rates from standard to actual conditions:

$$q_a = q_s(T_a/T_s) \qquad (3.8)$$

where $q_a =$ volumetric flow rate at actual conditions (ft^3/h)
$\quad\quad\ q_s =$ volumetric flow rate at standard conditions (ft^3/h)
$\quad\quad\ T_a =$ actual absolute temperature $(°R)$
$\quad\quad\ T_s =$ standard absolute temperature $(°R)$

The reader is again reminded that absolute temperatures and pressures must be employed in all ideal gas law calculations.

ILLUSTRATIVE EXAMPLE 3.6

Data from a thermal device indicate a volumetric flow rate of 30,000 scfm (60°F, 1 atm). If the operating temperature and pressure of the unit are 1100°F and 1 atm, respectively, calculate the flow rate in actual cubic feet per minute (acfm).

SOLUTION: Since the pressure remains constant, the standard cubic feet per minute using Charles' law is calculated as:

$$q_a = q_s \left(\frac{T_a}{T_s} \right) \qquad (3.2)$$

$$= 30,000 \left(\frac{1100 + 460}{60 + 460} \right)$$

$$= 90,000 \text{ acfm}$$

The reader is again cautioned on the use of acfm and/or scfm. Predicting the performance of and designing equipment should always be based on *actual* conditions.   ■

ILLUSTRATIVE EXAMPLE 3.7

The exhaust gas flow rate from a facility is 1000 scfm. All of the gas is vented through a small stack that has an inlet area of 1.0 ft^2. The exhaust gas temperature is 300°F. What is the velocity of the gas through the stack inlet in feet per second? Assume standard conditions to be 70°F and 1.0 atm. Neglect the pressure drop across the stack.

SOLUTION: The actual flow rate, in acfm, using Charles' law is calculated as:

$$q_a = q_s \left(\frac{T_a}{T_s} \right) = 1000 \left(\frac{460 + 300}{460 + 70} \right)$$

$$= 1434 \text{ acfm}$$

Note that since the gas in vented through the stack to the atmosphere, the pressure is 1.0 atm. The velocity of the gas is calculated as follows:

$$v = \frac{q_a}{A} = \frac{1434}{1.0}$$

$$= 1434 \, \text{ft/min} \qquad \blacksquare$$

PARTIAL PRESSURE AND PARTIAL VOLUME

Mixtures of gases are more often encountered than single or pure gases in engineering practice. The ideal gas law is based on the *number* of molecules present in the gas volume; the *kind* of molecules is not a significant factor, only the number. This law applies equally well to mixtures and pure gases alike. Dalton and Amagat both applied the ideal gas law to mixtures of gases. Since pressure is caused by gas molecules colliding with the walls of the container, it seems reasonable that the total pressure of a gas mixture is made up of pressure contributions due to each of the component gases. These pressure contributions are called *partial pressures*.

Dalton defined the partial pressure of a component as the pressure that would be exerted if the same mass of the component gas occupied the same total volume *alone* at the same temperature as the mixture. The sum of these partial pressures equals the total pressure:

$$P = p_A + p_B + p_C + \cdots + p_n = \sum_{i=1}^{n} p_i \qquad (3.9)$$

where P = total pressure
       n = number of components
       p_i = partial pressure of component i

Equation (3.9) is known as *Dalton's law*. Applying the ideal gas law to one component (*A*) only,

$$p_A V = n_A R T \qquad (3.10)$$

where n_A = number of moles of component *A*.
Eliminating R, T, and V between Equation (3.5) and (3.10) yields

$$\frac{p_A}{P} = \frac{n_A}{n} = y_A$$

or

$$p_A = y_A P \qquad (3.11)$$

where y_A = mole fraction of component *A*.

Amagat's law is similar to Dalton's. Instead of considering the total pressure to be made up of partial pressures where each component occupies the total container volume, Amagat considered the total volume to be made up of partial volumes in which each component is exerting the total pressure. The definition of the *partial volume* is therefore the volume occupied by a component gas alone at the same temperature and pressure as the mixture. For this case:

$$V = V_A + V_B + V_C + \cdots + V_n = \sum_{i=1}^{n} V_i \qquad (3.12)$$

Applying Equation (3.5), as before,

$$\frac{V_A}{V} = \frac{n_A}{n} = y_A \qquad (3.13)$$

where V_A = partial volume of component A.

It is common in engineering practice to describe low concentrations of components in gaseous mixtures in parts per million (ppm) by volume. Since partial volumes are proportional to mole fractions, it is only necessary to multiply the mole fraction of the component by 1 million to obtain the concentration in parts per million. [For liquids and solids, parts per million (ppm) is also used to express concentration, although it is usually on a *mass* basis rather than a *volume* basis. The terms ppmv and ppmw are sometimes used to distinguish between the volume and mass bases, respectively.]

ILLUSTRATIVE EXAMPLE 3.8

The exhaust to the atmosphere from a thermal device has a CO concentration expressed as 0.15 mm Hg partial pressure. Calculate the parts per million of CO in the exhaust.

SOLUTION: First calculate the mole fraction (y). By definition,

$$y_{CO} = p_{CO}/P \qquad (3.11)$$

(Strictly speaking, this equation only applies to ideal gas mixtures). Since the exhaust is discharged to the atmosphere, the atmospheric pressure (760 mm Hg) is the total pressure (P). Thus,

$$y_{CO} = (0.15)/(760) = 1.97 \times 10^{-4}$$

As noted above, partial pressures may be converted to ppm (parts per million) by multiplying by 10^6:

$$\begin{aligned} ppm &= (y_{CO})(10^6) = (1.97 \times 10^{-4})(10^6) \\ &= 197\ ppm \end{aligned}$$

Note: Since the concentration of a *gas* is involved, it is understood (unless otherwise specified) that ppm is on a volume basis. ∎

CRITICAL AND REDUCED PROPERTIES

There are some engineering calculations that require deviations from ideality to be included in the analysis. Many of the non-ideal correlations involve the critical temperature T_c, the critical pressure P_c, and a term defined as the acentric factor ω. An abbreviated list of these properties is presented in Table 3.3. These reduced quantities find wide application in thermodynamic analyses of non-ideal systems.

The critical temperature and pressure are employed in the calculation of the reduced temperature T_r and the reduced pressure P_r, as provided in Equations (3.14) and (3.15):

$$T_r = T/T_c \tag{3.14}$$

$$P_r = P/P_c \tag{3.15}$$

Both reduced properties are dimensionless.

Many physical and chemical properties of elements and compounds can be estimated from models (equations) that are based on the reduced temperature and pressure of the substance in question. These reduced properties have also served as the basis for many equations that are employed in practice to describe non-ideal gas (and liquid) behavior. Although a rigorous treatment of this material is beyond the scope of this text, information is available in the literature.[1–3] Highlights of this topic are presented in the next section.

ILLUSTRATIVE EXAMPLE 3.9

The following data is provided:

Chemical = chlorine gas

Temperature = 230°C

Pressure = 2500 psia

Calculate the reduced temperature and reduced pressure of the chlorine at this state.

SOLUTION: Refer to Table 3.3 to obtain the values of T_c, P_c, and ω:

$$T_c = 417 \text{K}$$
$$P_c = 76 \text{ atm}$$
$$\omega = 0.074$$

Calculate T_r:

$$T = 230 + 273$$
$$= 503 \text{K}$$
$$T_r = 503/417$$
$$= 1.21$$

Table 3.3 Critical Properties

Compound/ Element	Critical temperature, T_c, (K)	Critical pressure, P_c, (atm)	Acentric factor, ω
Acetylene	308.3	60.6	0.814
Ammonia	405.6	111.3	0.250
Benzene	562.1	48.3	0.210
Bromine	584.0	102.0	0.132
i-Butane	408.1	36.0	0.176
n-Butane	425.2	37.5	0.193
1-Butene	419.6	39.7	0.187
Carbon dioxide	304.2	72.8	0.225
Carbon monoxide	132.9	34.5	0.041
Carbon tetrachloride	556.4	45.0	0.193
Chlorine	417.0	76.0	0.074
Chloroform	536.4	54.0	0.214
Chlorobenzene	632.4	44.6	0.255
Ethane	305.4	48.2	0.091
Ethanol	516.2	63.0	0.635
Ethylene	282.4	49.7	0.086
Ethylene oxide	469.0	71.0	0.157
n-Hexane	407.4	29.3	0.296
n-Heptane	540.2	27.0	0.351
Hydrogen	33.2	12.8	0.000
Hydrogen chloride	324.6	82.0	0.266
Hydrogen cyanide	456.8	53.2	0.399
Hydrogen sulfide	373.2	88.2	0.100
Methane	190.6	45.4	0.007
Methanol	512.6	79.9	0.556
Methyl chloride	416.2	65.9	0.158
Nitric oxide (NO)	180.0	64.0	0.600
Nitrogen	126.2	33.5	0.040
Nitrous oxide (N_2O)	309.6	71.5	0.160
n-Octane	568.8	24.5	0.394
Oxygen	154.6	49.8	0.021
i-Pentane	460.4	33.4	0.227
n-Pentane	469.6	33.3	0.251
1-Pentene	464.7	40.0	0.245
Propane	369.8	41.9	0.145
Propylene	365.0	45.6	0.148

(*Continued*)

Table 3.3 *Continued*

Compound/ Element	Critical temperature, T_c, (K)	Critical pressure, P_c, (atm)	Acentric factor, ω
Sulfur	1314.0	116.0	0.070
Sulfur dioxide	430.8	77.8	0.273
Sulfur trioxide	491.0	81.0	0.510
Toluene	591.7	40.6	0.257
Water	647.1	217.6	0.348

Source: A. P. Kudchadker, G. H. Alani, and B. J. Zwolinski, *Chem. Rev.*, 68:659 (1968); J. F. Mathews, *Chem. Rev.*, 72:71 (1972); R. C. Reed and T. K. Sherwood, *The Properties of Liquids*, 2nd edition, McGraw-Hill, New York, 1966; C. A. Passut and R. P. Danner, *Ind. Eng. Chem. Process Des. Develop.*, 12:365 (1974).

Also calculate P_r:

$$P_r = 2500/(14.7)(76)$$
$$= 2.24 \qquad \blacksquare$$

NON-IDEAL GAS BEHAVIOR

No real gas conforms exactly to the ideal gas law, but it can be used as an excellent approximation for most gases at pressures about or less than 3 atm and near ambient temperatures. One of the non-ideal gas laws was that developed by van der Waals. His equation of state takes the form

$$P = [RT/(V - b)] - a/V^2 \qquad (3.16)$$

The two constants a and b are characteristics of the gas and are called *van der Waals constants*. This equation is rarely used today. It is shown here simply as an academic exercise, i.e., as a bridge between the simplest equation of state (ideal gas law) and the more complicated equations that will follow. The reader should note that the terms V and v are used interchangeably in the literature to represent either volume or specific molar volume.

One approach to account for deviations from ideality is to include a "correction factor", Z, which is defined as the *compressibility coefficient* or *compressibility factor*. The ideal gas law is then modified to the following form:

$$PV = ZnRT \qquad (3.17)$$

Note that Z approaches 1.0 as P approaches 0.0. For an ideal gas, Z is exactly unity. This equation may also be written as

$$Pv = ZRT \qquad (3.18)$$

where v is now the *specific* molar volume (not the total volume) with units of volume/ mole.

Equations of state can take many forms. The Virial equation is another equation of state that is used to describe gas behavior. The Virial equation is one of the most important of the non-ideal gas modelling correlations because it is the one upon which many other equations of state are based.[1] Its power series representation is:

$$Z = PV/RT = 1 + B/V + C/V^2 + D/V^3 + \cdots \qquad (3.19)$$

It can also be written as a function of pressure:

$$Z = PV/RT = 1 + B'P + C'P^2 + D'P^3 + \cdots \qquad (3.20)$$

where $B' = B/RT$
$C' = (C - B^2)/(RT)^2$
$\vdots$

In both cases above, B and C are temperature-dependent virial coefficients. Although this power series could be infinitely long, data is generally available only for the second virial coefficient. However, as one increases the number of terms, the accuracy of the equation improves. The two term equation can be used up to a pressure of about 15 atmospheres total pressure; with three terms, it can be used up to 50 atmospheres. Usually two or three terms are sufficient for accuracy.

ILLUSTRATIVE EXAMPLE 3.10

The first and second Virial coefficients for sulfur dioxide at 400K are:

$$B = -0.159 \text{ m}^3/\text{kgmol}$$
$$C = 0.009 \text{ (m}^3/\text{kgmol)}^2$$

Calculate the specific volume of SO_2 in L/gmol at 40 atm and 400K.

SOLUTION: Write the Virial equation. See Equation (3.19).

$$Z = PV/RT = 1 + B/V + C/V^2$$

Insert the appropriate values of the terms and coefficients. Use $R = 0.082$ L · atm/gmol · K = 82.06 cm³ · atm/gmol · K

$$(40)(V)/(0.082)(400) = 1 + (-0.159)/V + (0.009)/V^2$$
$$(1.22)(V) = 1 + (-0.159)/V + (0.009)/V^2$$

Note that the equation cannot simply be explicitly solved for V. A trial-and-error solution is required and any suitable numerical (or analytical) technique may be employed.

Solve for V iteratively.

For ideal gas conditions, $V = 1.0/1.22 = 0.820$ L/gmol. Guess $V = 0.820$ L/gmol, substitute into the right-hand side (RHS) of the equation and calculate an "updated" V from the left-hand side (LHS):

$$\text{for} \quad V_{guess} = 0.820, \quad V_{new} = 0.672$$
$$\text{for} \quad V_{guess} = 0.672, \quad V_{new} = 0.642$$
$$\text{for} \quad V_{guess} = 0.642, \quad V_{new} = 0.635$$

V is approximately 0.635 L/gmol. ∎

Pitzer's correlation[1] is another equation of state. As described earlier, the definition of an equation of state is one that relates pressure, molar or specific volume, and temperature for any pure fluid in an equilibrium state. For a real gas, molecular interactions do exist and the ideal gas law can no longer be applied to calculate an accurate result. One non-ideal correlation that takes this into account is the aforementioned compressibility factor, Z, as provided in Equation (3.19). This term is a function of the acentric factor and both the reduced temperature and pressure terms, which were defined earlier in Equations (3.14) and (3.15) as:

$$T_r = T/T_c$$

$$P_r = P/P_c$$

These two reduced properties and ω (acentric factor) are unique properties of each gas; numerical values were provided in Table 3.3.

The Pitzer correlation equation for the Z is based on the Virial equation discussed earlier:

$$Z = 1 + BP/RT$$
$$= 1 + (BP_c/RT_c)(P_r/T_r)$$

Part of the second term, BP_c/RT_c, is dimensionless and can be evaluated from the equation below:[2]

$$BP_c/RT_c = B^0 + \omega B^1 \tag{3.21}$$

where $B^0 = 0.083 - 0.422/T_r^{1.6}$
$\qquad\quad B^1 = 0.139 - 0.172/T_r^{4.2}$

The above calculational procedure for Z has been defined by one of the authors[3] as the "B" approach. The range of reduced temperatures and pressures for which the

"*B*" approach may be applied is somewhat limited but can be employed in most instances in lieu of other data and/or equations.

Pitzer's correlation, described above, allows for the calculation of the second Virial coefficient. Pitzer's generalized correlation for the compressibility factor is given by

$$Z = z^0 + \omega z^1 \tag{3.22}$$

where both z^0 and z^1 are functions of T_r and P_r. Lee and Kesler[4] provided tables of z^0 and z^1 as functions of T_r and P_r. This information also appeared in the Smith et al. text.[2] Due to the complexity of the tables, Van Vliet and Domato[5] converted the tabular results into equation form. These are provided in Table 3.4.

Table 3.4   Van Vliet and Domato Equations[5] for z^0 and z^1

A. Equations for z^0

If $P_r, T_r > 1$

$$z^0 = -0.168 + 0.00256(T_r)(P_r) + (1.130)\left(1 - \frac{1}{T_r^{1.5}P_r^{1.5}} + \frac{1}{T_r^{0.65}P_r^{0.65}} - \frac{1}{T_r^{0.95}P_r^{0.95}}\right)$$

$$- 0.00000846\left(1 - 0.5\frac{P_r}{T_r} + 0.3\frac{P_r^2}{T_r^2} - 0.5\frac{P_r^3}{T_r^3} + 0.05\frac{P_r^4}{T_r^4} - 0.2\frac{P_r^5}{T_r^5}\right)$$

If $T_r > 1, P_r < 1$

$$z^0 = 1.156 - 0.351e^{-T_r} - 0.0885e^{P_r}$$

If $T_r < 1, P_r > 1$

$$z^0 = -0.198(T_r)(P_r) + 0.137e^{-1/T_r} + 0.308e^{P_r}$$

If $P_r, T_r < 1$

$$z^0 = 0.996 - 0.276(T_r)(P_r) - 0.306((T_r)(P_r))^2$$

B. Equations for z^1

If $P_r, T_r > 1$

$$z^1 = 0.125 - 0.287/T_r^6 + 0.00907(T_r)(P_r) - 0.00000326\left(\frac{P_r}{T_r}\right)^5$$

If $T_r > 1, P_r < 1$

$$z^1 = -0.200 + 0.018(P_r)(T_r) + 0.2\left(1 - 0.2\frac{P_r}{T_r} + \frac{P_r^2}{T_r^2} - \frac{P_r^3}{T_r^3} + \frac{P_r^4}{T_r^4} - \frac{P_r^5}{T_r^5}\right)$$

If $T_r < 1, P_r > 1$

$$z^1 = 0.111(T_r)(P_r) - 0.275e^{-1/T_r} - 0.0989 P_r^{1.125}$$

If $P_r, T_r < 1$

$$z^1 = 0.00161 - 0.0525(T_r)(P_r) - 0.091((T_r)(P_r))^2$$

ILLUSTRATIVE EXAMPLE 3.11

The (P, V, T, n) values of a non-ideal gas are given at three given states:

$$P_1, \quad V_1, \quad T_1, \quad n_1$$
$$P_2, \quad V_2, \quad T_2, \quad n_2$$
$$P_3, \quad V_3, \quad T_3, \quad n_3$$

If it is known that the gas follows the "B" approach at each of the states, outline (do not solve) how to determine the identity of the gas. Is there a unique solution?

SOLUTION: Three pieces of information are required to solve the equations employing the "B" approach—T_r, P_r, and ω. Since three P, V, T values are specified, the problem, in principle, can ordinarily be solved. However, there are no guarantees that the solution will yield one set of T_r, P_r, and ω values that will satisfy the three values specified. Nor can one assume that the solution will be unique. ■

Another equation of state for gases is the Redlich–Kwong equation. It is one of the most widely used for engineering calculations. This is due to the reported accuracy it provides for many gases. The equation, developed in part from the earlier and less accurate Virial equation, is:

$$P = [RT/(V - b)] - a/[T^{1/2}V(V + b)] \tag{3.23}$$

where $\quad a = 0.42748R^2T_c^{2.5}/P_c$
$\quad\quad\quad b = 0.08664RT_c/P_c$

This equation has been verified from actual experimental PVT data.

ILLUSTRATIVE EXAMPLE 3.12

Calculate the molar volume of methane gas (ft^3/lbmol) at 373K and 10 atm. Employ the Redlich–Kwong equation.

SOLUTION: Obtain the critical properties T_c and P_c for methane from Table 3.3.

$$P_c = 45.4\,\text{atm}$$
$$T_c = 190.6\text{K}$$
$$= 343°\text{R}$$

In addition,

$$\omega = 0.007$$

Write the Redlich–Kwong equation in terms of a, b, and V.

$$P = [RT/(V - b)] - a/[T^{1/2}V(V + b)] \tag{3.23}$$
$$T = (373)(1.8)$$
$$= 671°R$$

$$10 = [(0.73)(671)/(V - b)] - a/[671^{1/2}V(V + b)]$$

Calculate the numerical values of the a and b constants.

$$a = (0.42748)(0.73)^2(343)^{2.5}/45.4$$
$$= 10,933$$
$$b = (0.08664)(0.73)(343)/45.4$$
$$= 0.478$$

Solve iteratively for the volume, V, in ft^3.

$$10 = [490/(V - 0.478)] - 10,933/(25.9)(V)(V + 0.478)$$

By trial-and-error,

$$V = 48.8 \text{ ft}^3 \qquad \blacksquare$$

Although the Redlich–Kwong and van der Waals' equations both deal with the critical properties of the gas, Redlich–Kwong takes into account the actual temperature in both terms of the equation. This correlation is popular because very little experimental data is needed to give a better representation of behavior than the ideal gas law.

The Redlich–Kwong equation can also be, and often is, written in an alternative form which is

$$Z = [1/(1 - h)] - (A/B)(h/[1 + h]) \tag{3.24}$$

where $h = b/V = BP/Z$
$B = b/RT$
$A/B = a/bRT$

The additional information needed for the application of an equation of state to non-ideal gas mixtures—beyond what is required for a pure gas—is the composition dependence of the constants in the equation of state. Theoretically, the dependence arises because of the difference between the force fields of unlike molecules. Due to the complexity of these fields, there are no exact theoretical equation(s) that describe these composition dependencies. There are essentially empirical "mixing rules" to estimate these relationships. This is discussed in the next section.

NON-IDEAL MIXTURES

The ideal gas law can be applied directly for ideal gas mixtures. However, the molecular weight of the mixture is based on a mole fraction average of the n components, i.e.,

$$\overline{MW} = \sum_{L=1}^{n} (y_i)(MW_i) \tag{3.25}$$

One approach to account for deviations from ideality is to assume the compressibility coefficient for the mixture, Z_m, is a linear mole fraction combination of the individual component Zs, i.e.,

$$Z_m = \sum y_i Z_i \tag{3.26}$$

Furthermore, Kay[6] has shown that the deviations arising in using this approach can be reduced by employing "pseudo-critical" values for T_c and P_c where

$$\overline{T}_c = \sum y_i T_{ci} \tag{3.27}$$

$$\overline{P}_c = \sum y_i P_{ci} \tag{3.28}$$

In lieu of other information, Theodore[7] suggests employing Equation (3.29) for the pseudo-critical value of ω:

$$\overline{\omega} = \sum y_i \omega_i \tag{3.29}$$

These pseudo-critical values—$\overline{T}_c$, $\overline{P}_c$, and $\overline{\omega}$—are then employed in the appropriate pure component equation of state. This approach has been defined by some as Kay's Rule.

ILLUSTRATIVE EXAMPLE 3.13

At 3000 ft^3/day, a process stream from a petroleum refinery is metered at 350 psia and 200°F. An analysis of the gas mixture gave the following results:

Methane	40% by mole
Ethane	10%
Ethylene	30%
Propane	7%
Propylene	7%
Butane	6%

Calculate the gas volumetric flow rate in ft^3/day at standard conditions (60°F, 1 atm). Assume ω for the mixture is 0.020 and Kay's rule to apply. Account for deviation from ideality using both the "B" approach and the equations provided by Van Vliet and Domato (see Table 3.4).

SOLUTION: The data and calculated results are presented in Table 3.5.

Table 3.5 Gas Mixture Results

	y_i	T_c, K	P_c, atm	$y_i T_{ci}$	$y_i P_{ci}$
C_1	0.4	191	45.8	76.4	18.32
C_2	0.1	306	48.2	30.6	4.82
C_2^-	0.3	282	50.0	84.6	15.00
C_3	0.07	373	42.0	26.1	2.94
C_3^-	0.07	365	45.8	25.6	3.21
C_4	0.06	425	37.5	25.5	2.25
				$\overline{T}_c = \sum = 268.8\text{K}$	$\overline{P}_c = \sum = 46.54\,\text{atm}$

The reduced properties are therefore

$$T_r = \frac{660}{(268.8)(1.8)} = 1.36$$

$$P_r = \frac{350}{(46.54)(14.7)} = 0.511$$

For standard conditions

$$T_r = 1.074$$
$$P_r = 0.021$$

so that $Z \simeq 1.0$.

Employing the "B" approach:

$$B^o = 0.083 - [0.422/(1.36)^{1.6}] = -0.193$$

$$B' = 0.139 - [0.172/(1.36)^{4.2}] = 0.0917$$

$$B^o + \omega B' = -0.193 + (0.2)(0.0917) = -0.175$$

$$Z_A = 1 - 0.175(0.511/1.36) = 0.941$$

Therefore,

$$q_S = q_A(Z_S/Z_A)(T_S/T_A)(P_A/P_S)$$
$$= 3000(1.0/0.941)(520/660)(350/14.7)$$
$$= 59,800\,\text{ft}^3/\text{day}$$

The problem can also be solved using the "Z" approach. First note that $T_r > 1, P_r < 1$. The following equations from Table 3.4 that are given below are to be employed to solve this problem.

$$z^0 = 1.156 - 0.351e^{-T_r} - 0.0885e^{P_r}$$

and

$$z^1 = -0.200 + 0.018(P_r)(T_r) + 0.2\left(1 - 0.2\frac{P_r}{T_r} + \frac{P_r^2}{T_r^2} - \frac{P_r^3}{T_r^3} + \frac{P_r^4}{T_r^4} - \frac{P_r^5}{T_r^5}\right)$$

For this approach,

$$Z = z^0 + \omega z^1$$

Substituting

$$z^0 = 1.156 - 0.351e^{-1.36} - 0.0885e^{0.511}$$

$$z^0 = 0.918$$

$$z^1 = -0.200 + 0.018(0.511)(1.36)$$

$$+ 0.2\left(1 - 0.2\frac{0.511}{1.36} + \frac{0.511^2}{1.36^2} - \frac{0.511^3}{1.36^3} + \frac{0.511^4}{1.36^4} - \frac{0.511^5}{1.36^5}\right)$$

$$z^1 = 0.018$$

$$Z = z^0 + \omega z^1$$

$$= 0.918 + (0.02)(0.018)$$

$$= 0.919$$

Therefore,

$$q_S = q_A(Z_S/Z_A)(T_S/T_A)(P_A/P_S)$$

$$= 3000(1.0/0.919)(520/660)(350/14.7)$$

$$= 62,200 \text{ ft}^3/\text{day}$$

The results of both approaches are in reasonable agreement.   ∎

REFERENCES

1. K. S. PITZER, "*Thermodynamics*," 3rd edition, McGraw-Hill, New York, 1995.
2. J. SMITH, H. VAN NESS, and M. ABBOTT, "*Chemical Engineering Thermodynamics*," 6th edition, McGraw-Hill, New York, 2001.

3. L. THEODORE; personal notes, Manhattan College, 1995.
4. B. LEE and M. KESLER, *AIChE J.*, Vol. 21, 1975.
5. T. VAN VLIET and D. DOMATO, communication to L. Theodore, Manhattan College, 2008.
6. W. KAY, "Density of hydrocarbon gases and vapors," *Incl. Eng. Chem.*, Vol. 28, p. 1014, 1936.
7. L. THEODORE; personal notes, Manhattan College, 1996.

NOTE: Additional problems for each chapter are available for all readers at www. These problems may be used for additional review or homework purposes.

Chapter **4**

Conservation Laws

Old Testament

TEKEL; Thou art weighed in the balances, and art found wanting.

—Daniel. V, 27

Sir Arthur Stanley Eddington [1882–1944]

It is one thing for the human mind to extract from the phenomena of nature the laws which it has itself put into them; it may be a far harder thing to extract laws over which it has no control. It is even possible that laws which have not their origin in the mind may be irrational, and we can never succeed in formulating them.

—Space, Time, and Gravitation [1920]

INTRODUCTION

In order to better understand thermodynamics, it is necessary to first understand the theory underlying this science. How can one predict what products will be emitted from effluent streams? At what temperature must a thermal device be operated? How much energy in the form of heat is given off during a combustion process? Is it economically feasible to recover this heat? Is the feed high enough in heating value, or must additional fuel be added to assist in the thermal process? If so, how much fuel must be added? The answers to these questions are rooted in the various theories of thermodynamics, including thermochemistry, phase equilibrium, and chemical reaction equilibrium.

One of the keys necessary to answer the above questions is often obtained via the application of one or more of the conservation laws, and the contents of this chapter deal with these laws. Topics covered include:

1 The Conservation Laws

2 The Conservation Law for Momentum

Thermodynamics for the Practicing Engineer. By L. Theodore, F. Ricci, and T. Van Vliet
Copyright © 2009 by John Wiley & Sons, Inc.

3 The Conservation Law for Mass

4 The Conservation Law for Energy

Obviously, the heart of the chapter is topic **4**, and to a lesser extent topic **3**. However, some momentum considerations come into play in the last part of this book.

Four important terms are defined below before proceeding to the conservation laws.

1 A *system* is any portion of the universe that is set aside for study.

2 Once a system has been chosen, the rest of the universe is referred to as the *surroundings*.

3 A *system* is described by specifying that it is in a certain state.

4 The *path*, or series of values certain variables assume in passing from one state to another, defines a process.

THE CONSERVATION LAWS

Momentum, energy, and mass are all conserved. As such, each quantity obeys the conservation law below as applied within a system.

$$\left\{ \begin{array}{c} \text{quantity} \\ \text{into} \\ \text{system} \end{array} \right\} - \left\{ \begin{array}{c} \text{quantity} \\ \text{out of} \\ \text{system} \end{array} \right\} + \left\{ \begin{array}{c} \text{quantity} \\ \text{generated} \\ \text{in system} \end{array} \right\} = \left\{ \begin{array}{c} \text{quantity} \\ \text{accumulated} \\ \text{in system} \end{array} \right\} \quad (4.1)$$

This equation may also be written on a *time* rate basis:

$$\left\{ \begin{array}{c} \text{rate of} \\ \text{quantity} \\ \text{into} \\ \text{system} \end{array} \right\} - \left\{ \begin{array}{c} \text{rate of} \\ \text{quantity} \\ \text{out of} \\ \text{system} \end{array} \right\} + \left\{ \begin{array}{c} \text{rate of} \\ \text{quantity} \\ \text{generated} \\ \text{in system} \end{array} \right\} = \left\{ \begin{array}{c} \text{rate of} \\ \text{quantity} \\ \text{accumulated} \\ \text{in system} \end{array} \right\} \quad (4.2)$$

The conservation law may be applied at the macroscopic, microscopic, or molecular level. One can best illustrate the differences in these methods with an example. Consider a system in which a fluid is flowing through a cylindrical tube (see Fig. 4.1). Define the system as the fluid contained within the tube between points 1 and 2 at any time. If one is interested in determining changes occurring at the inlet and outlet of the system, the conservation law is applied on a "macroscopic" level to the entire system. The resultant equation describes the overall changes occurring *to* the system without regard for internal variations *within* the system. This approach is usually applied by the practicing engineer. The microscopic

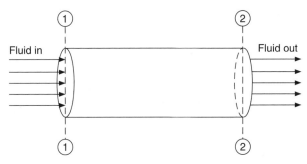

Figure 4.1 Conservation law example.

approach is employed when detailed information concerning the behavior *within* the system is required, and this is often requested of and by the engineer or scientist. The conservation law is then applied to a *differential* element within the system which is large compared to an individual molecule, but small compared to the entire system. The resultant equation is then expanded, via an integration, to describe the behavior of the entire system. This is defined by some as the *transport phenomena approach*. The molecular approach involves the application of the conservation law to individual molecules. This leads to a study of statistical and quantum mechanics—both of which are beyond the scope of this text. In any case, the description of individual molecules at the molecular level is of little value to the practicing engineer. However, the statistical averaging of molecular quantities in either a differential or finite element within a system leads to a more meaningful description of the behavior of a system. The macroscopic approach is adopted and applied in this text, and no further reference to microscopic or molecular analyses will be made. This chapter's aim, then, is to express the laws of conservation for momentum, energy, and mass in algebraic or finite difference form.

It should be noted that the applied mathematician has developed differential equations describing the detailed behavior of systems by applying the appropriate conservation law to a differential element or shell within the system. Equations were derived with each new application. The engineer later removed the need for these tedious and error-prone derivations by developing a general set of equations that could be used to describe systems. These came to be defined as the aforementioned *transport equations*.[1,2] Needless to say, these transport equations have proven an asset in describing the behavior of some systems, operations, and processes.

This text departs from the approach of developing these differential equations even though this method has a great deal to commend it. Experience has indicated that the engineer possessing a working knowledge of the conservation laws is likely to obtain a more integrated and unified picture of thermodynamics by developing the equations in algebraic form.

THE CONSERVATION LAW FOR MOMENTUM

The general conservation law for momentum on a rate basis is first applied to a volume element:

$$
\underbrace{\left\{\begin{array}{c}\text{rate of} \\ \text{momentum} \\ \text{in by} \\ \text{convection}\end{array}\right\}}_{(1)} - \underbrace{\left\{\begin{array}{c}\text{rate of} \\ \text{momentum} \\ \text{out by} \\ \text{convection}\end{array}\right\}}_{(2)} + \underbrace{\left\{\begin{array}{c}\text{rate of} \\ \text{momentum} \\ \text{in by} \\ \text{molecular} \\ \text{diffusion}\end{array}\right\}}_{(3)} - \underbrace{\left\{\begin{array}{c}\text{rate of} \\ \text{momentum} \\ \text{out by} \\ \text{molecular} \\ \text{diffusion}\end{array}\right\}}_{(4)}
$$

$$
\downarrow\downarrow\downarrow\downarrow\downarrow\downarrow\downarrow \qquad\qquad \downarrow\downarrow\downarrow\downarrow\downarrow\downarrow\downarrow
$$
$$
\text{bulk flow} \qquad\qquad \text{velocity gradients}
$$

$$
+ \underbrace{\left\{\begin{array}{c}\text{external forces} \\ \text{exerted on fluid}\end{array}\right\}}_{(5)} = \underbrace{\left\{\begin{array}{c}\text{rate of momentum} \\ \text{accumulation}\end{array}\right\}}_{(6)} \tag{4.3}
$$

$$
\downarrow\downarrow\downarrow\downarrow \qquad\qquad \downarrow\downarrow\downarrow\downarrow
$$
$$
\text{(a)\quad surface force} \qquad\qquad \text{inventory}
$$

$$
\text{(b)\quad body force}
$$

The above enumerates the rate of momentum and forces acting on the moving fluid in a volume element of concern at any time t. Each rate of momentum or force term in the above equation can be expressed in units of lb_f in order to maintain dimensional consistency. It is suggested that the reader refers to the literature for more information.[1,2]

The application of this conservation law finds extensive use in the field of fluid mechanics or fluid flow.[3] Applications in the field of thermodynamics are limited.

ILLUSTRATIVE EXAMPLE 4.1

A 10-cm-diameter horizontal line carries saturated steam at 420 m/s. Water is entrained by the steam at a rate of 0.15 kg/s. The line has a 90° bend. Calculate the force components in the horizontal and vertical directions required to hold the bend in place due to the entrained water.

SOLUTION: A line diagram of the system is provided in Fig. 4.2. Select the fluid in the bend as the system and apply the conservation law for mass (see next Section):

$$
\dot{m}_1 = \dot{m}_2
$$

Since the density and cross-sectional area are constant,

$$
v_1 = v_2
$$

where $\dot{m}_1, \dot{m}_2$ = mass flowrate at 1 and 2, respectively
v_1, v_2 = velocity at 1 and 2, respectively

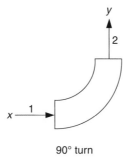

90° turn

Figure 4.2   Diagram for Illustrative Example 4.1.

A linear momentum ($\dot{M}$) balance in the horizontal direction provides the force applied by the channel wall on the fluid in the x-direction, F_x:

$$F_x g_c = \dot{M}_{x,\text{out}} - \dot{M}_{x,\text{in}}$$

$$= \frac{d}{dt}(mv)_{x,\text{out}} - \frac{d}{dt}(mv)_{x,\text{in}}$$

Note that the equation assumes that the pressure drop across the bend is negligible.
Since $v_{x,\text{out}} = 0$ and $dm/dt = \dot{m}$,

$$F_y g_c = 0 - \dot{m}v_{x,\text{in}} = -\frac{(0.15)(420)}{(1)} = -63\,\text{N} = -14.1\,\text{lb}_f$$

The x-direction supporting force acting on the 90° elbow is 14.1 lb$_f$ acting toward the left.
A linear momentum balance in the vertical direction results in

$$F_y g_c = \dot{M}_{y,\text{out}} - \dot{M}_{y,\text{in}}$$

$$= \dot{m}v_{y,\text{out}} - \dot{m}v_{y,\text{in}}$$

$$= \dot{m}v_2 - 0 = (0.15)(420) = 63\,\text{N} = 14.1\,\text{lb}_f$$

The y-direction supporting free force on the 90° elbow is 14.1 lb$_f$ acting upwards.   ■

ILLUSTRATIVE EXAMPLE 4.2

Refer to Illustrative Example 4.1. Calculate the magnitude and direction of the resultant force.

SOLUTION: The resultant supporting force is given by:

$$F_{\text{res}} = \sqrt{F_x^2 + F_y^2}$$

Substituting,

$$F_{\text{res}} = \sqrt{(-63)^2 + 63^2} = 89.1\,\text{N} = 19.1\,\text{lb}_f$$

The direction is given by

$$\tan \theta = \frac{F_y}{F_x} = \frac{63}{-63} = -1$$

$$\theta = 135°$$

where θ is the angle between the positive x axis and the direction of the force. The counter-clockwise rotation of the direction from the x axis is defined as positive.

The supporting force is therefore 19.1 lb_f acting in the "northwest" direction.   ∎

THE CONSERVATION LAW FOR MASS

The *conservation law* for mass can be applied to any process or system. The general form of this law is given as:

$$\text{mass in} - \text{mass out} + \text{mass generated} = \text{mass accumulated} \qquad (4.4)$$

or, on a time rate basis, by

$$\left\{ \begin{array}{c} \text{rate of} \\ \text{mass in} \end{array} \right\} - \left\{ \begin{array}{c} \text{rate of} \\ \text{mass out} \end{array} \right\} + \left\{ \begin{array}{c} \text{rate of mass} \\ \text{generated} \end{array} \right\} = \left\{ \begin{array}{c} \text{rate of mass} \\ \text{accumulated} \end{array} \right\} \qquad (4.5)$$

In thermodynamic-related processes, it is often necessary to obtain quantitative relationships by writing mass balances on the various elements in the system. This equation may be applied either to the total mass involved or to a particular species on either a mole or mass basis. This law can be applied to steady-state or unsteady-state (transient) processes and to batch or continuous systems. As noted earlier, in order to isolate a system for study, it is separated from the surroundings by a boundary or envelope. This boundary may be real (e.g., the walls of a thermal device) or imaginary. Mass crossing the boundary and entering the system is part of the *mass in* term in Equation (4.5), while that crossing the boundary and leaving the system is part of the *mass out* term. Equation (4.5) may be written for any compound whose quantity is not changed by chemical reaction and for any chemical element whether or not it has participated in a chemical reaction. (This is treated in more detail in the next chapter.) It may be written for one piece of equipment, around several pieces of equipment, or around an entire process. It may be used to calculate an unknown quantity directly, to check the validity of experimental data, or to express one or more of the independent relationships among the unknown quantities in a particular problem situation.

A *steady-state* process is one in which there is no change in conditions (pressure, temperature, composition, etc.) or rates of flow with time at any given point in the system. The accumulation term in Equation (4.5) is then zero. (If there is no chemical or nuclear reaction, the generation term is also zero.) All other processes are *unsteady state*.

In a *batch* process, a given quantity of reactants is placed in a container, and by chemical and/or physical means, a change is made to occur. At the end of the process, the container (or containers) to which material may have been transferred holds the product or products. In a *continuous* process, reactants are continuously fed to a piece of equipment or to several pieces in series, and products are continuously removed from one or more points. A continuous process may or may not be steady state. A coal-fired power plant, for example, operates continuously. However, because of the wide variation in power demand between peak and slack periods, there is an equally wide variation in the rate at which the coal is fired. For this reason, power plant problems may require the use of average data over long periods of time. However, most thermal operations are assumed to be steady state and continuous.

As indicated previously, Equation (4.5) may be applied to the total mass of each stream (referred to as an *overall* or *total material balance*) or to the individual component(s) of the stream (referred to as a *componential* or *component material balance*). The primary task in preparing a material balance in thermodynamics calculations is to often develop the quantitative relationships among the streams. The primary factors, therefore, are those that *tie* the streams together. An element, compound, or unreactive mass (e.g., ash) that enters or exits in a single stream or passes through a process unchanged is so convenient for this purpose that it may be considered a *key* to the calculations. If sufficient data are given about this component, it can be used in a component balance to determine the total masses of the entering and exiting streams. Such a component is sometimes referred to as a *key component*. Since a key component does not react in a process, it must retain its identity as it passes through the process. Obviously, except for nuclear reactions, elements may always be used as key components because they do not change identity even though they may undergo a chemical reaction. Thus CO (carbon monoxide) may be used as a key component only when it does not react, but C (carbon) may always be used as a key component. A component that enters the system in only one stream and leaves in only one stream is usually the most convenient choice for a key component.

Four important processing concepts are *bypass*, *recycle*, *purge*, and *makeup*. With *bypass*, part of the inlet stream is diverted around the equipment to rejoin the (main) stream after the unit (see Fig. 4.3). This stream effectively moves in parallel with the stream passing through the equipment. In *recycle*, part of the product stream is sent back to mix with the feed. If a small quantity of nonreactive material is present in the feed to a process that includes recycle, it may be necessary to remove the

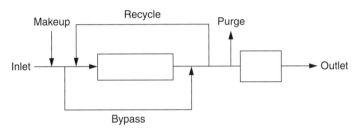

Figure 4.3   Recycle, bypass, and purge.

nonreactive material in a *purge* stream to prevent its building up above a maximum tolerable value. This can also occur in a process without recycle; if a nonreactive material is added in the feed and not totally removed in the products, it will accumulate until purged. The purging process is sometimes referred to as *blowdown*. *Makeup*, as its name implies, involves adding or making up part of a stream that has been removed from a process. Makeup may be thought of as the opposite of purge and/or blowdown.

ILLUSTRATIVE EXAMPLE 4.3

Fuel is fed into a boiler at a rate of 10,000 lb/h in the presence of 20,000 lb/h of air. Due to the low heating value of the fuel, 2000 lb/h of methane is added to assist in the combustion of the fuel. At what rate (lb/h) do the product gases exit the incinerator?

SOLUTION: Apply the conservation law for mass to the boiler. Assume steady-state conditions to apply:

$$\text{Rate of mass in } (\dot{m}_{in}) = \text{rate of mass out } (\dot{m}_{out})$$

Substituting

$$\dot{m}_{in} = (10{,}000 + 20{,}000 + 2000)$$
$$= 32{,}000 \, \text{lb/h}$$

Therefore,

$$\dot{m}_{out} = 32{,}000 \, \text{lb/h}. \qquad \blacksquare$$

ILLUSTRATIVE EXAMPLE 4.4

C_6H_5Cl is fed into a thermal oxidizer at a rate of 5000 scfm (60°F, 1 atm) and is combusted in the presence of air fed at a rate of 3000 scfm (60°F, 1 atm). Both streams enter the oxidizer at 70°F. The products are cooled from 2000°F and exit a cooler at 180°F. At what rate (lb/h) do the products exit the cooler? The molecular weight of C_6H_5Cl is 112.5; the molecular weight of air is 29.

SOLUTION: First convert scfm to acfm using Charles' law:

$$5000 \, \text{scfm} \left(\frac{460 + 70}{460 + 60} \right) = 5096 \, \text{acfm of } C_6H_5Cl$$

$$3000 \, \text{scfm} \left(\frac{460 + 70}{460 + 60} \right) = 3058 \, \text{acfm of air}$$

One pound · mole of any ideal gas occupies 387 ft³ at 70°F and 1 atm. Therefore, the molar flow rate ($\dot{n}$) may be calculated by dividing these results by 387:

$$\dot{n}(C_6H_5Cl) = \frac{5096}{387}$$
$$= 13.17 \text{ lbmol/min}$$

$$\dot{n}(\text{air}) = \frac{3058}{387}$$
$$= 7.90 \text{ lbmol/min}$$

The mass flow rate is obtained by multiplying these results by the molecular weight.

$$\dot{m}(C_6H_5Cl) = (13.17)(112.5)(60)$$
$$= 88{,}898 \text{ lb/h}$$
$$\dot{m}(\text{air}) = (7.90)(29)(60)$$
$$= 13{,}746 \text{ lb/h}$$

Since mass is conserved, $\dot{m}_{\text{in}}$ is equal to $\dot{m}_{\text{out}}$:

$$\dot{m}_{\text{in.total}} = 88{,}898 + 13{,}746$$
$$= 102{,}644 \text{ lb/h} \qquad\blacksquare$$

ILLUSTRATIVE EXAMPLE 4.5

A proposed absorber design requires that a packed column and a spray tower are to be used in series for the removal of HCl from a gas. The spray tower is operating at an efficiency of 65% and the packed column at an efficiency of 98%. Calculate the mass flow rate of HCl leaving the spray tower, the mass flow rate of HCl entering the packed tower, and the overall fractional efficiency of the removal system if 76.0 lb of HCl enters the system every hour.

SOLUTION: By definition

$$E = (\dot{m}_{\text{in}} - \dot{m}_{\text{out}})/\dot{m}_{\text{in}}$$
$$\dot{m}_{\text{out}} = (1 - E)(\dot{m}_{\text{in}})$$

For the spray tower:

$$\dot{m}_{\text{out}} = (1 - 0.65)(76.0)$$
$$= 26.6 \text{ lb/h HCl}$$

Note that the mass flow rate of HCl leaving spray tower equals the mass flow rate HCl entering packed column.

For the packed column:

$$\dot{m}_{out} = (1 - 0.98)(26.6)$$
$$= 0.532 \, lb/h \, HCl$$

The overall fractional efficiency is therefore

$$E = (\dot{m}_{in} - \dot{m}_{out})/\dot{m}_{in}$$
$$= (76.0 - 0.532)/76.0$$
$$= 0.993$$ ∎

ILLUSTRATIVE EXAMPLE 4.6

Refer to Illustrative Example 4.5. Calculate the overall efficiency on a percent basis.

SOLUTION: By definition

$$\%(percent) \, basis = (fraction \, basis)100$$

Thus the overall efficiency on a percent basis is

$$(0.993)100 = 99.3\%$$

∎

ILLUSTRATIVE EXAMPLE 4.7

Consider the flow diagram in Fig. 4.4 for a wastewater treatment system. The following flowrate data are given:

$$\dot{m}_1 = 1000 \, lb/min$$
$$\dot{m}_2 = 1000 \, lb/min$$
$$\dot{m}_4 = 200 \, lb/min$$

Find the amount of water lost by evaporation in the operation, $\dot{m}$.

SOLUTION: Apply a material balance around the treatment system to determine the value of $\dot{m}_5$. The value of $\dot{m}_5$ is given by:

$$\dot{m}_4 + \dot{m}_5 = \dot{m}_3$$
$$\dot{m}_4 + \dot{m}_5 = \dot{m}_1 + \dot{m}_2$$
$$200 + \dot{m}_5 = 1000 + 1000$$
$$\dot{m}_5 = 1800 \, lb/min$$

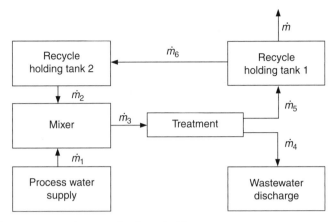

Figure 4.4   Flow diagram for Illustrative Example 4.7.

Similarly (for tank 2),

$$\dot{m}_6 = \dot{m}_2$$
$$\dot{m}_6 = 1000\,\text{lb}/\text{min}$$

Thus (for tank 1),

$$\dot{m}_5 + \dot{m} = \dot{m}_6$$
$$1800 - \dot{m} = 1000$$
$$\dot{m} = 800\,\text{lb}/\text{min}$$

One sees that 800 lb of water per minute are lost in the operation.   ■

ILLUSTRATIVE EXAMPLE 4.8

Consider the system shown in Fig. 4.5.

The following volumetric flowrate and phosphate concentration (volume basis) data have been provided by the plant manager. Are the data correct and/or consistent?

$$
\begin{array}{ll}
q_1 = 1000\ \text{gal}/\text{day} & C_1 = 4\ \text{ppm} \\
q_2 = 1000\ \text{gal}/\text{day} & C_2 = 0\ \text{ppm} \\
q_3 = 2000\ \text{gal}/\text{day} & C_3 = 2\ \text{ppm} \\
q_4 = 200\ \text{gal}/\text{day} & C_4 = 20\ \text{ppm} \\
q_5 = 1800\ \text{gal}/\text{day} & C_5 = 0\ \text{ppm} \\
q_6 = 1000\ \text{gal}/\text{day} & C_6 = 0\ \text{ppm}
\end{array}
$$

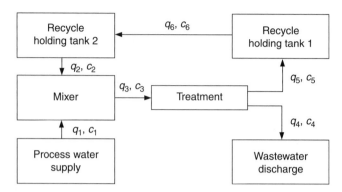

Figure 4.5   Flow diagram for Illustrative Example 4.8.

SOLUTION: A componential balance around the mixer (in lb) gives (with the conversion factor 120,000 gal/10^6 lb)

$$C_1 q_1 + C_2 q_2 = C_3 q_3$$

$$\left(\frac{4}{120,000}\right)(1000) + \left(\frac{0}{120,000}\right)(1000) = \left(\frac{2}{120,000}\right)(2000)$$

$$4000 = 4000 \quad \text{OK}$$

A balance around the treatment tank gives

$$C_3 q_3 = C_4 q_4 + C_5 q_5$$

$$\left(\frac{2}{120,000}\right)(2000) = \left(\frac{20}{120,000}\right)(200) + \left(\frac{0}{120,000}\right)(1800)$$

$$4000 = 4000 \quad \text{OK}$$

A balance around hold tank 1 gives

$$C_5 q_5 = C_6 q_6$$
$$(0)(1800) = (0)(1000)$$
$$0 = 0 \quad \text{OK}$$

A balance around hold tank 2 gives

$$C_2 q_2 = C_6 q_6$$
$$(0)(1000) = (0)(1000)$$
$$0 = 0 \quad \text{OK}$$

The data appear to be consistent.   ∎

THE CONSERVATION LAW FOR ENERGY

From the early recognition of energy, man has studied its effects upon objects, its transfer from object to object, and its transformation from one form to another. This field of study is an integral part of thermodynamics. Before proceeding to the first law of thermodynamics, often referred to as the conservation law for energy, certain important terms are defined as follows:

1 *Isothermal* means constant temperature.

2 *Isobaric* means constant pressure.

3 *Isochoric* is constant volume.

4 *Adiabatic* specifies no transfer of heat to or from systems.

The first law of thermodynamics specifies that energy is conserved. In effect, this law states that energy is neither created nor destroyed. Thus, the change in energy of a system is exactly equal to the negative of the change in the surroundings. For a system of constant mass (a closed system), the only way the system and surroundings may interchange energy is by work and heat. Work and heat are defined as energy in transit. They are not properties and cannot be stored in a system. Two common forms of work are expansion and electrical. Heat is energy in transit because of a temperature difference. This heat transfer may take place by conduction, convection, or radiation.

The energy balance makes use of the conservation law to account for all the energy in a chemical process, or in any other process for that matter. After a system is defined, the energy balance considers the energy entering the system across the boundary, the energy leaving the system across the boundary, and the accumulation of energy within the system. This may be written in a simplified equation form as

$$\text{Energy in} - \text{Energy out} = \text{Energy accumulation} \qquad (4.6)$$

This expression has the same form as the general law of conservation of mass as well as the conservation law for momentum. It may also be written on a rate basis.

All forms of energy must be included in an energy balance. In many processes, certain energy forms remain constant and changes in them may be neglected. However, these forms should be recognized and understood before their magnitude and constancy can be determined. Some forms of energy are easily recognized in everyday life: the energy of a moving object, the energy given off by a fire, and the energy content of a container of hot water. Other forms of energy are less easily recognized. However, the five key energy terms are kinetic, potential, internal, heat, and work. These are briefly described below.

1 *Kinetic energy.* The energy of a moving object is called "kinetic energy." A baseball thrown by a pitcher possesses a definite kinetic energy as it travels toward the catcher. A pound of flowing fluid possesses kinetic energy as it travels through a duct.

2 *Potential energy.* The energy possessed by a mass by virtue of its position in the Earth's gravitational field is called "potential energy." A boulder lying at the top

of a cliff possesses potential energy with reference to the bottom of the cliff. If the boulder is pushed off the cliff, its potential energy is transformed into kinetic energy as it falls. Similarly, a mass of fluid in a flowing system possesses a potential energy because of its height above an arbitrary reference level.

3 *Internal energy.* The component molecules of a substance are constantly moving within the substance. This motion imparts internal energy to the material. The molecules may rotate, vibrate, or migrate within the substance. The addition of heat to a material increases its molecular activity and hence, its internal energy. The temperature of a material is a measure of its internal energy.

4 *Heat.* As noted above, when energy if transferred between a system and its surroundings, it is transferred either as work or as heat. Thus, heat is energy in transit. This type of energy transfer occurs whenever a hot body is brought into contact with a cold body. Energy flows as heat from the hot body to the cold body until the temperature difference is dissipated—i.e., until thermal equilibrium is established. For this reason, heat may be considered as energy being transferred due to a temperature difference.

5 *Work.* Work is also energy in transit. Work is done whenever a force acts through a distance.

The first law of thermodynamics may be stated formally—as opposed to equation form—in many ways. One of these is as follows: although energy assumes many forms, the total quantity of energy is constant, and when energy disappears in one form, it must appear simultaneously in other forms.

As noted earlier, application of the conservation law for energy gives rise to the first law of thermodynamics. This law, in steady-state equation form for batch and flow processes, is presented here.

For *batch* processes:

$$\Delta U = Q + W \tag{4.7}$$

For flow processes:

$$\Delta H = Q + W_s \tag{4.8}$$

where potential, kinetic, and other energy effects have been neglected and

Q = energy in the form of heat transferred across the boundaries of the system

W = energy in the form of work transferred across the boundaries of the system

W_s = energy in the form of mechanical work transferred across the boundaries of the system

U = internal energy of the system

H = enthalpy of the system (defined below)

$\Delta U, \Delta H$ = changes in the internal energy and enthalpy, respectively, during the process

Contrary to an earlier convention, both Q and W (or W_s) are considered positive if added/transferred to the system. Also note that for a flow process

$$\Delta H = Q \tag{4.9}$$

if $W_s = 0$.

The internal energy and enthalpy in Equations (4.7) and (4.8), as well as the other equations in this section, may be on a *mass* basis (i.e., for 1 gal or 1 lb of material), on a *mole* basis (i.e., for 1 gmol or 1 lbmol of material), or represent the total internal energy and enthalpy of the entire system. As long as these equations are dimensionally consistent, it makes no difference. For the sake of clarity, the same convention i.e., used for the heat capacities will be employed throughout this text—uppercase letters (e.g., H, U, C_P) represent properties on a mole basis, while lowercase letters (e.g., h, e, c_P) represent properties on a mass basis. Properties for the entire system will rarely be used and therefore require no special symbols.

Perhaps the most important thermodynamic function the engineer works with is the aforementioned *enthalpy*. The enthalpy is defined by:

$$H = U + PV \tag{4.10}$$

where $P =$ pressure of the system
      $V =$ volume of the system

ILLUSTRATIVE EXAMPLE 4.9

If an 1800 kg car traveling at 40 km/h is stopped by applying a braking force of 5 kN, what is the distance the car will travel before it comes to a stop?

SOLUTION: Work may be defined as:

$$W = \int F \cdot ds = \int_0^s F \, ds$$

where $F =$ the force, e.g., N
      $ds =$ the differential displacement distance, e.g., m

If the force and the displacement are collinear (acting in the same direction), the above integral simplifies to force times distance. Furthermore, it they both act in the same direction, the term is positive; otherwise, it is negative (such as in the case of friction). Also note that a braking force is a friction force. Therefore, it will appear as a negative quantity when representing work in going from a moving state to a state of rest.

The kinetic energy at state 1 (before braking) is

$$KE_1 = \frac{1}{2} mv^2$$

$$KE_1 = \frac{1}{2}(1800 \, \text{kg}) \left[\left(\frac{40{,}000 \, \text{m}}{\text{hr}} \right) \left(\frac{1 \, \text{hr}}{3600 \, \text{s}} \right) \right]^2 = 111 \, \text{kJ}$$

The work in braking is

$$W_{1\to2} = -Fs$$
$$W_{1\to2} = -(5000)s$$

The total energy at state 2 is

$$KE_2 = 0$$

The total system energy balance is then given by

$$KE_1 - W_{1\to2} = KE_2$$

Substitute and solve for the distance

$$111 \times 10^3 - 5000s = 0$$
$$s = 22.2\,\text{m}$$ ■

ILLUSTRATIVE EXAMPLE 4.10

A process plant pumps 2000 lb of water to an elevation of 1200 ft above the turbogenerators. Determine the change in potential energy in Btu.

SOLUTION: By definition, potential energy change (ΔPE) is given by

$$\Delta PE = \frac{mg}{g_c}(\Delta Z) \tag{4.11}$$

where m = mass, lb
   g = acceleration due to gravity, 32.2 ft/s^2 at sea level
   g_c = gravitational constant, 32.2 lb · ft/(lb$_f$ · s^2)
   ΔZ = change in height

Substituting the data yields

$$PE = (2000\,\text{lb})\left[\frac{32.2\,\text{ft/s}^2}{32.2\,\text{lb} \cdot \text{ft}/(\text{lb}_f \cdot \text{s}^2)}\right]600\,\text{ft} = 1.2 \times 10^6 \text{ ft} \cdot \text{lb}_f$$

Since 1 Btu = 778.17 ft · lb$_f$,

$$PE = (1.2 \times 10^6 \text{ ft} \cdot \text{lb}_f)(1\,\text{Btu}/778.17\,\text{ft} \cdot \text{lb}_f) = 1543\,\text{Btu}$$ ■

ILLUSTRATIVE EXAMPLE 4.11

If 2000 lb of water has its velocity increased from 8 to 30 ft/s, calculate the change in kinetic energy of the water in Btu and ft · lb$_f$, and the minimum energy required to accomplish this change.

SOLUTION: By definition, the kinetic energy (KE) is given by

$$KE = \frac{mv^2}{2g_c} \tag{4.12}$$

where m = mass, lb
 v = velocity of water flow
 g_c = gravitational constant, lb · ft/lb$_f$ · s^2

This equation permits one to evaluate the energy possessed by a body of mass m, and having a velocity v, relative to a stationary reference; it is customary to use the Earth as the reference.
 The kinetic energy of the body initially is

$$KE_1 = \frac{(2000\,\text{lb})(8\,\text{ft/s})^2}{2[32.2\,\text{ft} \cdot \text{lb}/(\text{s}^2 \cdot \text{lb}_f)]} = 1989\,\text{ft} \cdot \text{lb}_f$$

The kinetic energy at its terminal velocity of 30 ft/s is

$$KE_2 = \frac{(2000\,\text{lb})(30\,\text{ft/s})^2}{2[32.2\,\text{ft} \cdot \text{lb}/(\text{s} \cdot \text{lb}_f)]} = 27,972\,\text{ft} \cdot \text{lb}_f$$

The kinetic energy change or difference, ΔKE, is then

$$\Delta KE = 1989 - 27,972 = -25,983\,\text{ft} \cdot \text{lb}_f$$

Converting the answer to Btu yields

$$\Delta KE = (-25,983\,\text{ft} \cdot \text{lb}_f)(1\,\text{Btu}/778.17\,\text{ft} \cdot \text{lb}_f) = -33.390\,\text{Btu} \qquad ∎$$

ILLUSTRATIVE EXAMPLE 4.12

A lake is located at the top of a mountain. A power plant has been constructed at the bottom of the mountain. The potential energy of the water traveling downhill can be used to spin turbines and generate electricity. This is the operating mode in the daytime during peak electrical demand. At night, when demand is reduced, the water is pumped back up the mountain. The operation is shown in Fig. 4.6.
 Using the method of power "production" described above, determine how much power (Watts) is generated by the lake located at an elevation of 3000 ft above the power plant. The flowrate of water is 500,000 gpm. The turbine efficiency is 30%. Neglect friction effects.

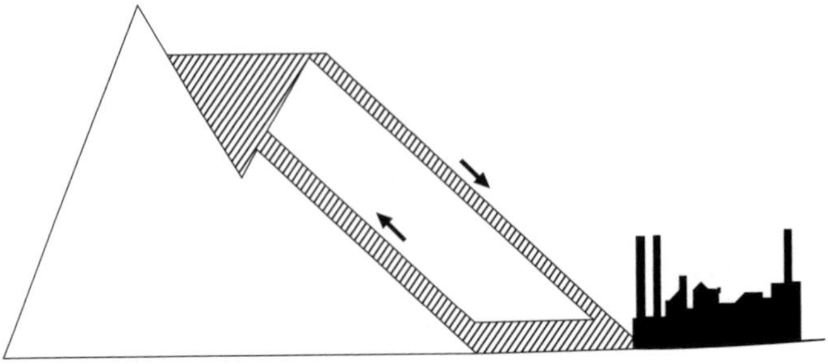

Figure 4.6   Schematic for Illustrative Example 4.12.

Note: This programmed-instructional problem is a modified and edited version (with permission) of an illustrative example prepared by Marie Gillman, a graduate mechanical engineering student at Manhattan College.

SOLUTION: First, convert height and flowrate to SI units in order to solve for the power in Watts:

$$(3000 \text{ ft})(0.3048 \text{ m/ft}) = 914.4 \text{ m}$$

$$(500{,}000 \text{ gal/min})(0.00378 \text{ m}^3/\text{gal}) = 1890 \text{ m}^3/\text{min}$$

The mass flow rate of the water in kilograms/second is

$$\frac{(1890 \text{ m}^3/\text{min})(1000 \text{ kg/m}^3)}{60 \text{ s/min}} = 31{,}500 \text{ kg/s}$$

The loss in potential energy, ΔPE, of the water flow is given by

$$\Delta \text{PE} = \frac{mg(\Delta Z)}{g_c} \tag{4.11}$$

Substituting yields

$$\Delta \text{PE} = (31{,}500 \text{ kg/s})(9.8 \text{ m/s}^2)(914.4 \text{ m})$$
$$= 2.82 \times 10^8 \text{ kg} \cdot \text{m/s}^3$$
$$= 2.82 \times 10^8 \text{ N/S}$$
$$= 282 \text{ MW}$$

Note that $g_c = 1$ in the SI system of units.

Assuming that the potential energy decrease is entirely converted to energy input to the turbine, the actual power output is

$$P = (0.30)(282) = 84.6 \text{ MW}$$

This is enough power for a small town. No pollutants or greenhouse gases are generated because no fossil fuel is required. The initial construction expense would be quite high, but the long-term cost of producing electricity would probably be very economical. ∎

REFERENCES

1. R. BIRD, W. STEWART, and E. LIGHTFOOT, "*Transport Phenomena*," 2nd edition, John Wiley & Sons, Hoboken, NJ, 2002.
2. L. THEODORE, "*Transport Phenomena for Engineers*," International Textbook Co., Scranton, PA, 1971.
3. P. ABULENCIA and L. THEODORE, "*Fluid Flow for the Practicing Engineer*," John Wiley & Sons, Hoboken, NJ, 2009.

NOTE: Additional problems for each chapter are available for all readers at www. These problems may be used for additional review or homework purposes.

Chapter **5**

Stoichiometry

Joseph Addison [1672–1719]

'Tis pride, rank pride, and haughtiness of soul;
I think the Romans call it stoicism.

—Cato. Act I, Sc. 4

INTRODUCTION

The term *stoichiometry* has come to mean different things to different people. In a loose sense, stoichiometry involves the balancing of an equation for a chemical reaction that provides a quantitative relationship among the reactants and products. In the simplest stoichiometric situation, exact quantities of pure reactants are available, and these quantities react completely to give the desired product(s). In an industrial process, the reactants usually are not pure, one reactant is usually in excess of what is needed for the reaction, and the desired reaction may not go to completion because of a host of other considerations.[1,2]

This chapter serves to introduce the general subject of stoichiometry, which in a very real sense, is an extension of the conservation law for mass discussed in the previous chapter. To simplify the presentation to follow, the textual matter and illustrative examples will focus on *combustion* reactions. Topics to be reviewed include:

1 Combustion of Methane

2 Excess and Limiting Reactant(s)

3 Combustion of Ethane

4 Combustion of Chlorobenzene

It should also be noted that much of this stoichiometric material is required in the treatment of "Chemical Reaction Enthalpy Effects" in Chapter 10.

COMBUSTION OF METHANE

The complete combustion of pure hydrocarbons yields carbon dioxide and water as the reaction products. Consider the combustion of methane in oxygen:

$$CH_4 + O_2 \longrightarrow CO_2 + H_2O \tag{5.1}$$

In order to balance this reaction, two molecules of oxygen are needed. This requires that there be four oxygen atoms on the right side of the reaction. This is satisfied by producing two molecules of water as product. The final balanced reaction becomes:

$$CH_4 + 2O_2 \longrightarrow CO_2 + 2H_2O \tag{5.2}$$

Thus, two molecules (or moles) of oxygen are required to completely combust one molecule (or mole) of methane to yield one molecule (or mole) of carbon dioxide and two molecules (or moles) of water. Note that the numbers of carbon, oxygen, and hydrogen atoms on the right-hand side of this reaction are equal to those on the left-hand side. The reader should verify that the total mass (obtained by multiplying the number of each molecule by its molecular weight and summing) on each side of the reaction is the same. [The term *moles* here may refer to either gram · moles (gmol) or pound · moles (lbmol); it makes no difference.]

ILLUSTRATIVE EXAMPLE 5.1

The reaction equation (not balanced) for the combustion of butane is

$$C_4H_{10} + O_2 \longrightarrow CO_2 + H_2O$$

Determine the mole ratio of reactants to products.

SOLUTION: A chemical equation provides a variety of qualitative and quantitative information essential for the calculation of the quantity of reactants reacted and products formed in a chemical process. A balanced chemical equation, as noted above, must have the same number of atoms of each type in the reactants and products. Thus, the balanced equation for butane is

$$C_4H_{10} + \left(\tfrac{13}{2}\right)O_2 \longrightarrow 4CO_2 + 5H_2O$$

Note that:

   number of carbons in reactants = number of carbons in products = 4

   number of oxygens in reactants = number of oxygens in products = 13

   number of hydrogens in reactants = number of hydrogens in products = 10

   number of moles of reactants is 1 mol C_4H_{10} + 6.5 mol O_2 = 7.5 mol total

   number of moles of products is 4 mol CO_2 + 5 mol H_2O = 9 mol total

The reader should note that although the number of moles on both sides of the equation do *not* balance, the masses of reactants and products (in line with the conservation law for mass) *must* balance.   ∎

EXCESS AND LIMITING REACTANT(S)[1,2]

The terms used to describe a reaction that do not involve stoichiometric ratios of reactants must be carefully defined in order to avoid confusion. If the reactants are not present in formula or stoichiometric ratio, one reactant is said to be *limiting*; the others are said to be in *excess*. Consider the following reaction:

$$CO + \tfrac{1}{2}O_2 \longrightarrow CO_2 \tag{5.3}$$

If the starting amounts are 1 mol of CO and 3 mol of O_2, CO is the limiting reactant, with O_2 present in excess. There are 2.5 mol of excess O_2, because only 0.5 mol is required to combine with the CO. Thus there is 500% excess oxygen present. The percentage of excess must be defined in relation to the amount of the reactant necessary to react completely with the limiting reactant. Thus, if for some reason only part of the CO actually reacts, this does not alter the fact that the oxygen is in excess by 500%. However, there are often several possible products. For instance, the reactions

$$C + O_2 \longrightarrow CO_2 \quad \text{and} \quad C + \tfrac{1}{2}O_2 \longrightarrow CO \tag{5.4}$$

can occur simultaneously. In this case, if there are 3 mol of oxygen present per mole of carbon, the oxygen is in excess. The extent of this excess, however, cannot be definitely fixed. It is customary to choose one product (e.g., the desired one) and specify the excess reactant in terms of this product. For this case, there is 200% excess oxygen for the reaction going to CO_2, and there is 500% excess oxygen for the reaction going to CO. The discussion on excess oxygen can be extended to excess air using the same approach. Stoichiometric or theoretical oxygen (or air) is defined as 0% excess oxygen (or air). This is an important concept since thermal combustion units operate with excess air. For example, approximately 25–50% excess air is employed with liquid injection incinerators, waste-fired boilers and process furnaces. Minimum excess air requirements for rotary kilns are approximately 50–80% if only bulk solids are burned and 80–120% if containerized solids are incinerated.

ILLUSTRATIVE EXAMPLE 5.2

Complete combustion of carbon disulfide results in combustion products of CO_2 and SO_2 according to the reaction

$$CS_2 + O_2 \longrightarrow CO_2 + SO_2$$

Balance this reaction equation.

SOLUTION: The balanced equation is

$$CS_2 + 3O_2 \longrightarrow CO_2 + 2SO_2$$

■

ILLUSTRATIVE EXAMPLE 5.3

Refer to Illustrative Example 5.2. If 500 lb of CS_2 is combusted with 225 lb of oxygen, which is the limiting reactant?
Data: MW of $CS_2 = 76.14$; MW of $SO_2 = 64.07$; MW of $CO_2 = 44$

SOLUTION: The initial molar amounts of each reactant is

$$(500\,\text{lb}\,CS_2)(1\,\text{lbmol}\,CS_2/76.14\,\text{lb}\,CS_2) = 6.57\,\text{lbmol}\,CS_2$$
$$(225\,\text{lb}\,O_2)(1\,\text{lbmol}\,O_2/32\,\text{lb}\,O_2) = 7.03\,\text{lbmol}\,O_2$$

The amount of O_2 needed to consume *all* the CS_2, i.e., the stoichiometric amount, is then

$$(6.57\,\text{lbmol})(3\,\text{lbmol}/1\,\text{lbmol}) = 19.71\,\text{lbmol}\,O_2$$

Therefore, O_2 is the limiting reactant since 19.7 mol of O_2 are required for complete combustion but only 7.03 mol of O_2 are available.   ■

ILLUSTRATIVE EXAMPLE 5.4

Refer to Illustrative Example 5.3. How much of each product is formed (lb)?

SOLUTION: The limiting reactant is used to calculate the amount of product formed. First note that

$$(7.03\,\text{lbmol}\,O_2)(1\,\text{lbmol}\,CS_2/3\,\text{lbmol}\,O_2) = 2.34\,\text{lbmol}\,CS_2$$
$$(2.34\,\text{lbmol}\,CS_2)(76.14\,\text{lb}/1\,\text{lbmol}\,CS_2) = 322\,\text{lb}\,CS_2\,\text{reacted}$$

Therefore,

$$(7.03\,\text{lbmol}\,O_2)(1\,\text{lbmol}\,CO_2/3\,\text{lbmol}\,O_2) = 2.34\,\text{lbmol}\,CO_2$$
$$(2.34\,\text{lbmol}\,CO_2)(44\,\text{lb}\,CO_2/1\,\text{lbmol}\,CO_2) = 103\,\text{lb}\,CO_2\,\text{produced}$$

$$(7.03\,\text{lbmol}\,O_2)(2\,\text{lbmol}\,SO_2/3\,\text{lbmol}\,O_2) = 4.68\,\text{lbmol}\,SO_2$$
$$(4.68\,\text{lbmol}\,SO_2)(64.07\,\text{lb}/1\,\text{lbmol}\,SO_2) = 300\,\text{lb}\,SO_2\,\text{produced}$$

■

COMBUSTION OF ETHANE

Consider the combustion of 1 mol of ethane. The reaction for complete combustion may be written

$$C_2H_6 + 3.5O_2 \longrightarrow 2CO_2 + 3H_2O \tag{5.5}$$

Thus, 2 mol of CO_2 and 3 mol of H_2O will be formed from the complete combustion of 1 mol of C_2H_6. The oxygen required is 3.5 mol. If 60% excess is used, an additional 2.1 mol, or a total of 5.6 mol of oxygen is required.

Equation (5.5) describes a gas-phase reaction. In accordance with Charles' law, one may also interpret this equation as follows: When 1 ft^3 of C_2H_6 reacts with 3.5 ft^3 of O_2, 2 ft^3 of CO_2 and 3 ft^3 of H_2O will form. Similar balanced stoichiometric reactions may be written for other hydrocarbons and organics to determine oxygen (or air) requirements and the products of combustion. However, this somewhat tedious calculation may be bypassed by use of Table 5.1. The reader should carefully review this table before proceeding to the remaining Illustrative Examples in this chapter. This table is employed again in Chapter 10.

This discussion may now be extended to reaction systems that involve the combustion of carbon and/or carbonaceous compounds and fuels. To simplify matters, it will be assumed that both the air and fuel mixture are dry. Throughout this text, air is assumed to contain 21% O_2 by volume. This is perhaps a bit high by a few hundredths of a percent. The remaining 79% is assumed to be inert and to consist of nitrogen (and a trace of the noble gases). The usual combustion products of hydrocarbons are CO_2 and H_2O. The combustion products of other organics may contain additional compounds; e.g., an organic chloride will also produce chlorine and/or hydrochloric acid. Combustion products from certain fuels will yield sulfur dioxide and nitrogen.

The so-called complete combustion of an organic or a fuel involves conversion of all the elemental carbon to carbon dioxide, hydrogen to water, sulfur to sulfur dioxide, and nitrogen to its elemental form of N_2. Thus, the *theoretical* or *stoichiometric* oxygen described for a combustion reaction is the amount of oxygen required to burn all the carbon to carbon dioxide, all the hydrogen to water, etc.; excess oxygen (or excess air) is the oxygen furnished in excess of the theoretical oxygen required for combustion. Since combustion calculations assume dry air to contain 21% oxygen and 79% nitrogen on a mole or volume basis, 4.76 mol of air consists of 1.0 mol of oxygen and 3.76 mol of nitrogen. One mole of O_2 equals 32 lb; 3.76 mol of N_2 equals 105.28 lb. Therefore, on a weight basis, dry air contains 23.3% O_2 by weight (32/137.28) and 77% N_2 by weight (105.28/137.28).

The temperature of the products that results from complete combustion under adiabatic conditions is defined as the *adiabatic flame temperature*; this too will be discussed in more detail in Chapter 10.

ILLUSTRATIVE EXAMPLE 5.5

The offensive odor of butanol can be removed from stack gases by its complete combustion to carbon dioxide and water. It is of interest that the incomplete combustion of butanol actually results in a more serious odor pollution problem than the original one. Write the equations showing the two intermediate malodorous products formed if butanol undergoes incomplete combustion.

Table 5.1 Combustion Constants[3]

Compound	Density lb/ft³	Specific volume ft³/lb	Heat of combustion (Btu/ft³) Gross (high)	Heat of combustion (Btu/ft³) Net (low)	Heat of combustion (Btu/lb) Gross (high)	Heat of combustion (Btu/lb) Net (low)	For 100% total air** (mol/mol or ft³/ft³ of combustible) — Required for combustion O₂	N₂	Air	Flue products CO₂	H₂O	N₂	For 100% total air** (lb/lb of combustible) — Required for combustion O₂	N₂	Air	Flue products CO₂	H₂O	N₂	Flammability limit (% by volume) Lower	Upper
Carbon, C*	—	—	—	—	14,093	14,093	1.0	3.76	4.76	1.0	—	3.76	2.66	8.86	11.53	3.66	—	8.86	—	—
Hydrogen, H₂	0.0053	187.723	325	275	61,100	51,623	0.5	1.88	2.38	—	1.0	1.88	7.94	26.41	34.34	—	8.94	26.41	4.00	74.20
Oxygen, O₂	0.0846	11.819	—	—	—	—	—	—	—	—	—	—	—	—	—	—	—	—	—	—
Nitrogen, N₂	0.0744	13.443	—	—	—	—	—	—	—	—	—	—	—	—	—	—	—	—	—	—
Carbon monoxide, CO	0.0740	13.506	322	322	4347	4347	0.5	1.88	2.38	1.0	—	1.88	0.57	1.90	2.47	1.57	—	1.90	12.50	74.20
Carbon dioxide, CO₂	0.1170	8.548	—	—	—	—	—	—	—	—	—	—	—	—	—	—	—	—	—	—
Paraffin series																				
Methane, CH₄	0.0424	23.565	1013	913	23,879	21,520	2.0	7.53	9.53	1.0	2.0	7.53	3.99	13.28	17.27	2.74	2.25	13.28	5.00	15.00
Ethane, C₂H₆	0.0803	12.455	1792	1641	22,320	20,432	3.5	13.18	16.68	2.0	3.0	13.18	3.73	12.39	16.12	2.93	1.80	12.39	3.00	12.50
Propane, C₃H₈	0.1196	8.365	2590	2385	21,661	19,944	5.0	18.82	23.82	3.0	4.0	18.82	3.63	12.07	15.70	2.99	1.68	12.07	2.12	9.35
n-Butane, C₄H₁₀	0.1582	6.321	3370	3113	21,308	19,680	6.5	24.47	30.97	4.0	5.0	24.47	3.58	11.91	15.49	3.03	1.55	11.91	1.86	8.41
Isobutane, C₄H₁₀	0.1582	6.321	3363	3105	21,257	19,629	6.5	24.47	30.97	4.0	5.0	24.47	3.58	11.91	15.49	3.03	1.55	11.91	1.80	8.44
n-Pentane, C₅H₁₂	0.1904	5.252	4016	3709	21,091	19,517	8.0	30.11	38.11	5.0	6.0	30.11	3.55	11.81	15.35	3.05	1.50	11.81	—	—
Isopentane, C₅H₁₂	0.1904	5.252	4008	3716	21,052	19,478	8.0	30.11	38.11	5.0	6.0	30.11	3.55	11.81	15.35	3.05	1.50	11.81	—	—
Neopentane, C₅H₁₂	0.1904	5.252	3993	3693	20,970	19,396	8.0	30.11	38.11	5.0	6.0	30.11	3.55	11.81	15.35	3.05	1.50	11.81	—	—
n-Hexane, C₆H₁₄	0.2274	4.398	4762	4412	20,940	19,403	9.5	35.76	45.26	6.0	7.0	35.76	3.53	11.74	15.27	3.06	1.46	11.74	1.18	7.40
Olefin series																				
Ethylene, C₂H₄	0.0746	13.412	1614	1513	21,644	20,295	3.0	11.29	14.29	2.0	2.0	11.29	3.42	11.39	14.81	3.14	1.29	11.39	2.75	28.60
Propylene, C₃H₆	0.1110	9.007	2336	2186	21,041	19,691	4.5	16.94	21.44	3.0	3.0	16.94	3.42	11.39	14.81	3.14	1.29	11.39	2.00	11.10

n-Butene, C_4H_8	0.1480	6.756	3084	2885	20,840	19,496	6.0	22.59	28.59	4.0	4.0	22.59	3.42	11.39	14.81	3.14	1.29	11.39	1.75	9.70
Isobutene, C_4H_8	0.1480	6.756	3068	2869	20,730	19,382	6.0	22.59	28.59	4.0	4.0	22.59	3.42	11.39	14.81	3.14	1.29	11.39	—	—
n-Pentene, C_5H_{10}	0.1852	5.400	3836	3586	20,712	19,363	7.5	28.23	35.73	5.0	5.0	28.23	3.42	11.39	14.81	3.14	1.29	11.39	—	—
Aromatic series																				
Benzene, C_6H_6	0.2060	4.852	3751	3601	18,210	17,480	7.5	28.23	35.73	6.0	3.0	28.23	3.07	10.22	13.30	3.38	0.69	10.22	1.40	7.10
Toluene, C_7H_8	0.2431	4.113	4484	4284	18,440	17,620	9.0	33.88	42.88	7.0	4.0	33.88	3.13	10.40	13.53	3.34	0.78	10.40	1.27	6.75
Xylene, C_8H_{10}	0.2803	3.567	5230	4980	18,650	17,760	10.5	39.52	50.02	8.0	5.0	39.52	3.17	10.53	13.70	3.32	0.85	10.53	1.00	6.00
Miscellaneous gases																				
Acetylene, C_2H_2	0.0697	14.344	1499	1448	21,500	20,776	2.5	9.41	11.91	2.0	1.0	9.41	3.07	10.22	13.30	3.38	0.69	10.22	—	—
Naphthalene, $C_{10}H_8$	0.3384	2.955	5854	5654	17,298	16,708	12.0	45.17	57.17	10.0	4.0	45.17	3.00	9.97	12.96	3.43	0.56	9.97	—	—
Methyl alcohol, CH_3OH	0.0846	11.820	868	768	10,259	9078	1.5	5.65	7.15	1.0	2.0	5.65	1.50	4.98	6.48	1.37	1.13	4.98	6.72	36.50
Ethyl alcohol, C_2H_5OH	0.1216	8.221	1600	1451	13,161	11,929	3.0	11.29	14.29	2.0	3.0	11.29	2.08	6.93	9.02	1.92	1.17	6.93	3.28	18.95
Ammonia, NH_3	0.0456	21.914	441	365	9668	8001	0.75	2.82	3.57	—	1.5	3.32	1.41	4.69	6.10	—	1.59	5.51	15.50	27.00
Sulfur, S*	—	—	—	—	3983	3983	1.0	3.76	4.76	1.0 (SO_2)	—	3.76	1.00	3.29	4.29	2.00 (SO_2)	—	3.29	—	—
Hydrogen sulfide, H_2S	0.0911	10.979	647	596	7100	6545	1.5	5.65	7.15	1.0 (SO_2)	1.0	5.65	1.41	4.69	6.10	1.88 (SO_2)	0.53	4.69	4.30	45.50
Sulfur dioxide, SO_2	0.1733	5.770	—	—	—	—	—	—	—	—	—	—	—	—	—	—	—	—	—	—
Water vapor, H_2O	0.0476	21.017	—	—	—	—	—	—	—	—	—	—	—	—	—	—	—	—	—	—
Air	0.0766	13.063	—	—	—	—	—	—	—	—	—	—	—	—	—	—	—	—	—	—
Gasoline	—	—	—	—	—	—	—	—	—	—	—	—	—	—	—	—	—	—	1.40	7.60

*Carbon and sulfur are considered as gases for molal calculations only.

**100% total air in this table refers to stoichiometric air or zero percent excess air.

Adapted from *Fuel Flue Gases*, American Gas Association, *Combustion Flame and Explosions of Gases*, 1951.

SOLUTION: The malodorous products are butyraldehyde (C_4H_8O) and butyric acid (C_3H_7COOH), which can be formed sequentially as follows:

$$C_4H_9OH + \tfrac{1}{2}O_2 \longrightarrow C_4H_8O + H_2O$$
$$C_4H_8O + \tfrac{1}{2}O_2 \longrightarrow C_3H_7COOH$$

or the acid can be formed directly as follows:

$$C_4H_9OH + O_2 \longrightarrow C_3H_7COOH + H_2O$$

For complete combustion:

$$C_4H_9OH + 6O_2 \longrightarrow 4CO_2 + 5H_2O \qquad \blacksquare$$

COMBUSTION OF CHLOROBENZENE[1,2]

The balanced reaction for the combustion of 1.0 lbmol of chlorobenzene (C_6H_5Cl) in stoichiometric oxygen is

$$C_6H_5Cl + 7O_2 \longrightarrow 6CO_2 + 2H_2O + HCl \qquad (5.6)$$

Since air, not oxygen, is employed in combustion processes, this reaction with air becomes

$$C_6H_5Cl + 7O_2 + [26.3N_2] \longrightarrow 6CO_2 + 2H_2O + HCl + [26.3N_2] \qquad (5.7)$$

where the nitrogen in the air has been retained in brackets on both sides of the equation since it does not participate in the combustion reaction. The moles and masses involved in this reaction, based on the stoichiometric combustion of 1.0 lbmol of C_6H_5Cl, are given here:

$$C_6H_5Cl + 7O_2 + [26.3N_2] \longrightarrow 6CO_2 + 2H_2O + HCl + [26.3N_2]$$

Moles	1	7	26.3	6	2	1	26.3
MW	112.5	32	28	44	18	36.5	28
Mass	112.5	224	736.4	264	36	36.5	736.4

$$(5.8)$$

Initial mass = 1072.9   Initial number of moles = 34.3
Final mass = 1072.9   Final number of moles = 35.3

Note that, in accordance with the conservation law for mass, the initial and final masses balance. The number of moles, as is typically the case in chemical reaction/combustion calculations, do not balance. The concentrations of the various species may also be calculated. For example,

$$\%CO_2 \text{ by weight} = (264/1072.9)100\% = 24.61\%$$
$$\%CO_2 \text{ by mol (or volume)} = (6/35.3)100\% = 17.0\%$$
$$\%CO_2 \text{ by weight (dry basis)} = (264/1036.9)100\% = 25.46\%$$
$$\%CO_2 \text{ by mol (dry basis)} = (6/33.3)100\% = 18.0\%$$

The air requirement for this reaction is 33.3 lbmol. This is stoichiometric or 0% excess air (EA). For 100% EA (100% above stoichiometric), one would use (33.3)(2.0) or 66.6 lbmol of air. For 50% EA one would use (33.3)(1.5) or 50 lbmol of air; for this condition, 16.7 lbmol excess or additional air is employed.

If 100% excess air is employed in the combustion of 1.0 lbmol of C_6H_5Cl, the combustion reaction would become

$$C_6H_5Cl + 14O_2 + 52.6N_2 \longrightarrow 6CO_2 + 2H_2O + HCl + 7O_2 + 52.6N_2 \quad (5.9)$$

ILLUSTRATIVE EXAMPLE 5.6

If chlorobenzene is combusted in 100% excess air at 1.0 atm, calculate the partial pressures of O_2, HCl, and H_2O.

SOLUTION: Assume 1.0 lbmol chlorobenzene as a basis. Based on the development above, the total final number of moles = 68.6. Therefore

$$\%O_2 \text{ by mol (or volume)} = (7/68.6)100\% = 10.2\%$$
$$\%HCl \text{ by mol} = (1/68.6)100\% = 1.46\%$$
$$\%H_2O \text{ by mol} = (2/68.6)100\% = 2.92\%$$

Since this operation is conducted at atmospheric pressure, then

$$\text{Partial pressure } O_2 = 0.102 \text{ atm}$$
$$\text{Partial pressure } HCl = 0.0146 \text{ atm}$$
$$\text{Partial pressure } H_2O = 0.0292 \text{ atm} \qquad \blacksquare$$

ILLUSTRATIVE EXAMPLE 5.7

If the chlorobenzene in the previous example contains 0.5% sulfur (S) by mass, estimate the partial pressure of the SO_2 in the flue gas.

SOLUTION

$$\text{Weight of S} = (0.005)(112.5) = 0.5625 \text{ lb}$$
$$\text{Number of lbmol of S} = 0.5625/32 = 0.0176 \text{ lbmol}$$
$$\text{Number of lbmol of } SO_2 \text{ formed} = \text{number of lbmol of S} = 0.0176 \text{ lbmol}$$

For this condition (approximately),

$$\%SO_2 \text{ by mol} = (0.0176/68.6)100\% = 0.0257\%$$
$$\text{Partial pressure } SO_2(p_{SO_2}) = 2.57 \times 10^{-4} \text{ atm}$$
$$= 257 \text{ ppm(v)} \qquad \blacksquare$$

ILLUSTRATIVE EXAMPLE 5.8

Refer to Illustrative Example 5.7. Calculate the concentration of SO_2 in ppm(v).

SOLUTION: By definition

$$ppm = (p)10^6 = (y)10^6$$

Substituting

$$ppm\,(SO_2) = (2.57 \times 10^{-4})(10^6)$$
$$= 257 \qquad \blacksquare$$

ILLUSTRATIVE EXAMPLE 5.9

A separation unit produces a pure hydrocarbon compound with a concentration not high enough to economically justify recovering and recycling it. The engineering division of the company has made a decision that it would be worthwhile to combust the compound and recover the heat generated as a makeup heat source for the separation unit.

If this hydrocarbon compound contains three atoms of carbon, determine its chemical formula if the flue gas composition on a dry basis is:

$$CO_2: 7.5\% \qquad CO: 1.3\% \qquad O_2: 8.1\% \qquad N_2: 83.1\%$$

SOLUTION: Assume a basis 100 mol of dry flue gas. The moles of each component in the dry product gas is then

CO_2	7.5 mol
CO	1.3 mol
O_2	8.1 mol
N_2	83.1 mol

Determine the amount of oxygen fed for combustion. Since nitrogen does not react (*key component*), using the ratio of oxygen to nitrogen in air will provide the amount of oxygen fed:

$$O_{2,\text{fed}} = \left(\frac{21}{79}\right)(83.1) = 22.1 \text{ mol}$$

A balanced equation for the combustion of the hydrocarbon in terms of N moles of the hydrocarbon and n hydrogen atoms in the hydrocarbon yields

$$NC_3H_n + 22.1O_2 \longrightarrow 7.5CO_2 + 1.3CO + 8.1O_2 + N(n/2)H_2O$$

The moles of hydrocarbon, N, is obtained by performing an elemental carbon balance:

$$3N = 7.5 + 1.3$$
$$N = 8.8/3 = 2.93$$

Similarly, the moles of water formed is obtained by performing an elemental oxygen balance:

$$2(22.1) = 2(7.5) + 1.3 + 2(8.1) + N(n/2)$$
$$N(n/2) = 44.2 - 15 - 1.3 - 16.2$$
$$= 11.7$$

The number of hydrogen atoms, n, in the hydrocarbon is then

$$n = 2(11.7)/N$$
$$= 23.4/2.93$$
$$= 7.99 \approx 8$$

Since $n = 8$, the hydrocarbon is C_3H_8, propane. ∎

ILLUSTRATIVE EXAMPLE 5.10

The general formula for alkyl dichlorobenzenes is

$$C_nH_{2n-8}Cl_2 \qquad \text{where } n > 6$$

Write a balanced, general chemical equation for the decomposition of alkyl dichlorobenzenes in the presence of oxygen.

SOLUTION: The general balanced equation for the complete combustion of the alkyl dichlorobenzenes is as follows:

$$C_nH_{2n-8}Cl_2 + (1.5n - 2.5)O_2 \longrightarrow nCO_2 + 2HCl + (n - 5)H_2O$$

where $n > 6$. ∎

ILLUSTRATIVE EXAMPLE 5.11

A state incinerator emissions limit requires 99% HCl control and allows a maximum particulate discharge of 0.07 gr/dscf (grains/dry standard cubic feet) at 68°F, corrected to 50% EA. An incinerator is to burn 5 tons/h hazardous sludge waste containing 2% Cl, 80% C, 5% inerts, and the balance H_2O by weight. Calculate the maximum mass emission rate of equivalent HCl in lb/h that may be emitted.

SOLUTION: The incinerator receives 5 tons/h of waste with 2% Cl content. Assuming all chlorine is converted to HCl, the amount of HCl formed is given as:

$$\text{Cl in feed} = (5\,\text{ton/h})(2000\,\text{lb/ton})(0.02\,\text{lb Cl/lb waste})$$
$$= 200\,\text{lb Cl/h}$$

$$\text{HCl formed} = [(200\,\text{lb Cl/h})(36.5\,\text{lb HCl})]/(35.5\,\text{lb Cl})$$
$$= 205.6\,\text{lb HCl/h}$$

The maximum permissible mass emission rate of HCl at 99% control is

$$(205.6\,\text{lb HCl/h})(1 - 0.99) = 2.06\,\text{lb HCl/h emitted} \qquad \blacksquare$$

ILLUSTRATIVE EXAMPLE 5.12

Assume perfect combustion of a contaminated fuel oil with stoichiometric air. The gravimetric weight percentage analysis of a sample of this fuel oil is

88.52% carbon	0.10% nitrogen
10.87% hydrogen	0.06% oxygen
0.40% sulfur	0.05% ash

Calculate the gravimetric analysis of the flue gas.

SOLUTION: Convert the gravimetric percent of each component to lbmol by selecting 100 lb of fuel as a basis.

C	$88.52/12 = 7.38\,\text{lbmol}$
H	$10.87/2 = 5.44\,\text{lbmol}$
S	$0.40/32 = 0.0125\,\text{lbmol}$
O_2	$0.06/32 = 0.00188\,\text{lbmol}$
N_2	$0.10/28 = 0.00357\,\text{lbmol}$

Since the molecular weight (MW) of the ash is unknown and its content is small, it is neglected in the calculation.

Write the combustion reaction for each component in the fuel:

$$C + O_2 \longrightarrow CO_2$$
$$H_2 + \tfrac{1}{2}O_2 \longrightarrow H_2O$$
$$S + O_2 \longrightarrow SO_2$$

Determine the amount of oxygen required for each component. Use the previous results to complete the last column:

Component	lbmol in fuel	lbmol O_2 required
C	7.38	7.38
H_2	5.44	2.72
S	0.0125	0.0125
O_2	0.00188	−0.00188
		Total = 10.1

Note: Oxygen in the fuel reduces air requirements. Therefore, the amount of oxygen in the fuel has been subtracted from the amount of O_2 required.

Calculate the amount of nitrogen from the oxygen requirement:

$$\text{Amount of } N_2 = (\text{amount of } O_2)\frac{0.79 \text{ lbmol } N_2}{0.21 \text{ lbmol } O_2}$$
$$= (10.1)(0.79/0.21)$$
$$= 38.0 \text{ lbmol } N_2$$

The lbmol of N_2 required for each component may also be calculated.

Component	lbmol N_2 required
C	27.76
H_2	0.23
S	0.047
O_2	0.0
	Total = 38.03

Determine the gravimetric composition (percentage) of the flue gas:

Component	lbmol	MW	lb	wt %
CO_2	7.38	44	324.7	21.8
H_2O	5.44	18	97.9	6.6
SO_2	0.00125	64	0.08	0.005
O_2	0.0	32	0.0	0.0
N_2	38.03	28	1064.3	71.6
	Total = 50.85		Total = 1487.0	

■

ILLUSTRATIVE EXAMPLE 5.13

Refer to Illustrative Example 5.12. Calculate the total volume of flue gas (at 500°F and 1 atm) per pound of oil burned.

SOLUTION: Noting that 100 lb of fuel was used as a basis, the total lbmol of flue gas produced per pound of oil burned is:

$$
\text{lbmol of flue gas} = \frac{\text{total lbmol from above}}{100}
$$

$$
= \frac{50.85}{100}
$$

$$
= 0.5085 \text{ lbmol gas/lb oil}
$$

Calculate the total volume of flue gas at 500°F and 1 atm using the ideal gas law:

$$
V = \frac{nRT}{P} = \frac{(0.5085 \text{ lbmol flue gas/lb oil})(0.7302)(500 + 460)}{1.0}
$$

$$
= 356.1 \text{ ft}^3/\text{lb oil} \qquad \blacksquare
$$

ILLUSTRATIVE EXAMPLE 5.14

Refer to Illustrative Examples (5.12) and (5.13). Calculate the volume percent of CO_2 in the *dry* flue gas, i.e., on a dry basis.

SOLUTION: One can now complete the following:

Component of dry flue gas	lbmol
CO_2	7.38
SO_2	0.00125
N_2	38.03
	Total = 45.41

The volume percentage of CO_2 in the dry flue gas is

$$
\%CO_2 = \frac{\text{lbmol } CO_2}{\text{total lbmol dry flue gas}}(100) = \frac{7.38}{45.41}(100)
$$

$$
= 16.25\% \qquad \blacksquare
$$

ILLUSTRATIVE EXAMPLE 5.15

An incinerator has been burning a certain mass of dichlorobenzene ($C_6H_4Cl_2$) per hour, and the HCl produced was neutralized with solid soda ash (Na_2CO_3). If the incinerator switches to burning an equal mass of mixed tetrachlorobiphenyls ($C_{12}H_6Cl_4$), by what factor will the consumption of soda ash be increased?

SOLUTION: The balanced stoichiometric reaction for the oxidation of dichlorobenzene is shown below:

$$C_6H_4Cl_2 + 6.5O_2 \longrightarrow 6CO_2 + H_2O + 2HCl$$

Therefore, for 1 lb of dichlorobenzene (DCB), the following mass of HCl is produced:

$$\left(\frac{1\,\text{lb}}{147\,\text{lb/lbmol DCB}}\right)\left(\frac{2\,\text{mol HCl}}{\text{mol DCB}}\right) = 0.0136\,\text{lbmol HCl produced}$$

The balanced stoichiometric reaction for the oxidation of tetrachlorobiphenyl is as shown below:

$$C_{12}H_6Cl_4 + 12.5O_2 \longrightarrow 12CO_2 + H_2O + 4HCl$$

Therefore, for 1 lb of tetrachlorobiphenyl (TCB), the following mass of HCl is produced:

$$\left(\frac{1\,\text{lb}}{290\,\text{lb/lbmol TCB}}\right)\left(\frac{4\,\text{mol HCl}}{\text{mol DCB}}\right) = 0.0138\,\text{lbmol HCl produced}$$

Thus, the amount of acid produced does not change significantly ($\approx 1.5\%$), and neither will the amount of base required for neutralization. ∎

REFERENCES

1. J. REYNOLDS, J. JERIS, and L. THEODORE, "*Handbook of Chemical and Environmental Engineering Calculations,*" John Wiley & Sons, Hoboken, NJ, 2004.
2. J. SANTOLERI, J. REYNOLDS, and L. THEODORE, "*Introduction to Hazardous Waste Incineration,*" 2nd edition, John Wiley & Sons, Hoboken, NJ, 2000.
3. Adapted from: "*Fuel Flue Gases,*" American Gas Association, 1951.

NOTE: Additional problems for each chapter are available for all readers at www. These problems may be used for additional review or homework purposes.

Chapter **6**

The Second Law of Thermodynamics

Plato [427–347 b.c.]

Democracy, which is a charming form of government, full of variety and disorder, and dispensing a sort of equality to equals and unequals alike.

—*The Republic. Book VIII, 558–C*

INTRODUCTION

Historically, the study of the second law of thermodynamics was developed by individuals such as Carnot (a French engineer), Clausius, and Kelvin in the middle of nineteenth century. This development was made purely on a macroscopic scale and is referred to as the "classical approach to the second law," which does not require the existence of an atomic or molecular theory.

The law of conservation of energy has already been defined in Chapter 4 as the first law of thermodynamics. Its application allows calculations of energy relationships associated with all kinds of processes. The "limiting" law is called the "second law of thermodynamics." Applications to be addressed later in the text involve calculations for maximum power outputs from a power plant and equilibrium yields in chemical reactions. In principle, this law states that water cannot flow uphill and heat cannot flow from a cold to a hot body of its own accord. Other defining statements for this law that have appeared in the literature are provided below:

1 Any process whose sole net result is the transfer of heat from a lower temperature level to a higher one is impossible.

2 No apparatus, equipment, or process can operate in such a way that its only effect (in system and surroundings) is to convert heat taken in completely into work.

3 It is impossible to convert the heat taken into a system completely into work in a cyclical process.

Thermodynamics for the Practicing Engineer. By L. Theodore, F. Ricci, and T. Van Vliet

Both a qualitative and quantitative review of entropy (to be defined shortly) and the second law are presented in this chapter. This is followed by a section titled "The Heat Exchanger Dilemma" that examines the interrelationship of entropy and heat exchanger design optimization. This, in turn, is followed by a discussion of energy conservation measures that can be implemented in chemical plant and process applications. The chapter concludes with a short discussion on the Third Law of Thermodynamics—a law that defines the value of entropy at absolute zero temperature.

QUALITATIVE REVIEW OF THE SECOND LAW

As described above, the first law of thermodynamics is a conservation law for energy transformations. Regardless of the types of energy involved in processes—thermal, mechanical, electrical, elastic, magnetic, etc.—the change in the energy of a system is equal to the difference between energy input and energy output. The first law also allows free convertibility from one form of energy to another, as long as the over-all quantity is conserved. Thus, this law places no restriction on the conversion of work into heat, or on its counterpart—the conversion of heat into work.

The unrestricted conversion of work into heat is well known to most technical individuals. Frictional effects are frequently associated with mechanical forms of work that result in a temperature rise of the bodies in contact. However, the transform-ation of heat into work is of greater concern. In nations with a partially developed or developing technological society, the ability to produce energy in the form of work takes on prime importance. Work transformations are necessary to transport people and goods, drive machinery, pump liquids, compress gases, and provide energy input to so many other processes that are taken for granted in highly developed societies. Much of the work input in such societies is available in the form of electrical energy, which is then converted to rotational mechanical work. Although some of this electrical energy (work) is produced by hydroelectric power plants, by far the greatest part of it is obtained from the combustion of fossil fuels or nuclear fuels. These fuels allow the engineer to produce a relatively high-temperature gas or liquid stream that acts as a thermal (heat) source for the production of work. Hence the study of the con-version of heat to work is extremely important—especially in the light of developing shortages of fossil and nuclear fuels, along with the accompanying environmental pro-blems, particularly with global warming.

The brief discussion of energy-conversion devices above leads to an important second-law consideration—i.e., that energy has "quality" as well as quantity. Becuase work is 100% convertible to heat whereas the reverse situation is not true, work is a more valuable form of energy than heat. Although it is not as obvious, it can also be shown through second-law arguments that heat has "quality" in terms of the temperature at which it is discharged from a system. The higher the temperature at which heat transfer occurs, the greater the possible energy transformation into work. Thus, thermal energy stored at high temperatures is generally more useful to society than that available at lower temperatures. While there is an immense quantity of energy stored in the oceans, for example, its present availability to society for performing useful tasks is quite low. This implies, in turn, that thermal energy loses some of its "quality" or is degraded when it is transferred by means of heat transfer from one temperature to a lower one. Other forms of energy degradation include energy transformations due to frictional effects and electric resistance. Such effects are highly undesirable if the use of energy for practical purposes is to be maximized. The second law provides some means of measuring this energy degradation[1] through a term referred to as *entropy* and it is the second law (of thermodynamics) that serves to define this important thermodynamic property. Entropy is normally designated as S with units of energy/absolute temperature, e.g., Btu/°R or cal/K, and like enthalpy and internal energy is a point function. At the heart of the "quality energy" discussion is the relatively recent concept of exergy, which is a direct measure of a system's potential to do useful work (a measure of its quality energy). For a more detailed discussion on the concept of exergy, refer to Chapter 19.

Entropy calculations[2,3] can provide quantitative information on the "quality" of energy and energy degradation. Detailed second law equations and calculations are presented in the next section.

In line with the discussion regarding the "quality" of energy, individuals at home and in the workplace are often instructed to "conserve energy." However, this comment, if taken literally, is a misnomer because energy is automatically conserved by the provisions of the first law. (The reader is referred to the presentation in Chapter 4 on energy conservation.) In reality, the comment "conserve energy" addresses only the concern associated with the "quality" of energy. If the light in a room is not turned off, "quality" energy is degraded although energy is conserved, i.e., the electrical energy is converted to internal energy (which heats up the room). Note, however, that this energy transformation will produce a token rise in temperature of the room from which little, if any, "quality" energy can be recovered and used again (for lighting or other useful purposes).[1]

There are a number of other phenomena that cannot be explained by the law of conservation of energy. It is the second law of thermodynamics that provides guidelines to the understanding and analysis of these diverse effects. Among other considerations, the second law can:

1 Provide the means of measuring the aforementioned "quality" of energy.

2 Establish the criteria for the "ideal" performance of engineered devices.

3 Determine the direction of change for processes.

4 Establish the final equilibrium state for spontaneous processes.

QUANTITATIVE REVIEW OF THE SECOND LAW

It can be shown that the change in entropy for a reversible adiabatic process is always zero, i.e.,

$$\Delta S = 0 \qquad (6.1)$$

For liquids and solids, the entropy change for a system undergoing a temperature change from T_1 to T_2 is given by

$$\Delta S = C_p \ln(T_2/T_1); \quad \text{Btu/lbmol} \cdot {}^\circ\text{R} \tag{6.2}$$

ILLUSTRATIVE EXAMPLE 6.1

A copper block with a mass of 20 lb is originally at 100°C. It is placed in a pail containing 6 gallons of water at 25°C. Neglecting heat losses, calculate the change in the entropy of the copper block. The average heat capacity at constant pressure, c_p, for copper and water is approximately 0.092 and 1.0 Btu/lb · °F, respectively.

SOLUTION: First write the equation describing the energy change of the copper in cooling from 100°C to T.

$$Q_c = m_c c_{p,c}(100 - T)$$

where Q_c = heat lost from the copper block.
    Write the equation describing the energy change of the water in being heated from 25°C to T.

$$Q_w = m_w c_{p,w}(T - 25)$$

Calculate the final temperature T in °C.

$$m_c c_{p,c}(100 - T) = m_w c_{p,w}(T - 25)$$
$$(20)(0.092)(100 - T) = (6)(8.33)(1.0)(T - 25)$$
$$T = 27.66°C$$
$$= 300.66\text{K}$$

Calculate the entropy change of the copper in Btu/°F by employing Equation (6.2).

$$\Delta S_c = m_c c_{p,c} \ln(T/373)$$
$$= (20)(0.092) \ln(300.66/373)$$
$$= -0.397 \text{ Btu/°F}$$

Although the units in this equation are mixed, they do satisfy the equation dimensionally since the temperature ratio is in absolute units.   ∎

ILLUSTRATIVE EXAMPLE 6.2

Refer to Illustrative Example 6.1. Calculate the change in the entropy of the water.

SOLUTION: Calculate the entropy change of the water by once again employing Equation (6.2).

$$\Delta S_w = m_w c_{p,w} \ln(T/298)$$
$$= (6)(8.33)(1.0) \ln(300.66/298)$$
$$= +0.444 \, \text{Btu}/^\circ\text{F}$$

■

ILLUSTRATIVE EXAMPLE 6.3

Refer to Illustrative Example (6.1) and (6.2). Neglecting heat losses, calculate the change in the entropy of the entire system.

SOLUTION: Calculate the overall entropy change of the process. Use the results from the two previous examples,

$$\Delta S_T = \Delta S_c + \Delta S_w$$
$$= -0.397 + 0.444$$
$$= +0.047 \, \text{Btu}/^\circ\text{F}$$

Note that the overall entropy change is positive.   ■

ILLUSTRATIVE EXAMPLE 6.4

Comment on the results of the previous example.

SOLUTION: If ΔS_{syst} and ΔS_{surr} represent the entropy change of a system and the surroundings, respectively, it can be shown that, for a particular process and as a consequence of the second law, the total entropy change ΔS_{tot} is given by

$$\Delta S_{tot} = \Delta S_{syst} + \Delta S_{surr} \geq 0 \tag{6.3}$$

In effect, the second law requires that for every process, the total entropy change is positive; the only exception is if the process is reversible and then

$$(\Delta S_{tot})_{rev} = 0 \tag{6.4}$$

Thus, no real process can occur for which the total entropy change is zero or negative. The fundamental facts relative to the entropy concept are that the entropy change of a system may be positive (+), negative (−), or zero (0); the entropy change of the surroundings during this process may likewise be positive, negative, or zero. However, the total entropy change, ΔS_{tot}, must be $\Delta S_{tot} \geq 0$. Once again, the equality sign applies if the change occurs reversibly and adiabatically.   ■

Consider now the entropy change of gases. The entropy change of an ideal gas undergoing a change of state from P_1 to P_2 at a constant temperature T is given by

$$\Delta S_T = R \ln(P_1/P_2); \quad \text{Btu/lbmol} \cdot {}^\circ \text{R} \tag{6.5}$$

The entropy change of an ideal gas undergoing a change of state from T_1 to T_2 at a constant pressure is given by

$$\Delta S_P = C_p \ln(T_2/T_1); \quad C_p = \text{constant} \tag{6.6}$$

Correspondingly, the entropy change for an ideal gas undergoing a change from (P_1, T_1) to (P_2, T_2) is

$$\Delta S = R \ln(P_1/P_2) + C_p \ln(T_2/T_1) \tag{6.7}$$

Regarding phase changes, the entropy change is given by

$$\Delta S_{\text{phase}} = \Delta H_{\text{phase}}/T \tag{6.8}$$

where ΔH_{phase} is the energy change associated with the phase change and T is once again the absolute temperature.

ILLUSTRATIVE EXAMPLE 6.5

Calculate the change in entropy, ΔS, of 5 lbmol of an ideal gas as it changes state in an irreversible process from 100°F and 1 atm to 400°F and 10 atm. The heat capacity of the ideal gas may be taken as 5.0 Btu/lbmol · °F.

SOLUTION: Convert the temperatures to an absolute scale:

$$T_1 = 100 + 460$$
$$= 560°\text{R}$$
$$T_2 = 400 + 460$$
$$= 860°\text{R}$$

Write the equation describing the entropy change of the gas:

$$\Delta S = nR \ln(P_1/P_2) + nC_p \ln(T_2/T_1) \tag{6.7}$$

Calculate the entropy for the irreversible process in Btu/°R:

$$\Delta S = (5)[(1.987) \ln(1/10) + (5.0) \ln(860/560)]$$
$$= (5)[-4.575 + 2.145]$$
$$= -12.15 \, \text{Btu/}^\circ\text{R}$$

■

ILLUSTRATIVE EXAMPLE 6.6

Refer to Illustrative Example 6.5. Calculate ΔS if the change occurs reversibly.

SOLUTION: Employ units of Btu/°R. Since entropy is a point function, ΔS is not a function of the path, i.e., it does not matter whether the process is reversible or irreversible. Thus,

$$\Delta S = -12.15\,\text{Btu}/°\text{R}$$

∎

ILLUSTRATIVE EXAMPLE 6.7

Refer to Illustrative Example 6.6. Calculate the entropy change of the surroundings.

SOLUTION: Since the process is reversible,

$$\Delta S_{\text{tot}} = 0$$

In line with Equation (6.4), the ΔS for the surroundings is therefore $+12.15$ Btu/°R.   ∎

ILLUSTRATIVE EXAMPLE 6.8

Refer to Fig. 6.1(I) where a $-$, $+$, or 0 sign for the entropy change of a system for irreversible processes A, B, and C have been added. Fill in the remaining six boxes

SOLUTION: Since the process is irreversible, the total entropy change for any process—and this includes process A, B, or C—must be positive, that is, $+$. This is noted in the RHS of Fig. 6(II). The completed figure is as shown in Fig. 6.1(III) and the following conclusions can be drawn.

1. For process A, the entropy change of the surroundings must be large enough, i.e., greater in absolute magnitude than the entropy change of the system, for the total entropy change to be ($+$).

2. For process B, the entropy change must be ($+$) since the total entropy change is ($+$).

3. For process C, the entropy change can be (0) or ($+$); however, it can also be ($-$) provided the total entropy change is ($+$).   ∎

IDEAL WORK AND LOST WORK

To reexamine the concept of "quality" energy, consider the insulated space pictured in Fig. 6.2(A) and (B).

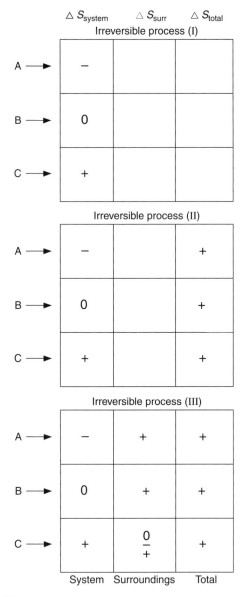

Figure 6.1 Irreversible process.

Since the space is insulated, and is a closed system with no work term, one can conclude from the first law that

$$U_A = U_B \tag{6.9}$$

Although the energy levels are the same, note that system A has the capability of doing useful work (because of the high-temperature high-pressure steam) while system B

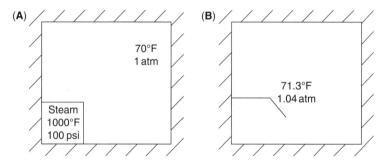

Figure 6.2   Entropy application.

does not. If an entropy analysis is performed on both A and B, one would deduce that (as demonstrated earlier)

$$S_A < S_B \tag{6.10}$$

In effect, the entropy level has increased for the system that has lost its ability to do useful work. It is in this manner that the concept of entropy can be used to determine a system's ability to either do useful work or lose its ability to do useful work.

This development can now be extended to introduce the concept of ideal work, W_{ideal}, and lost work, W_{lost}. As noted in the literature,[2] "in any steady-state flow process requiring work, there is an absolute minimum amount which must be expended to accomplish the desired change of state of the fluid flowing through the system. In a process producing work, there is an absolute maximum amount which may be accomplished as the result of a given change of state of the fluid flowing through the system. For either case, the limiting value for work is obtained when the change of state associated with the process is accomplished *completely reversibly*; for such a process, the entropy generation is zero," which is in line with Equation (6.1).

For real-world, or actual, or irreversible processes, the actual work W_s (or $\dot{W}_s$ on a rate basis) can be compared with the ideal work. If W_{ideal} (or $\dot{W}_{ideal}$ on a rate basis) is positive, it is the *minimum work required* and is smaller than $\dot{W}_s$. The thermodynamic efficiency E_t can then be defined as the ratio of the ideal work to the actual work:

$$E_t \text{ (work required)} = \frac{W_{ideal}}{W_s} = \frac{\dot{W}_{ideal}}{\dot{W}_s} \tag{6.11}$$

When W_{ideal} (or $\dot{W}_{ideal}$) is negative, the ideal work is the *maximum work obtainable* and is larger than the actual work. The thermodynamic efficiency is then defined as:

$$E_t \text{ (work produced)} = \frac{W_s}{W_{ideal}} = \frac{\dot{W}_s}{\dot{W}_{ideal}} \tag{6.12}$$

Work that is lost because the process is irreversible is defined as the *lost work*, W_{lost}, and represents the difference between the actual work and the ideal work, i.e.,

$$W_{lost} = W_s - W_{ideal} \qquad (6.13)$$

Equation (6.13) may also be written on a time rate basis

$$\dot{W}_{lost} = \dot{W}_s - \dot{W}_{ideal} \qquad (6.14)$$

This lost work may be viewed as the decrease in the aforementioned "quality" energy of the system. In any event, the second law leads to the conclusion that the greater the irreversibility of a process, the greater the (rate of) entropy increase and the greater the amount of energy that becomes unavailable for work, i.e., the lost quality work.

ILLUSTRATIVE EXAMPLE 6.9

A patent has been submitted that claims a newly developed energy conversion process has unique capabilities. The patent claim indicates that the process can take steam at 212°F and 1 atm and provide heat to a constant temperature reservoir at 300°C while discharging heat to a low-temperature reservoir at 60°F. Pertinent data is provided in the flow diagram of Fig. 6.3. Is this claim possible?

Note: Of all the intellectual property rights, the most pertinent for inventions are patents. A patent can protect, for example, a composition of matter, an article of manufacture, or a method of doing something. Patent rights are private property rights. Infringement of a patent is a civil offense, not criminal. If necessary, the patent owner must come to his or her own defense through litigation.

SOLUTION: For any energy-producing process to be theoretically possible, it must meet the requirements of both the first and second laws of thermodynamics. As with the first law, the second law has been verified by experiments; its validity cannot be questioned at the macrolevel.

The first law was discussed earlier. The second law states that every naturally occurring process produces a total entropy change that is positive; the limiting value of zero is attained only for a reversible process. Thus, no process is possible for which the total entropy decreases.

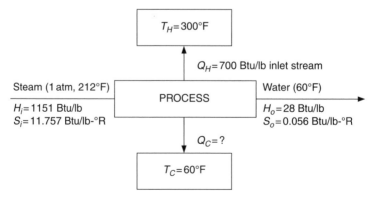

Figure 6.3   Flow diagram for Illustrative Example 6.9.

To satisfy the first law for the flow process above (assuming a basis of 1 lb of steam)

$$\Delta H - \sum Q_i = 0$$
$$1151 - 28 - 700 - Q_C = 0$$
$$1123 - 700 - Q_C = 0$$
$$Q_C = 423 \, \text{Btu/lb}$$

Regarding the second law, the entropy change of the steam is

$$\Delta S_S = 0.056 - 1.757$$
$$= -1.701 \, \text{Btu/lb} \cdot {}^\circ \text{R}$$

The entropy gained by the reservoir is given by

$$\Delta S_i = \frac{Q_i}{T_i}$$

For the hot reservoir,

$$\Delta S_H = \frac{Q_H}{T_H} = \frac{700}{300 + 460}$$
$$= 0.921 \, \text{Btu/lb} \cdot {}^\circ \text{R}$$

For the discharge to the cold reservoir,

$$\Delta S_C = \frac{Q_C}{T_C} = \frac{423}{60 + 460}$$
$$= 0.813$$

The total entropy change is therefore

$$\Delta S_T = \Delta S_S + \Delta S_H + \Delta S_C$$
$$= -1.701 + 0.921 + 0.813$$
$$= +0.033 \, \text{Btu/lb} \cdot {}^\circ \text{R}$$

The calculations performed above automatically satisfies the first law since the energy lost by the conversion of the steam to water, i.e., 1123 Btu/lb, was set equal to the energy transferred to the hot and cold reservoir, i.e., 700 and 423 Btu/lb, respectively. Interestingly, the second law is also satisfied, but only marginally. This suggests that the claim may not be possible. ∎

THE HEAT EXCHANGER DILEMMA[4]

Environmental concerns involving conservation of energy issues gained increasing prominence during and immediately after the Arab oil embargo of 1973. In addition, global population growth has led to an ever-increasing demand for energy. The use of energy has resulted in great benefits; however, the environmental and human health impacts of this energy use has become a concern. One of the keys to reducing and/or eliminating these problems will be achieved through energy conservation.

One of the areas where energy conservation can be realized is in the design and specification of process (operating) conditions for a heat exchanger. This can be best accomplished by the inclusion of second law principles in the analysis. The quantity of heat recovered in an exchanger is not alone in influencing its size and cost. As the temperature difference driving force in the exchanger approaches zero, the "quality" heat recovery approaches 100 percent.

Regarding the design of heat exchangers, if Q represents the rate of heat transfer between a hot and cold fluid flowing in the heat exchanger, application of the conservation law for energy gives

$$Q_H = \dot{m}_H c_{P,H}(T_{HI} - T_{HO}) \tag{6.15}$$

and

$$Q_C = \dot{m}_C c_{P,C}(T_{CO} - T_{CI}) \tag{6.16}$$

where the subscripts H and C refer to the hot and cold fluids, respectively, I and O refer to the fluid inlet and outlet temperature, respectively, $\dot{m}$ represents the mass flow rate, and c_P is once again the heat capacity at constant pressure (assumed constant). In addition, if there is no heat loss between the exchanger and the surroundings,

$$Q_H = Q_C \tag{6.17}$$

The following important equation relates Q to the average temperature difference between the hot and cold fluids.

$$Q = UA\Delta T_{LM} \tag{6.18}$$

This is often referred to as the heat exchanger design equation. The terms, U, A, and ΔT_{LM} represent the overall heat transfer coefficient (a function of the resistance to heat transfer), the area for heat transfer, and the log mean temperature difference driving force (TDDF), respectively. For some exchangers, the latter term is given by

$$\Delta T_{LM} = TDDF = \frac{\Delta T_2 - \Delta T_1}{\ln(\Delta T_2/\Delta T_1)} \tag{6.19}$$

where ΔT_2 and ΔT_1 represents the temperature difference between the hot and cold fluid at each end of the exchanger, respectively. For purposes of the analysis to follow, Equation (6.18) is rearranged in the form

$$\frac{Q}{U\Delta T_{LM}} = A \tag{6.20}$$

An individual designing a heat exchanger is faced with two choices. He or she may decide on a design with a large temperature difference driving force that would result in a compact (smaller area) design but with a larger entropy change that is accompanied by the loss of "quality" energy. Alternatively, a design with a smaller driving force would result in a larger heat exchanger, but with a smaller entropy change with a larger recovery of "quality" energy.

Most heat exchangers are designed with the requirement/specification that the temperature difference between the hot and cold fluid should be positive at all times and at least 20°F.[5] This temperature difference or driving force is referred to as the approach temperature. Obviously, and as will be demonstrated in the Illustrative Examples to follow, the corresponding entropy change is proportional to the driving force. Thus, large temperature difference driving forces result in large irreversibilities and the associated large entropy change.

Regarding the cooling medium for a given heat transfer duty, it is possible to circulate a large quantity with a small temperature change or a small quantity with a large temperature change. The temperature change (or range) of the coolant naturally affects the TDDF. If a large quantity is used, the TDDF is larger and less heat transfer area is required as a result of the larger TDDF. Although this will reduce the original investment and fixed charges (capital and operating costs), the amount of "quality" energy recovered will be smaller owing to the greater quantity of coolant employed. It is therefore apparent that there must be an optimum between the two conditions: much coolant, smaller surface, and the recovery of less "quality" energy or little coolant, larger surface, and the recovery of more quality energy. In the limit, as the TDDF $\rightarrow$ 0, the area requirement $A \rightarrow \infty$, the entropy change $\Delta S \rightarrow 0$, and the "quality" energy remains unchanged. Clearly, cost must be minimized, but just as clearly, the "quality" of energy recovered must be included in the analysis. This dilemma is addressed below.[4]

Consider first the model of operation for the three heat exchangers pictured in Figure 6.4. (*Note:* For purposes of analysis, set $m_C = m_H = 1.0$ lb and $c_P = 1.0$ Btu/lb · °F.)

For operation (A), the entropy change for the hot fluid is

$$\Delta S_H = m_H c_P \ln \frac{T_2}{T_1}$$

$$= (1)(1) \ln \left(\frac{300 + 460}{540 + 460} \right)$$

$$= \ln \left(\frac{760}{1000} \right) = -0.2744 \text{ Btu}/°\text{R}$$

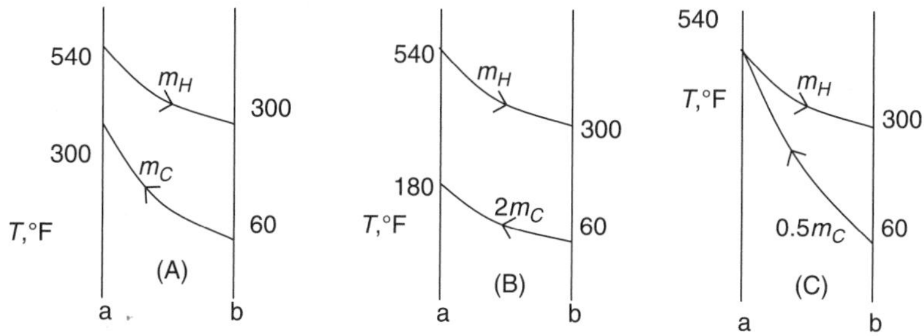

Figure 6.4   Heat Exchanger Operation.

and

$$\Delta S_C = m_C c_P \ln \left(\frac{300 + 460}{60 + 460} \right)$$

$$= \ln \left(\frac{760}{520} \right)$$

$$= 0.3795 \, \text{Btu/}^\circ\text{R}$$

The total entropy increase is therefore

$$\Delta S_{T,A} = -0.2744 + 0.3795$$

$$= 0.1054 \, \text{Btu/}^\circ\text{R}$$

In addition,

$$\Delta T_{LM,A} = 540 - 300 = 300 - 60$$

$$= 240^\circ\text{F}$$

For operation (B), the entropy change can be calculated in a manner similar to that of operation (A).

$$\Delta S_H = -0.2744 \, \text{Btu/}^\circ\text{R}$$

and

$$\Delta S_C = (2)(1) \ln \left(\frac{180 + 460}{60 + 460} \right)$$

$$= 2 \ln \left(\frac{640}{520} \right)$$

$$= 0.4153 \, \text{Btu/}^\circ\text{R}$$

The total entropy increase for operation (B) is therefore

$$\Delta S_{T,B} = -0.2744 + 0.4153$$

$$= 0.1409 \, \text{Btu/}^\circ\text{R}$$

In addition,

$$\Delta T_{LM,B} = 360 - 240 / \ln(360/240)$$

$$= 296^\circ\text{F}$$

For operation (C), the entropy change for the hot fluid is again

$$\Delta S_H = -0.2744 \text{ Btu}/^\circ R$$

while

$$\Delta S_C = (0.5)(1)\ln\left(\frac{540 + 460}{60 + 460}\right)$$

$$= 2\ln\left(\frac{1000}{520}\right)$$

$$= 0.3270 \text{ Btu}/^\circ R$$

The total entropy increase for operation (C) is therefore

$$\Delta S_{T,C} = -0.2744 + 0.3270$$

$$= 0.0526 \text{ Btu}/^\circ R$$

In addition,

$$\Delta T_{LM,C} = 240/\infty$$

$$= 0^\circ F$$

A summary of the results for operations A, B, and C plus the heat exchanger area requirement (A) and quality energy (QE) available are provided in Table 6.1.

One concludes that as the ΔT_{LM} (or TDDF) increases, the area requirement decreases; howerver, the QE available correspondingly decreases. Alternatively, if ΔT_{LM} decreases both A and QE increase.

Consider now the operation of heat exchangers A and B, as provided in Fig. 6.5. For Case I, note that

$$TDDF_A = TDDF_B$$

$$A_A = A_B; \quad A_A = A_B = A$$

$$\Delta S_A = \Delta S_B$$

End Result: Two $\dot{m}_C$ streams ($\dot{m}_{C,A}$ and $\dot{m}_{C,B}$) result at $300^\circ F$ where $\dot{m}_{C,A} = \dot{m}_{C,B} = \dot{m}_C$.

Table 6.1   Heat Exchanger-Entropy Analysis Results

	ΔT_{LM}, $^\circ F$	ΔS_T, Btu$/^\circ R$	A	QE
Operation A	0.1054	240	moderate	moderate
Operation B	0.1409	296	lower	lower
Operation C	0.0526	0	∞	higher

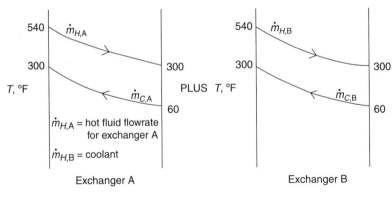

Figure 6.5   Heat exchanger comparison: Case I.

In Fig. 6.6, consider Case II. Here the coolant discharge from C serves as the inlet coolant to exchanger D. For Case II, note that

$$\text{TDDF}_C > \text{TDDF}_D; \quad \text{TDDF}_D = 0$$
$$A_C < A_D; \quad A_D = \infty$$
$$\Delta S_C > \Delta S_D; \quad \Delta S_D = 0$$

End Result: One $\dot{m}_C$ ($\dot{m}_{C,C}$) results at 540°F.

Case I/Case II Comparison: One $\dot{m}_C$ stream at 540°F (II) vs. two $\dot{m}_C$ at 300°F (I)

$$A = D \text{ (II) vs. } 2A \text{ (I)}$$
$$\Delta S \text{ (II)} < \Delta S \text{ (I)}$$

Thus, for Case II, Stream II can heat (for example) another fluid to 520°F while Stream I cannot.

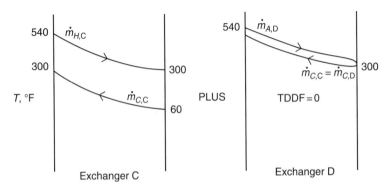

Figure 6.6   Heat exchanger comparison: Case II.

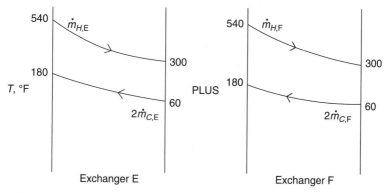

Figure 6.7   Heat exchanger comparison: Case III.

Now consider Case III (see Fig. 6.7). Here twice the coolant is employed in each exchanger. For Case III, note that

TDDF has increased

A has decreased

ΔS has increased

End Result: Four m_C at 180°F, A has decreased and ΔS has increased.

Case I/Case II/Case III Comparison: See next four Illustrative Examples.

ILLUSTRATIVE EXAMPLE 6.10

Refer to Case I, in Fig. 6.5. Calculate the entropy change of exchanger A. Assume $c_p = 1.0$ and $m_H = m_C = 1.0$, consistent units.

SOLUTION: For the hot fluid,

$$\Delta S_H = mc_p \ln\left(\frac{300 + 460}{540 + 460}\right)$$

$$= (1)(1)(-0.2744)$$

$$= -0.2744$$

For the cold fluid,

$$\Delta S_C = mc_p \ln\left(\frac{300 + 460}{60 + 460}\right)$$

$$= (1)(1)\ln\left(\frac{70}{520}\right)$$

$$= 0.3795$$

Therefore, for one exchanger—say A,

$$\Delta S_A = \Delta S_H + \Delta S_C$$
$$= -0.2744 + 0.3795$$
$$= +0.1051$$

Since there are two similar exchangers,

$$\Delta S_A = \Delta S_B$$

and

$$\Delta S_{tot,I} = (2)(0.1051)$$
$$= +0.2102 \, \text{Btu}/^\circ\text{R}$$

As expected, there is a positive entropy change.   ■

ILLUSTRATIVE EXAMPLE 6.11

Refer to Case II in Fig. 6.6. Calculate the total entropy change.

SOLUTION: Consider exchanger C first.

$$\Delta S_C = \Delta S_{H,C} + \Delta S_{C,C}$$

This is given by the result in the previous Illustrative Example, i.e.,

$$\Delta S_C = 0.1051$$

Now consider exchanger D. Since the TDDF is zero, the operation is reversible. Therefore,

$$\Delta S_D = 0$$

The total entropy change is then

$$\Delta S_{tot,II} = \Delta S_C + \Delta S_D$$
$$= 0.1051 + 0.0$$
$$= 0.1051 \, \text{Btu}/^\circ\text{R}$$ ■

ILLUSTRATIVE EXAMPLE 6.12

Refer to Case III in Fig. 6.7. Calculate the total entropy change.

SOLUTION: Consider exchanger E:

$$\Delta S_E = \Delta S_{H,E} + \Delta S_{C,E}$$

$$= -0.2744 + (2)(1) \ln\left(\frac{180 + 460}{60 + 460}\right)$$

$$= -0.2744 + 0.4153$$

$$= 0.1409$$

Since there are two similar exchangers,

$$\Delta S_E = \Delta S_F$$

Therefore

$$\Delta S_{tot, III} = \Delta S_E + \Delta S_F$$

$$= (2)(0.1409)$$

$$= 0.2818 \, \text{Btu}/°\text{R}$$ ∎

ILLUSTRATIVE EXAMPLE 6.13

Refer to Illustrative Examples (6.10)–(6.12). Comment on the results.

SOLUTION: The calculated results of Illustrative Examples 6.10–6.12 paint a clear picture. As the TDDF decreases, the area cost requirement ($ per ft^2 of heat exchanger area) increases, the entropy change decreases, and the "quality" of energy increases. Thus, from a "conservation of energy" perspective, second law considerations mandates its inclusion in heat exchanger applications with appropriate economic considerations, i.e., both the cost of the exchanger and the economic factors associated with the quality of the recovered energy must be considered in the analysis. Note that pressure drop, materials for construction, etc., are not included in the analysis.

The reader is again reminded that the subject of "quality" energy will be revisited in Chapter 19, where the more commonly used term "exergy" will be introduced along with example problems.   ∎

CHEMICAL PLANT AND PROCESS APPLICATIONS[1,6]

There are numerous general energy conservation practices that can be instituted at chemical plants. Ten of the simpler ones are detailed below:

1 Lubricate fans.

2 Lubricate pumps.

3 Lubricate compressors.

4 Repair steam and compressed air leaks.

5 Insulate bare steam lines.

6 Inspect and repair steam traps.

7 Increase condensate return.

8 Minimize boiler blowdown.

9 Maintain and inspect temperature-measuring devices.

10 Maintain and inspect pressure-measuring devices.

Some energy conservation practices applicable to specific chemical operations are also provided below:

1 Recover energy from hot gases.

2 Recover energy from hot liquids.

3 Reduce reflux ratio in distillation columns.

4 Reuse hot wash water.

5 Add effects to existing evaporators.

6 Use liquefied gases as refrigerants.

7 Recompress vapor(s) for low-pressure steam.

8 Generate low-pressure steam from flash operations.

9 Use waste heat for absorption refrigeration.

10 Cover tanks of hot liquids to reduce heat loss.

Recycling can also be achieved in the recovery of energy through the use of waste as a fuel supplement or fuel substitute. Waste may be processed in fossil-fuel-fired plants or in incinerators equipped with an energy recovery system. Usually, a variety of high-Btu wastes with different compositions are blended to produce a fuel with a certain specification.

For the purposes of implementing an energy conservation strategy, process changes and/or designs can be divided into four phases, each presenting different opportunities for implementing energy conservation measures:

1 Product conception.

2 Laboratory research.

3 Process development (pilot plant).

4 Mechanical (physical) design.

Energy conservation "training" measures that can be taken in the chemical process industry include the following:

1 Implement a sound operation, maintenance, and inspection (OM&I) program.

2 Implement a pollution prevention program.[1,6,7]

3 Institute a formal training program for all employees.

The following energy conservation practices are recommended at the plant's "office" level:[8]

1 Maintain air-conditioner efficiency, and reduce heated and cooled spaces.
2 Maintain boiler efficiency.
3 Use natural ventilation whenever and wherever possible, reduce air infiltration, and seal leaks in pipes and ducts.
4 Raise office temperatures in summer.
5 Lower office temperatures in winter.
6 Use shading efficiently.
7 Close windows and doors when and where applicable.
8 Fix broken windows and other air leaks.
9 Do not use lights unnecessarily.
10 Turn off office equipment that is not in use.

THE THIRD LAW OF THERMODYNAMICS

Based on extensive experimental data, Walther Hermann Nernst (1864–1941), a German physicist, postulated a "heat theorem" that the entropy change for any chemical reaction at absolute zero temperature, i.e., $0°R$ or $0K$, was zero. Thus,

$$\Delta S = 0 \quad \text{at} \quad T = 0 \qquad (6.15)$$

Planck, another German physicist, further proposed that at absolute zero temperature

$$S = 0 \qquad (6.16)$$

Experimental data has verified these two conclusions. The third law of thermodynamics is based on Equation (6.16), i.e., third law of thermodynamics is concerned with the absolute values of entropy. Thus, by definition, the entropy of all pure crystalline materials at absolute zero temperature is exactly zero and in line with Equation (6.16),

$$S_{-273°C} = S_{0K} = 0$$

Note, however, that engineers are usually concerned with *changes* in thermodynamic properties. The third law allows one to calculate the entropy at another temperature via the following equation:

$$\Delta S = \int_{0}^{T_f} \frac{(C_P)_s}{T} dT + \frac{\Delta H_f}{T_f} + \int_{T_f}^{T_v} \frac{(C_P)_l}{T} dT + \frac{\Delta H_v}{T_v} + \int_{T_v}^{T} \frac{(C_P)_g}{T} dT \qquad (6.17)$$

where the first, third, and fifth terms on the right hand side (RHS) represent the entropy change associated with temperature changes for the solid, liquid, and gas phases, respectively, and the second and fourth RHS terms represent the entropy change associated with fusion and vaporization, respectively. Since the entropy is zero at absolute zero

$$\Delta S = S_T - S_0 = S_T \qquad (6.18)$$

The above equation provides the entropy at the temperature in question, T.

The calculation of entropy changes for reactions will come into play in Chapter 13. The entropy change for a reaction in a reference or standard state—usually 298K—will allow one to perform equilibrium calculations.

ILLUSTRATIVE EXAMPLE 6.14

The following experimental data at a standard condition of 298K is provided for the reaction

$$A + 3B = 2C$$

$$S_A^\circ = 26.3 \, \text{Btu/lbmol} \cdot {}^\circ\text{R}$$
$$S_B^\circ = 21.0 \, \text{Btu/lbmol} \cdot {}^\circ\text{R}$$
$$S_C^\circ = 29.9 \, \text{Btu/lbmol} \cdot {}^\circ\text{R}$$

Calculate the entropy change at 298K.

SOLUTION: By definition, the entropy change for this reaction is given by the entropy of the product(s) multiplied by their stoichiometric coefficients *minus* the entropy of the reaction(s) multiplied by their stoichiometric coefficients. Thus

$$\Delta S^\circ = 2S_C^\circ - 3S_B^\circ - S_A^\circ$$

Substituting

$$\Delta S^\circ = (2)(29.9) - (3)(21.0) - 26.3$$
$$= -29.5 \, \text{Btu/lbmol} \cdot {}^\circ\text{R}$$
$$= -29.5 \, \text{cal/gmol} \cdot \text{K} \qquad \blacksquare$$

REFERENCES

1. M. K. THEODORE and L. THEODORE, "*Introduction to Environmental Management*," 2nd edition, CRC Press/Taylor & Francis, Boca Raton, FL, 2009.
2. J. SMITH, H. VAN NESS, and M. ABBOTT, "*Chemical Engineering Thermodynamics*," 6th edition, McGraw-Hill, New York, 2001.

3. L. THEODORE and J. REYNOLDS, *"Thermodynamics,"* A Theodore Tutorial, East Williston, NY, 1994.
4. L. THEODORE: unpublished paper, Manhattan College, 2006.
5. L. THEODORE: personal notes, Manhattan College, 1988.
6. G. BURKE, B. SINGH, and L. THEODORE, *"Handbook of Environmental Management and Technology,"* 2nd edition, John Wiley & Sons, Hoboken, NJ, 2000.
7. R. DUPONT, L. THEODORE, and K. GANESAN, *"Pollution Prevention,"* Lewis/CRC Press, Boca Raton, FL, 2000.
8. M.K. THEODORE and L. THEODORE, *"Environmental Management,"* CRC Press/Taylor & Francis Group, Boca Raton, FL, 2009.

NOTE: Additional problems for each chapter are available for all readers at www. These problems may be used for additional review or homework purposes.

Part II

Enthalpy Effects

Benedict (Baruch) Spinoza [1632–1677]

Nothing exists from whose nature some effect does not follow.

—*Ethics [1677]. Part I, Prop. XXXVI*

This Part reviews enthalpy effects, particularly those associated with a chemical reaction. To introduce this subject, the reader is reminded that the engineer and applied scientist are rarely concerned with the *magnitude* or *amount* of the energy in a system; the primary concern is with *changes* in the amount of energy. In measuring energy changes for systems, the enthalpy (H) has been found to be the most convenient term to work with.

There are many different types of enthalpy effects; these include:

1 Sensible (temperature)
2 Latent (phase)
3 Dilution (with water), e.g., HCl with H_2O
4 Solution (nonaqueous), e.g., HCl with a solvent other than H_2O
5 Reaction (chemical)

Chapter 7 is concerned with sensible enthalpy effects while Chapter 8 is concerned with latent enthalpy effects. Chapter 9 addresses enthaply effects associated with mixing processes (3) and (4). This Part concludes with a fairly extensive presentation on chemical reaction effects in Chapter 10.

Chapter 7

Sensible Enthalpy Effects

Edith Nesbit [1858–1924] and Ogden Nash [1902–1992]

Only the gamefish swims upstream,
But the sensible fish swims down.

—When You Say That, Smile

INTRODUCTION

Sensible enthalpy effects are associated with temperature. This introductory chapter to Part II provides methods that can be employed to calculate these changes. These methods include the use of:

1 Enthalpy values

2 Average heat capacity values

3 Heat capacity as a function of temperature

Two classes of heat capacity are considered in topic **3**—employing a, b, c constants and α, β, γ constants. In addition to an introduction to the Gibbs Phase Rule, this chapter presents various approaches that can be used to estimate heat capacity values from basic principles.

THE GIBBS PHASE RULE (GPR)

The reader is introduced to the term "degrees of freedom" before proceeding to the (GPR).[1] The *degrees of freedom F* or the *variance* of a system is defined as the *smallest number* of independent variables (such as pressure, temperature, concentration) that must be specified in order to completely define (the remaining variables of) the system. The significance of the degrees of freedom of a system may be drawn from the following examples. In order to specify the

---

Thermodynamics for the Practicing Engineer. By L. Theodore, F. Ricci, and T. Van Vliet
Copyright © 2009 by John Wiley & Sons, Inc.

density of a gaseous (vapor) steam, it is necessary to state both the temperature and pressure to which this density corresponds. For example, the density of steam has a particular value at 150°C and 1 atm pressure. A statement of this density at 150°C without mention of pressure does not clearly define the state of the steam, for at 150°C the steam may exist at many other possible pressures. Similarly, mention of the pressure without the temperature leaves ambiguity. Therefore, for the complete description of the state of the steam, two variables must be given, and this phase, when present alone in a system, possesses two degrees of freedom, or the system is said to be *bivariant*. When liquid water and steam exist in equilibrium, however, the temperature and the densities of the phases are determined only by the pressure, and a statement of some arbitrary value of the latter is sufficient to define all the other variables. The same applies to the choice of temperature as the independent variable. At each arbitrarily chosen temperature (within the range of existence of the two phases), equilibrium is possible only at a given pressure, and once again the system is defined in terms of one variable. Under these conditions, the system possesses only one degree of freedom or it is *monovariant*.

There is a definite relation in a system between the number of degrees of freedom, the number of components, and the number of phases present. This relationship was first established by J. Willard Gibbs in 1876. This relation, known as the *Gibbs Phase Rule*, is a principle of the widest generality. It is one of the most often used rules in thermodynamic analyses, particularly in the representation of equilibrium conditions existing in heterogeneous systems.

To arrive at a mathematical description of the phase rule, consider a system of C components in which P phases are present. The problem is to determine the total number of variables upon which such a system depends. First, the state of the system will depend upon the pressure and the temperature. Again, in order to define the composition of each phase, it is necessary to specify the concentration of ($C - 1$) constituents of the phase, the concentration of the remaining component being determined by difference. Since there are P phases, the total number of concentration variables will be $P(C - 1)$, and these along with the temperature and pressure constitute a total of $[P(C - 1) + 2]$ variables.

Recall from algebra that when a system possesses n independent variables, n independent equations are necessary in order to solve for the value of each of these. Similarly, in order to define the $[P(C - 1) + 2]$ variables of a system, this number of equations relating these variables would have to be available. The next question is then: How many equations involving these variables can possibly be written from the conditions describing the system? To answer this, recourse must be had to thermodynamics. Equilibrium thermodynamic principles makes it possible to write *one equation* among the variables *for each constituent distributed between any two phases*. When P phases are present, ($P - 1$) equations are available for each constituent, and for C constituents a total of $C(P - 1)$ equations. If this number of equations is equal to the number of variables, the system is completely defined. However, this will generally not be the case, and the number of variables will exceed the number of equations by F, where

$$F = \text{number of variables} - \text{number of equations}$$
$$= [P(C - 1) + 2] - [C(P - 1)]$$
$$= C - P + 2 \tag{7.1}$$

Equation (7.1) is the celebrated phase rule of Gibbs. F Term is the number of degrees of freedom of a system and provides the number of variables whose values must be specified (arbitrarily) before the state of the system can be completely and unambiguously characterized. According to the phase rule, the number of degrees of freedom of a system is therefore given by the difference between the number of components and the number of phases present, i.e., by $(C - P)$.

It is assumed in the above derivation that each component is present in every phase. If a component is missing from a particular phase, however, the number of concentration variables is decreased by one. But at the same time, the number of possible equations is also decreased by one. Hence the value of $(C - P)$, and therefore F, remains the same whether each constituent is present in every phase or not. This effectively means that the phase rule is not restricted by this assumption, and is generally valid under all conditions of distribution provided that equilibrium exists in the system.

The simplest case of GPR is one in which only a single component in a single phase is present, as with ice or steam. When more than one component and/or phase is present in a system, the number of degrees of freedom correspondingly increases in accordance with Equation (7.1). Most (but not all) of the applications in this chapter involve one-component one-phase systems.

ILLUSTRATIVE EXAMPLE 7.1

Calculate the number of degrees of freedom for a one-component, one-phase system.

SOLUTION: Refer to Equation 7.1.

$$F = C - P + 2$$

Since $C = 1$ and $P = 1$,

$$F = 1 - 1 + 2$$
$$= 2$$

Thus, two independent variables must be specified to completely define the system. ∎

ENTHALPY VALUES

The term enthalpy was introduced in Chapter 4 in Equation (4.10).

$$H = U + PV$$

The terms H, U, and S (as well as free energy G, to be discussed later in Chapters 13 and 14), are *state* or *point* functions. As noted in the previous section, by fixing a certain number of variables on which the function depends, the numerical value of the function is automatically fixed, i.e., it is single-valued. For example, fixing the temperature and pressure of a one-component single-phase system immediately specifies the enthalpy and internal energy.

The change in enthalpy as it undergoes a change in state from (T_1, P_1) to (T_2, P_2) is given by

$$\Delta H = H_2 - H_1 \tag{7.2}$$

Note that enthalpy has two key properties:

1 Enthalpy is a point function, i.e., the enthalpy change from one state (say 200°F, 1 atm) to another state (say 400°F, 1 atm) is a function only of the two states and not the path of the process associated with the change.

2 Absolute values of enthalpy are not important. The enthalpy of water at 60°F, 1 atm, as recorded in some steam tables is zero Btu/lbmol. Another table indicates a value of 224.2 Btu/lbmol. Both are correct! Note that changing the temperature of water from 60°F to 100°F results in the same change in enthalpy using either table.

Also note that enthalpy (changes) may be described with (English) units of Btu, Btu/lb, Btu/lbmol, Btu/scf, or Btu/time depending on the available data and calculation required.

Most industrial facilities operate in a steady-state flow mode with no significant mechanical or shaft work added (or withdrawn) from the system. For this condition, the First Law reduces to

$$Q = \Delta H \tag{7.3}$$

This equation is routinely used in many engineering calculations. If a unit or system is operated *adiabatically*, $Q = 0$ and the above equation becomes

$$\Delta H = 0 \tag{7.4}$$

Enthalpy values and enthalpy changes may be calculated directly from enthalpy data. These are available in the literature. Typical tabulated values are provided in Table 7.1[2] Additional enthalpy data (on a mole basis) is located in Table 7.2.

Table 7.1 Enthalpy of Combustion Gases[2]

Temp (°F)	O$_2$	N$_2$	Air	CO	CO$_2$	SO$_2$	H$_2$	CH$_4$	H$_2$O^a	HCl	HF	HBr
						Btu/lb						
60	0.00	0.00	0.00	0.00	0.00	0.00	0.00	0.00	1060.00	0.00	0.00	0.00
100	8.80	9.90	9.60	10.00	8.00	5.90	137.00	21.00	1077.70	7.60	13.20	3.40
200	30.90	34.80	33.60	54.90	29.30	21.40	484.00	76.10	1122.40	26.70	45.20	12.00
300	53.30	59.90	57.70	59.90	52.00	37.50	832.00	136.40	1168.00	45.80	79.50	20.70
400	76.20	85.00	81.80	85.00	75.30	54.40	1182.00	202.10	1213.00	64.90	112.70	29.30
500	99.40	110.30	106.00	110.60	99.80	71.80	1532.00	272.60	1260.00	84.10	153.00	38.00
600	123.10	136.10	130.20	136.30	125.10	89.80	1882.00	347.80	1307.00	103.00	188.00	46.80
800	171.70	187.70	178.90	188.70	177.80	127.00	2584.00	511.20	1404.00	142.00	258.00	64.50
1000	221.70	240.70	235.00	242.70	233.60	165.50	3291.00	691.10	1504.00	182.00	329.00	82.90
1200	272.50	294.70	288.50	297.80	290.90	205.10	4007.00	886.20	1608.00	222.00	400.00	101.00
1400	324.30	350.80	343.00	354.30	349.70	245.40	4729.00	1094.00	1715.00	264.00	465.00	119.00
1600	377.30	407.30	398.00	407.50	416.30	286.40	5460.00	1313.00	1827.00	302.00	538.00	138.00
1800	430.70	465.00	455.00	465.30	470.90	327.80	6198.00	1542.00	1947.00	349.00	613.00	157.00
2000	484.00	523.80	513.00	523.80	532.80	369.10	6952.00		2060.00	388.00	688.00	177.00
2200	539.30	583.20	570.70	583.30	596.10	411.10	7717.00		2186.00	433.00	764.00	198.00
2400	594.40	642.30	628.50	643.00	659.20	452.70	8490.00		2303.00	477.00	841.00	218.00
2600	649.00	702.80	687.30	703.20	723.20	495.20	9272.00		2430.00	523.00	920.00	239.00
2800	702.80	763.10	746.60	771.30	787.40	537.50	10,060.00		2570.00	568.00	998.00	260.00
3000	758.60	824.10	806.30	832.60	852.00	580.00	10,870.00		2700.00	615.00	1085.00	283.00
3200	816.40	885.80	866.00	894.00	916.70	622.50	11,680.00		2830.00	665.00	1167.00	304.00
3400	873.40	947.60	925.90	956.00	981.60	665.00	12,510.00		2960.00	712.00	1249.00	326.00
3600	931.00	1010.00	986.10	1018.00	1047.00	707.50	13,330.00		3100.00	759.00	1331.00	347.00
3800	988.00	1070.00	1046.00	1081.00	1112.00	760.00			3240.00	806.00	1409.00	369.00
4000	1045.00	1132.00	1102.00	1138.00	1177.00	804.00			3380.00	854.00	1499.00	391.00

^{a}Water vapor enthalpy includes latent enthalpy (heat) of vaporization.

Table 7.2 Molar Enthalpies of Combustion Gases,[3] Btu/lbmol[a]

T (°F)	N_2	Air (MW 28.97)	CO_2	H_2O
32	0	0	0	0
60	194.9	194.6	243.1	224.2
77	312.2	312.7	392.2	360.5
100	473.3	472.7	597.9	545.3
200	1170	1170	1527	1353
300	1868	1870	2509	2171
400	2570	2576	3537	3001
500	3277	3289	4607	3842
600	3991	4010	5714	4700
700	4713	4740	6855	5572
800	5443	5479	8026	6460
900	6182	6227	9224	7364
1000	6929	6984	10,447	8284
1200	8452	8524	12,960	10,176
1500	10,799	10,895	16,860	13,140
2000	14,840	14,970	23,630	18,380
2500	19,020	19,170	30,620	23,950
3000	23,280	23,460	37,750	29,780

[a]To convert values to enthalpy in Btu/lb, divide by the molecular weight.

ILLUSTRATIVE EXAMPLE 7.2

If 20,000 scfm (60°F, 1 atm) of an air stream is heated from 200°F to 2000°F, calculate the heat transfer rate required to bring about this change of temperature. Use the following enthalpy data:

$$H_{200°F} = 1170 \, \text{Btu/lbmol}$$
$$H_{2000°F} = 14,970 \, \text{Btu/lbmol}$$

SOLUTION: Convert the scfm of air to a molar flow rate, $\dot{n}$. Note that 1 lbmol of an ideal gas occupies $379 \, \text{ft}^3$ at 60°F and 1 atm:

$$\dot{n} = (\text{scfm})(1.0 \, \text{lbmol}/379 \, \text{scf}) = 20,000/379$$
$$= 52.8 \, \text{lbmol/min}$$

Set the given enthalpies of air at 200 and 2000°F.

$$H_{200°F} = 1170 \, \text{Btu/lbmol}$$
$$H_{2000°F} = 14,970 \, \text{Btu/lbmol}$$

Calculate the heat transfer rate, Q, using this enthalpy data. Apply Equation (7.3),

$$Q = \dot{n}\Delta H$$
$$= \dot{n}(H_{2000°F} - H_{200°F}) = (52.8)(14{,}970 - 1170)$$
$$= 7.29 \times 10^5 \text{ Btu/min} \qquad\blacksquare$$

This approach does not directly account for temperature variations/effects on enthalpy calculations. These are considered in the next section.

Extensive data and information on enthalpies are provided in the literature.[3]

HEAT CAPACITY VALUES

For a mathematical representation of enthalpy for a one-component one-phase system, the GPR allows one to write

$$H = H(T, P) \qquad (7.5)$$

By the rules of partial differentiation, a differential change in H is given by

$$dH = \left(\frac{\partial H}{\partial T}\right)_P dT + \left(\frac{\partial H}{\partial P}\right)_T dP \qquad (7.6)$$

The term $(\partial H/\partial P)_T$ is assumed to be negligible in most engineering applications. It is exactly zero for an ideal gas and is small for solids, liquids, and gases near ambient conditions. The term $(\partial H/\partial T)_P$ is defined as the heat capacity at constant pressure (see Chapter 2 for additional details):

$$C_P = \left(\frac{\partial H}{\partial T}\right)_P \qquad (7.7)$$

Equation (7.6) may now be written as

$$dH = C_P \, dT \qquad (7.8)$$

If average molar heat capacity data are available, this equation may be integrated to yield

$$\Delta H = \overline{C}_P \Delta T \qquad (7.9)$$

where $\overline{C}_P$ = average molar value of C_P in the temperature range ΔT.

Average molar heat capacity data are provided in Table 7.3. Thus, the calculation of enthalpy change(s) associated with a temperature change may be accomplished through the application of either Equation (7.3) or (7.9).

Table 7.3 Mean Molar Heat Capacities of Gases for the Temperature Range 0°F to T [4]

$\overline{C}_P = $ Btu/lbmol · °F

T (°F)	N_2	O_2	H_2O	CO_2	H_2	CO	CH_4	SO_2	NH_3	HCl	NO
0	6.94	6.92	7.93	8.50	6.86	6.92	8.25	9.9	8.80	6.92	7.1
200	6.96	7.03	8.04	9.00	6.89	6.96	8.42	10.0	8.85	6.96	7.2
400	6.98	7.14	8.13	9.52	6.93	7.00	9.33	10.3	9.05	7.01	7.2
600	7.02	7.26	8.25	9.97	6.95	7.05	10.00	10.6	9.40	7.05	7.3
800	7.08	7.39	8.39	10.37	6.97	7.13	10.72	10.9	9.75	7.10	7.3
1000	7.15	7.51	8.54	10.72	6.98	7.21	11.45	11.2	10.06	7.15	7.4
1200	7.23	7.62	8.69	11.02	7.01	7.30	12.13	11.4	10.43	7.19	7.5
1400	7.31	7.71	8.85	11.29	7.03	7.38	12.78	11.7	10.77	7.24	7.6
1600	7.39	7.80	9.01	11.53	7.07	7.47	13.38	11.8		7.29	7.7
1800	7.46	7.88	9.17	11.75	7.10	7.55		12.0		7.33	7.7
2000	7.53	7.96	9.33	11.94	7.15	7.62		12.1		7.38	7.8
2200	7.60	8.02	9.48	12.12	7.20	7.68		12.2		7.43	7.8
2400	7.66	8.08	9.64	12.28	7.24	7.75		12.3		7.47	7.9
2600	7.72	8.14	9.79	12.42	7.28	7.80		12.4		7.52	8.0
2800	7.78	8.19	9.93	12.55	7.33	7.86		12.5		7.57	8.0
3000	7.83	8.24	10.07	12.67	7.38	7.91		12.5		7.61	8.1
3200	7.87	8.29	10.20	12.79	7.43	7.95					
3400	7.92	8.34	10.32	12.89	7.48	8.00					
3600	7.96	8.38	10.44	12.98	7.53	8.04					
3800	8.00	8.42	10.56	13.08	7.57	8.08					
4000	8.04	8.46	10.67	13.16	7.62	8.11					
4200	8.07	8.50	10.78	13.23	7.66	8.14					
4400	8.10	8.54	10.88	13.31	7.70	8.18					
4600	8.13	8.58	10.97	13.38	7.75	8.20					
4800	8.16	8.62	11.08	13.44	7.79	8.23					

ILLUSTRATIVE EXAMPLE 7.3

Refer to Illustrative Example 7.2. Use the following average heat capacity data to perform the calculations:

$$\overline{C}_{P,\text{av}} \text{ (over } 200-2000°F \text{ range)} = 7.53 \text{ Btu/lbmol} \cdot °F$$

SOLUTION: Calculate the heat transfer rate using the average heat capacity data provided:

$$Q = \Delta H$$
$$Q = \dot{n}\overline{C}_P\Delta T$$
$$= (52.8)(7.53)(2000 - 200)$$
$$= 7.16 \times 10^5 \text{ Btu/min}$$

This result is in excellent agreement with the result obtained in the previous example.   ∎

ILLUSTRATIVE EXAMPLE 7.4

Given the mass flow rate of a gas stream and its heat capacity, determine the required heat rate to change the gas stream from one temperature to another. Data are provided below.

Mass flow rate of gas stream, $\dot{m} = 1200$ lb/min
Average heat capacity of gas, $\overline{c}_P = 0.26$ Btu/lb · °F
Initial temperature, $T_1 = 200°F$
Final temperature, $T_2 = 1200°F$

SOLUTION: Write the equation describing the required heat rate, Q:

$$Q = \Delta H$$
$$Q = \dot{m}\overline{c}_P\Delta T$$
$$= \dot{m}\overline{c}_P(T_2 - T_1)$$

Calculate the required heat rate, Q, in Btu/min employing the data provided:

$$Q = \dot{m}\overline{c}_P(T_2 - T_1)$$
$$= (1200)(0.26)(1200 - 200)$$
$$= 3.12 \times 10^5 \text{ Btu/min}$$ ∎

ILLUSTRATIVE EXAMPLE 7.5

A total of 18.7×10^6 Btu/h of heat is transferred from the flue gas in a boiler. Calculate the outlet temperature of the gas stream using the following information:

  Average heat capacity, $\bar{c}_P$, of gas $= 0.26$ Btu/lb · °F

  Gas mass flow rate, $\dot{m} = 72{,}000$ lb/h

  Gas inlet temperature, $T_1 = 1200$°F

SOLUTION: Solve the conservation law for energy for the gas outlet temperature:

$$Q = \Delta H = \dot{m}\bar{c}_P\Delta T$$
$$= \dot{m}\bar{c}_P(T_2 - T_1) \qquad (7.9)$$

Rearranging,

$$T_2 = [Q/(\dot{m}\bar{c}_P)] + T_1$$

Substitute and calculate the gas outlet temperature:

$$T_2 = [-18.7 \times 10^6/(72{,}000)(0.26)] + 1200$$
$$= 200°F$$

The above calculation is based on adiabatic conditions, i.e., the entire heat load is transferred from the flowing gas. The unit is assumed to be perfectly insulated so that no heat is transferred to the surroundings. A 2–10% heat loss is often assumed in many engineering calculations—depending on the size of the unit, the refractory (insulation) construction and the local environment (outdoors, rain, wind, etc.).   ∎

A more rigorous approach to enthalpy calculations can be provided if heat capacity variation with temperature is available. If the heat capacity is a function of the temperature, the enthalpy change (as written) in Equation (7.8) applies. If the temperature variation of the heat capacity is given by

$$C_P = \alpha + \beta T + \gamma T^2 \qquad (7.10)$$

Equation (7.8) may be integrated directly between some reference or standard temperature (T_0) and the final temperature (T_1)

$$\Delta H = H_1 - H_0$$
$$= \alpha(T_1 - T_0) + (\beta/2)(T_1^2 - T_0^2) + (\gamma/3)(T_1^3 - T_0^3) \qquad (7.11)$$

Equation (7.8) may also be integrated if the heat capacity is a function of temperature of the form

$$C_P = a + bT + cT^{-2} \qquad (7.12)$$

The enthalpy change is then given by

$$\Delta H = a(T_1 - T_0) + (b/2)(T_1^2 - T_0^2) - c(T_1^{-1} - T_0^{-1}) \qquad (7.13)$$

Tabulated values of a, b, c and α, β, γ for a host of compounds (including some chlorinated organics), drawn from the literature[5-7] are provided in Tables 7.4 and 7.5. The reader is cautioned that the use of Equations (7.11) and (7.13) requires the use of the absolute temperature in Kelvin.

ILLUSTRATIVE EXAMPLE 7.6

Determine the heat rate required (in Btu/h) to raise the temperature of 1000 lb/h of carbon dioxide from 200°F to 3200°F. Perform the calculation assuming

$$C_P = 10.57 + 2.10 \times 10^{-3}T - 2.06 \times 10^5/T^2; \; Btu/lbmol \cdot °R$$

SOLUTION: Convert the flow rate from lb/h to lbmol/h:

$$\dot{n} = 1000/44$$
$$= 22.73 \, lbmol/h$$

Convert the two temperatures from °F to K:

$$T_0 = 200°F = 200 + 460 = 660°R = 660/1.8$$
$$= 366.48K$$
$$T_1 = 3200°F = 3200 + 460 = 3660°R = 3660/1.8$$
$$= 2033.15K$$

Assign values to a, b, and c:

$$a = 10.57$$
$$b = 2.1 \times 10^{-3}$$
$$c = -2.06 \times 10^5$$

Write the equation for Q in terms of a, b, and c. See Equation (7.13).

$$Q = \Delta H = \dot{n}[a(T_1 - T_0) + (b/2)(T_1^2 - T_0^2) - c(T_0 - T_1)/T_0 T_1]$$

Table 7.4   Molar Heat Capacities[5-7]

Compound	Formula	Temperature range (K)	a	$b \times 10^3$	$c \times 10^{-5}$
Inorganic gases					
Ammonia	NH_3	298–1800	7.11	6.00	−0.37
Bromine	Br_2	298–3000	8.92	0.12	−0.30
Carbon monoxide	CO	298–2500	6.79	0.98	−0.11
Carbon dioxide	CO_2	298–2500	10.57	2.10	−2.06
Carbon disulfide	CS_2	298–1800	12.45	1.60	−1.80
Chlorine	Cl_2	298–3000	8.85	0.16	−0.68
Hydrogen	H_2	298–3000	6.52	0.78	+0.12
Hydrogen sulfide	H_2S	298–2300	7.81	2.96	−0.46
Hydrogen chloride	HCl	298–2000	6.27	1.24	+0.30
Hydrogen cyanide	HCN	298–2500	9.41	2.70	−1.44
Nitrogen	N_2	298–3000	6.83	0.90	−0.12
Nitrous oxide	N_2O	298–2000	10.92	2.06	−2.04
Nitric oxide	NO	298–2500	7.03	0.92	−0.14
Nitrogen dioxide	NO_2	298–2000	10.07	2.28	−1.67
Nitrogen tetroxide	N_2O_4	298–1000	20.05	9.50	−3.56
Oxygen	O_2	298–3000	7.16	1.00	−0.40
Sulfur dioxide	SO_2	298–2000	11.04	1.88	−1.84
Sulfur trioxide	SO_3	298–1500	13.90	6.10	−3.22
Water	H_2O	298–2750	7.30	2.46	0.00
Chlorinated organic liquids					
Methyl chloride	CH_3Cl	298–1800	3.44	23.52	−0.81
Dichloromethane	CH_2Cl_2	298–1800	5.34	27.00	−1.16
Chloroform	$CHCl_3$	298–1800	8.91	27.18	−1.34
Carbon tetrachloride	CCl_4	298–1800	1.39	24.97	−1.36
Ethyl chloride	C_2H_5Cl	298–1800	2.69	46.81	−1.81
1,1-Dichloroethane	$C_2H_4Cl_2$	298–1800	5.99	47.31	−2.00
1,1,2,2-Tetrachloroethane	$C_2H_2Cl_4$	298–1800	1.15	49.83	−2.38
n-Propyl chloride	C_3H_7Cl	298–1800	2.88	66.18	−2.56
1,3-Dichloropropane	$C_3H_6Cl_2$	298–1800	7.35	63.07	−2.51
n-Butyl chloride	C_4H_9Cl	298–1800	3.22	85.65	−3.32
1-Chloropentane	$C_5H_{11}Cl$	298–1800	3.57	10.51	−4.07
1-Chloroethylene	C_2H_3Cl	298–1800	3.70	35.30	−1.48
trans-1,2-Dichloroethylene	$C_2H_6Cl_2$	298–1800	7.13	34.46	−1.55
Trichloroethylene	C_2HCl_3	298–1800	10.76	33.59	−1.64
Tetrachloroethylene	C_2Cl_4	298–1800	14.99	31.13	−1.63
3-Chloro-l-propene	C_3H_5Cl	298–1800	4.26	52.61	−2.07
Chlorobenzene	C_6H_5Cl	298–1800	−1.12	95.63	−4.24
p-Dichlorobenzene	$C_6H_5Cl_2$	298–1800	3.49	92.98	−4.24
Hexachlorobenzene	C_6Cl_6	298–1800	21.53	81.64	−4.13

Constants for the equation $C_P = a + bT + CT^{-2}$, where T is in K and C_P is in Btu/lbmol · °F or cal/gmol · °C.

Table 7.5 Molar Heat Capacities[5-7]

Compound	Formula	Constants		
		α	$\beta \times 10^3$	$\gamma \times 10^6$
Normal paraffins (gases)				
Methane	CH_4	3.381	18.044	-4.300
Ethane	C_2H_6	2.247	38.201	-11.049
Propane	C_3H_8	2.410	57.195	-17.533
n-Butane	C_4H_{10}	3.844	73.350	-22.655
n-Pentane	C_5H_{12}	4.895	90.113	-28.039
n-Hexane	C_6H_{14}	6.011	106.746	-33.363
n-Heptane	C_7H_{16}	7.094	123.447	-38.719
n-Octane	C_8H_{18}	8.163	140.217	-44.127
Increment per C atom above C_8	—	1.097	16.667	-5.338
Normal monoolefins (gases) (1-alkenes)				
Ethylene	C_2H_4	2.830	28.601	-8.726
Propylene	C_3H_6	3.253	45.116	-13.740
1-Butene	C_4H_8	3.909	62.848	-19.617
1-Pentene	C_5H_{10}	5.347	78.990	-24.733
1-Hexene	C_6H_{12}	6.399	95.752	-30.116
1-Heptene	C_7H_{14}	7.488	112.440	-35.462
1-Octene	C_8H_{16}	8.592	129.076	-40.775
Increment per C atom above C_8	—	1.097	16.667	-5.338
Miscellaneous gases				
Acetaldehyde	C_2H_4O	3.364	35.722	-12.236
Acetylene	C_2H_2	7.331	12.622	-3.889
Ammonia	NH_3	6.086	8.812	-1.506
Benzene	C_6H_6	-0.409	77.621	-26.429
1,3-Butadiene	C_4H_6	5.432	53.224	-17.649
Carbon dioxide	CO_2	6.214	10.396	-3.545
Carbon monoxide	CO	6.420	1.665	-0.196
Chlorine	Cl_2	7.576	2.424	-0.965
Cyclohexane	C_6H_{12}	-7.701	125.675	-41.584
Ethyl alcohol	C_3H_6O	6.990	39.741	-11.926
Hydrogen	H_2	6.947	-0.200	0.481
Hydrogen chloride	HCl	6.732	0.433	0.370
Hydrogen sulfide	H_2S	6.662	5.134	-0.854
Methyl alcohol	CH_4O	4.394	24.274	-6.855
Nitric oxide	NO	7.020	-0.370	2.546
Nitrogen	N_2	6.524	1.250	-0.001
Oxygen	O_2	6.148	3.102	-0.923
Sulfur dioxide	SO_2	7.116	9.512	3.511
Sulfur trioxide	SO_3	6.077	23.537	-0.687

(*Continued*)

TABLE 7.5 *Continued*

Compound	Formula	α	β × 10³	γ × 10⁶
			Constants	
Toluene	C_7H_8	0.576	93.493	−31.227
Water	H_2O	7.256	2.298	0.283
Chlorinated organic liquids				
Methyl chloride	CH_3Cl	8.994	10.280	−21.218
Dichloromethane	CH_2Cl_2	13.672	7.616	−33.272
Chloroform	$CHCl_3$	18.453	4.932	−37.923
Carbon tetrachloride	CCl_4	23.777	2.224	−39.558
Ethyl chloride	C_2H_5Cl	15.267	17.060	−48.798
1,1-Dichloroethane	$C_2H_4Cl_2$	20.014	14.315	−54.917
1,1,2,2-Tetrachloroethane	$C_2H_2Cl_4$	28.397	10.338	−66.739
n-Propyl chloride	C_3H_7Cl	20.703	24.070	−69.351
1,3-Dichloropropane	$C_3H_6Cl_2$	24.840	21.810	−68.214
n-Butyl chloride	C_4H_9Cl	26.258	31.208	−89.482
1-Chloropentane	$C_5H_{11}Cl$	31.812	38.345	−109.612
1-Chloroethylene	C_2H_3Cl	14.058	10.939	−40.586
trans-1,2-Dichloroethylene	$C_2H_6Cl_2$	18.050	8.845	−42.981
Trichloroethylene	C_2HCl_3	22.410	6.394	−46.257
Tetrachloroethylene	C_2Cl_4	26.595	4.106	−46.208
3-Chloro-1-propene	C_3H_5Cl	18.728	18.514	−56.495
Chlorobenzene	C_6H_5Cl	28.558	25.786	−116.150
p-Dichlorobenzene	$C_6H_5Cl_2$	33.265	23.025	−116.848
Hexachlorobenzene	C_6Cl_6	50.759	13.286	−115.620

Constants for the equation $C_P = \alpha + \beta T + \gamma T^2$, where T is in K and C_P is in Btu/lbmol · °F or cal/gmol · °C; applicable range: 298–1000K.

Calculate Q in Btu/h.

$$Q = (22.73)(1.8)[10.57(1666.67)$$
$$+ (2.10 \times 10^{-3}/2)(2033.15^2 - 366.48^2)$$
$$- (-2.06 \times 10^5)(366.48 - 2033.15)/(366.48)(2033.15)]$$
$$= 873,600 \, \text{Btu/h} \qquad \blacksquare$$

ILLUSTRATIVE EXAMPLE 7.7

Refer to Illustrative Example 7.6. Perform the calculation assuming

$$C_P = 6.214 + 10.396 \times 10^{-3} \, T - 3.545 \times 10^{-6} \, T^2; \; \text{Btu} \cdot \text{lbmol}/°\text{R}$$

SOLUTION: Assign values to α, β, and γ.

$$\alpha = 6.214$$
$$\beta = 10.396 \times 10^{-3}$$
$$\gamma = -3.545 \times 10^{-6}$$

Write the equation for Q in terms of α, β, and γ. See Equation (7.71).

$$Q = \dot{n}[\alpha(T_1 - T_0) + (\beta/2)(T_1^2 - T_0^2) + (\gamma/3)(T_1^3 - T_0^3)]$$

Calculate Q.

$$Q = (22.73)(1.8)\,[(6.214)(2033.15 - 366.48) + (10.396 \times 10^{-3}/2)(2033.15^2 - 366.48^2)$$
$$- (3.545 \times 10^{-6}/3)(2033.15^3 - 366.48^3)]$$
$$= 870{,}300 \text{ Btu/h}$$

As expected, the results of this and the previous Illustrative Example compare favorably.   ∎

ILLUSTRATIVE EXAMPLE 7.8

Determine the percentage of a river stream's flow available to an industry for cooling such that the river temperature does not increase more than $10°F$. Fifty percent of the industrial withdrawal is lost by evaporation and the industrial water returned to the river is $60°F$ warmer than the river.

Note: This problem is a modified and edited version (with permission) of an illustrative example prepared by Ms. Marie Gillman, a graduate mechanical engineering student at Manhattan College.

SOLUTION: Draw a flow diagram representing the process. See Fig. 7.1. Express the volumetric flow lost by evaporation from the process in terms of that entering the process:

$$q_{\text{lost}} = 0.5 q_{\text{in}}$$

Express the process outlet temperature and the maximum river temperature in terms of the upstream temperature:

$$T_{\text{out}} = T_{\text{up}} + 60°F$$
$$T_{\text{mix}} = T_{\text{up}} + 10°F$$

Using the conservation law for mass, express the process outlet flow in terms of the process inlet flow. Also express the flow bypassing the process in terms of the upstream and process

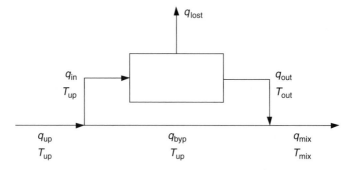

Figure 7.1 Flow diagram for Illustrative Example 7.8.

inlet flows:

$$q_{out} = 0.5q_{in}$$
$$q_{byp} = q_{up} - q_{in}$$
$$q_{mix} = q_{up} - 0.5q_{in}$$

The flow diagram with the expressions developed above are shown in Fig. 7.2.

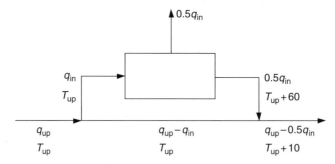

Figure 7.2 Flow diagram for Illustrative Example 7.8 after applying mass balances.

Noting that the enthalpy of any stream can be represented by $qc_p\rho(T - T_{ref})$, an energy balance around the downstream mixing point leads to

$$(q_{up} - q_{in})c_p\rho(T_{up} - T_{ref}) + 0.5q_{in}c_p\rho(T_{up} + 60 - T_{ref}) = (q_{up} - 0.5q_{in})c_p$$
$$\times \rho(T_{up} + 10 - T_{ref})$$

Note that T_{ref} may be arbitrarily assigned a value and thus indirectly defines a basis for the enthalpy. Setting $T_{ref} = 0$ and assuming that density and heat capacity are constant yields

$$(q_{up} - q_{in})T_{up} + 0.5q_{in}(T_{up} + 60) = (q_{up} - 0.5q_{in})(T_{up} + 10)$$

The equation may now be solved for the inlet volumetric flow to the process in terms of the upstream flow.

$$q_{up}T_{up} - q_{in}T_{up} + 0.5q_{in}T_{up} + 30q_{in}$$
$$= q_{up}T_{up} + 10q_{up} - 0.5q_{in}T_{up} - 5q_{in}$$

Canceling terms produces

$$35q_{in} = 10q_{up}$$
$$q_{in} = 0.286q_{up}$$

Therefore, 28.6% of the river flow, q_{up}, is available for cooling.

Note that the problem can also be solved by setting $T_{ref} = T_{up}$. The solution to the problem is greatly simplified, since for this condition, $T_{ref} - T_{up} = 0$.   ■

ILLUSTRATIVE EXAMPLE 7.9

A plant has three streams to be heated (see Table 7.6) and three streams to be cooled (see Table 7.7). Cooling water (90°F supply, 155°F return) and steam (saturated at 250 psia) are available. Note that the saturated steam at 250 psia has a temperature of 401°F. Calculate the heating and cooling duties and indicate what utility (or utilities) should be employed.

SOLUTION: The sensible heating duties for all streams are first calculated. The results are shown in Table 7.8.

Table 7.6 Streams to be Heated in Illustrative Example 7.9

Stream	Flowrate (lb/h)	c_P [Btu/(lb · °F)]	T_{in} (°F)	T_{out} (°F)
1	50,000	0.65	70	300
2	60,000	0.58	120	310
3	80,000	0.78	90	250

Table 7.7 Streams to be Cooled in Illustrative Example 7.9

Stream	Flowrate (lb/h)	c_P [Btu/(lb · °F)]	T_{in} (°F)	T_{out} (°F)
4	60,000	0.70	420	120
5	40,000	0.52	300	100
6	35,000	0.60	240	90

Table 7.8 Duty Requirements in Illustrative Example 7.9

Stream	Duty, Btu/h
1	7,745,000
2	6,612,000
3	9,984,000
4	12,600,000
5	4,160,000
6	3,150,000

The total heating and cooling duties can now be compared.

$$\text{Heating: } 7,745,000 + 6,612,000 + 9,984,000 = 24,341,000 \text{ Btu/h}$$
$$\text{Cooling: } 12,600,000 + 4,160,000 + 3,150,000 = 19,910,000 \text{ Btu/h}$$
$$\text{Heating} - \text{Cooling} = 24,341,000 - 19,910,000 = 4,310,000 \text{ Btu/h}$$

As a minimum, 4,431,000 Btu/h will have to be supplied by steam or another hot medium. ∎

ILLUSTRATIVE EXAMPLE 7.10

Refer to Illustrative Example 7.9. Devise a network of heat exchangers that will make full use of heating and cooling streams against each other, using utilities only if necessary.

SOLUTION: The reader should note that this is an open-ended problem (see Chapter 16).

Figure 7.3 represents a system of heat exchangers that will transfer heat from the hot streams to the cold ones in the amounts desired. It is important to note that this is but one of

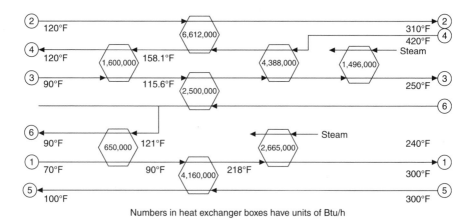

Numbers in heat exchanger boxes have units of Btu/h

Figure 7.3 Flow diagram for Illustrative Example 7.10.

many possible schemes. The optimum system would require a trial-and-error procedure that would examine a host of different schemes. Obviously, the economics would come into play.

It should also be noted that in many chemical and petrochemical plants there are cold streams that must be heated and hot steams that must be cooled. Rather than use steam to do all the heating and cooling water to do all the cooling, it is often advantageous, as demonstrated in this problem, to have some of the hot streams heat the cold ones. The problem of optimum heat exchanger networks has been extensively studied and is available in the literature. This problem gives one simple illustration.

Finally, highly interconnected networks of exchangers can save a great deal of "quality" energy (referred to in Chapter 6) in a chemical plant. The more interconnected they are, however, the harder the plant is to control, start-up, and shut down. Often auxiliary heat sources and cooling sources must be included in the plant design in order to ensure that the plant can operate smoothly. ∎

ILLUSTRATIVE EXAMPLE 7.11

Calcium oxide is the choice of reagent for most flue gas desulfurization (FGD) operations. Heat capacity values for this solid are provided below:

$$a = 11.67$$

$$b = 1.08 \times 10^{-3}$$

$$c = -1.56 \times 10^5$$

Estimate the additional heat that must be added to a FGD system to increase 1200 lb/hr of CaO from 42°F to 68°F.

SOLUTION: Refer to Equation (7.13)

$$Q = \Delta H = \dot{n}[a(T_2 - T_1) + (b/2)(T_2^2 - T_1^2) - c(T_1 - T_2)/T_1 T_2]$$

Convert the temperatures to K:

$$T_1 = 42°F = 500°R = 500/1.8$$
$$= 278K$$
$$T_2 = 68°F = 528°R = 528/1.8$$
$$= 293K$$

Substitute values.

$$Q = (1200/56)[11.67(293 - 278) + 1.08 \times 10^{-3}(293^2 - 278^2)$$
$$+ 1.56 \times 10^5(278 - 293)/(278)(293)](1.8)$$
$$= 21.43[175 + 9.25 - 28.7](1.8)$$
$$= 6000 \, \text{Btu/h}$$
$$= 1510 \, \text{kcal/hr}$$

∎

In dealing with mixtures, the authors recommend employing either a mass-average or mole-average value of the components in the mixture, i.e.,

$$\bar{C}_P = \sum_{i=1}^{n} w_i \, C_{Pi}; \quad \text{mass basis} \tag{7.10}$$

where C_P = average heat capacity; mass basis
$\quad w_i$ = mass fraction of component i
$\quad C_{Pi}$ = heat capacity of component i; mass basis
$\quad n$ = number of components
On a mole basis,

$$\bar{C}_P = \sum_{i=1}^{n} y_i \, C_{Pi}; \quad \text{mole basis} \tag{7.11}$$

where y_i = mole fraction of component i.
Thus, the average heat capacity on a mole basis for a two component mixture is given by

$$\bar{C}_P = y_1 C_{P1} + y_2 C_{P2} \tag{7.12}$$

Finally, it should be noted that heat capacity values for solids are available in the literature.[9] Values for other gases and liquids are also available. Much of the data is provided in terms of a, b, c and α, β, γ values.

PREDICTIVE METHODS FOR HEAT CAPACITY

Methods for predicting (as opposed to experimental) heat capacity values are available in the literature.[10] A procedure to calculate heat capacities for ideal gases was developed by Rihani and Doraiswamy[11] that is based on a group contribution method. The describing equation is given by

$$C_P^0 = \sum_i N_i a_i + \sum_i N_i b_i T + \sum_i N_i c_i T^2 + \sum_i N_i d_i T^3 \tag{7.13}$$

where N_i is the number of groups of type i, T is the temperature in Kelvins, and a_i, b_i, c_i, and d_i are additive group parameters.[11,12] For example, the structure of 2-methyl-1,3-butadiene is

$$H_2C{=}CH{-}C{=}CH_2$$
$$\underset{\displaystyle CH_3}{\vert}$$

and the groups for this compound are:

$$- CH_3 \qquad \overset{H}{\underset{/}{\diagdown}} C{=}CH_2 \qquad and \qquad \overset{\diagdown}{\underset{/}{}} C{=}CH_2$$

Group contribution values for this chemical are provided Table 7.9. Procedures to calculate real gas heat capacities are provided by Reid et al.[12]

Table 7.9   Group Contribution Values

	N	a	$b \times 10^2$	$c \times 10^4$	$d \times 10^6$
2-Methyl-1,3-butadiene:					
$-CH_3$	1	0.6087	2.1433	−0.0852	0.01135
$HC{=}CH_2$	1	0.2773	3.4580	−0.1918	0.004130
$C{=}CH_2$	1	−0.4173	3.8857	−0.2783	0.007364
$\sum (N)$(group parameter)		0.4687	9.4870	−0.5553	0.02284

ILLUSTRATIVE EXAMPLE 7.12

Write the ideal gas law equation for the heat capacity of 2-methyl-1,3-butadiene and calculate its value at 1000K. Refer to Table 7.9 and employ Equation (7.13).

SOLUTION: As noted above, the compound's structure is

$$H_2C{=}CH{-}\underset{\underset{CH_3}{|}}{C}{=}CH_2$$

with contributing groups

$$- CH_3 \qquad \overset{H}{\underset{/}{\diagdown}} C{=}CH_2 \qquad \overset{\diagdown}{\underset{/}{}} C{=}CH_2$$

The equation takes the form:

$$C_P^0 = 0.4687 + 9.4870 \times 10^{-2}(T) + (-0.5553 \times 10^{-4})(T)^2$$
$$+ 0.02284 \times 10^{-6}(T)^3$$

Substituting $T = 1000\text{K}$,

$$C_P^0 = 0.4687 + 9.4870 \times 10^{-2}(1000)$$
$$+ (-0.5553 \times 10^{-4})(1000)^2 + 0.02284 \times 10^{-6}(1000)^3$$
$$= 0.4687 + 94.9 - 55.5 + 22.84$$
$$= 62.71 \, \text{cal/gmol} \cdot \text{K}$$
$$= 62.71 \, \text{Btu/lbmol} \cdot {}^\circ\text{R}$$ ∎

REFERENCES

1. Adapted from: source unknown, date unknown.
2. J. SANTOLERI: private communication to L. Theodore, 1998.
3. K. KOBE and E. LONG, "Thermochemistry for the Petroleum Industry," *Pet. Refiner.*, 28(11), 127–132, 1949.
4. E. WILLIAMS and R. JOHNSON, "*Stoichiometry for Chemical Engineers*," McGraw-Hill, New York, 1958.
5. J. SANTOLERI, J. REYNOLDS, and L. THEODORE, "*Introduction to Hazardous Waste Incineration*," 2nd edition, John Wiley & Sons, Hoboken, NJ, 2000.
6. K. KELLEY, *U.S. Bur. Mines Bull.*, 584, 1960.
7. L. THEODORE and J. REYNOLDS: personal notes, Manhattan College, 1986.
8. H. M. SPENCER, J. JUSTICE, and G. FLANAGAN, *J. Am. Chem. Soc.*, 56, 2311, 1934; 64, 2511, 1942; 67, 1859, 1945; *Ind. Eng. Chem.* 40, 2152, 1948.
9. D. GREEN and R. PERRY (editors), "*Perry's Chemical Engineers Handbook*," 8th edition, McGraw-Hill, New York, 2008.
10. N. P. CHOPEY, "*Handbook of Chemical Engineering Calculations*," 2nd edition, McGraw-Hill, New York, 1994.
11. D. RIHANI and L. DORAISWAMY, *Ind. Eng. Chem. Fund.*, 4:17, ACS, Washington DC, 1965.
12. R. REID, J. PRAUSNITZ, and T. SHERWOOD, "*Properties of Gases and Liquids*," 3rd edition, McGraw-Hill, New York, 1977.

NOTE: Additional problems for each chapter are available for all readers at www. These problems may be used for additional review or homework purposes.

Chapter **8**

Latent Enthalpy Effects

Henri-Frédéric Amiei [1821–1881]

A man without passion is only a latent force, only a possibility, like a stone waiting for the blow from the iron to give forth sparks.

—Journal [1883]

INTRODUCTION

The term phase for a pure substance refers to a state of matter that is gas, liquid, or solid. Latent enthalpy effects are associated with phase changes. These phase changes involve no change in temperature but there is a transfer of energy to/from the substance. There are three possible latent effects, as detailed below:

1 vapor–liquid

2 liquid–solid

3 vapor–solid

Vapor–liquid changes are referred to as condensation when the vapor is condensing and vaporization when liquid is vaporizing. Liquid–solid changes are referred to as melting when the solid melts to liquid and freezing when a liquid solidifies. Vapor–solid changes are referred to as sublimation. One should also note that there are enthalpy effects associated with a phase change of a solid to another solid form; however, this enthalpy effect is small compared to the other effects mentioned above. Finally, vapor–liquid changes are highlighted in this chapter since they are of primary interest to the practicing engineer.

Latent enthalpy effects can be measured directly in a calorimeter for vapor–liquid changes. However, these effects can be predicted from semi-theoretical equations provided in the literature. This chapter discusses these methods and their application under the following section titles:

1 The Clausius–Clapeyron (C–C) Equation

2 Predictive Methods: Normal Boiling Point

3 Predictive Methods: Other Temperatures

4 Industrial Applications

A discussion of vapor pressure estimation is included in the C–C presentation since the describing equation is based on the temperature variation of vapor pressure. Procedures to estimate the derivative of the vapor pressure (p') variation with temperature, i.e., dp'/dT, is included in the Illustrative Examples. The Industrial Applications section provides five Illustrative Examples, three of which are power plant (utility) oriented.

THE CLAUSIUS–CLAPEYRON (C–C) EQUATION

The C–C equation can be derived from basic principles. The original form of the Clausius–Clapeyron equation is given by

$$\frac{dp'}{dT} = \frac{\Delta H}{T(v_g - v_l)} \tag{8.1}$$

where
- p' = vapor pressure
- T = temperature
- ΔH = enthalpy of vaporization
- v = specific volume

At a pressure significantly below the critical pressure

$$v_g \ggg v_l \tag{8.2}$$

so that

$$\frac{dp'}{dT} = \frac{\Delta H}{Tv_g} = \frac{\Delta H}{Tv} \tag{8.3}$$

Applying the ideal gas law to v, i.e., $v = RT/p'$

$$\frac{dp'}{dT} = \left(\frac{\Delta H}{RT^2}\right)p' \tag{8.4}$$

If ΔH is assumed constant, Equation (8.4) may be rewritten as

$$\frac{dp'}{p'} = \frac{\Delta H}{R}\frac{dT}{T^2} \tag{8.5}$$

and integrated to give

$$\ln\left(\frac{p_2'}{p_1'}\right) = -\frac{\Delta H(\text{at } T_{av})}{R}\left(\frac{1}{T_2} - \frac{1}{T_1}\right) \tag{8.6}$$

where $T_{av} = (T_2 + T_1)/2$.

Examining any of the above equations, one notes that vapor pressure data is required in order to use the C–C equation to calculate the enthalpy of vaporization. In addition, the derivative dp'/dT must be evaluated.

Vapor pressure data is available in the literature.[1] However, there are two equations that can be used in lieu of vapor pressure information—the Clapeyron equation and the Antoine equation. The two constant (A, B) Clapeyron equation is given by

$$\ln p' = A - (B/T) \tag{8.7}$$

where p' and T are the vapor pressure and temperature, respectively. The three constant (A, B, C) Antoine equation is given by

$$\ln p' = A - [B/(T + C)] \tag{8.8}$$

Note that for both equations, the units of p' and T must be specified for given values of A, B, and (possibly) C. Values for the Clapeyron equation coefficients—A and B—are provided in Table 8.1 for some compounds. Some Antoine equation coefficients—A, B, and C—are listed in Table 8.2. Additional values for the coefficients for both equations are available in the literature.[1,2]

Table 8.1   Appropriate Clapeyron Equation Coefficients*

	A	B
Acetaldehyde	18.0	3.32×10^3
Acetic anhydride	20.0	5.47×10^3
Ammonium chloride	23.0	10.0×10^3
Ammonium cyanide	22.9	11.5×10^3
Benzyl alcohol	21.9	7.14×10^3
Hydrogen peroxide	20.4	5.82×10^3
Nitrobenzene	18.8	5.87×10^3
Nitromethane	18.5	4.43×10^3
Phenol	19.8	5.96×10^3
Tetrachloroethane	17.5	4.38×10^3

*T in K, p' in mm Hg.

Table 8.2   Antoine Equation Coefficients*

	A	B	C
Acetone	14.3916	2795.82	230.00
Benzene	13.8594	2773.78	220.07
n-Pentane	13.8183	2477.07	233.21
Ethanol	16.6758	3674.49	226.45
n-Heptane	13.8587	2911.32	216.64
Methanol	16.5938	3644.30	239.76
n-Butane	13.6608	2154.70	238.79
Toluene	14.0098	3103.01	219.79
Water	16.2620	3799.89	226.35

*T in °C, p' in kPa.

ILLUSTRATIVE EXAMPLE 8.1

The Clapeyron equation coefficients have been experimentally determined for acetone to be

$$A = 15.03$$
$$B = 2817$$

with p' and T in mm Hg and K, respectively. Estimate the vapor pressure at 0°C.

SOLUTION: The Clapeyron equation is given by

$$\ln p' = A - (B/T) \tag{8.7}$$

Employing the correct units, substitute to give the vapor pressure, p', of acetone at 0°C.

$$\ln p' = 15.03 - 2817/(0 + 273)$$
$$= 4.7113$$
$$p' = 111.2 \, \text{mm Hg} \qquad \blacksquare$$

ILLUSTRATIVE EXAMPLE 8.2

The Antoine coefficients for acetone are:

$$A = 16.65$$
$$B = 2940$$
$$C = -35.93$$

with p' and T in mm Hg and K, respectively. Use the Antoine equation to estimate the vapor pressure of acetone at 0°C.

SOLUTION: The Antoine equation is given by

$$\ln p' = A - B/(T + C) \qquad (8.8)$$

Substituting, the vapor pressure of acetone at $0°C$ predicted by the Antoine equation is

$$\ln p' = 16.65 - 2940/(273 - 35.93)$$
$$= 4.2486$$
$$p' = 70.01 \text{ mm Hg} \qquad \blacksquare$$

ILLUSTRATIVE EXAMPLE 8.3

Comment on the results of Illustrative Examples 8.1 and 8.2.

SOLUTION: The reader should note that the Clapeyron equation generally overpredicts the vapor pressure at or near ambient conditions. The Antoine equation is widely used in industry and usually provides excellent results. Also note that, contrary to statements appearing in the *Federal Register* and some *Environmental Protection Agency* (*EPA*) publications, vapor pressure is not a function of pressure. $\blacksquare$

In order to employ Equation (8.6), the derivative must be evaluated in order to determine the enthalpy of vaporization. One method to accomplish this is to analytically form the derivation dp'/dT for either Equation (8.7) and (8.8). For the Clapeyron equation,

$$\frac{dp'}{dT} = \frac{d(e^{[A-(B/T)]})}{dT}$$
$$= \frac{(B)(e^{[A-(B/T)]})}{T^2} \qquad (8.9)$$

where T is in K.
    For the Antoine equation

$$\frac{dp'}{dT} = \frac{d(e^{[A-(B/(T+C))]})}{dT}$$
$$= (e^{A-[B/(T+C)]})\left[\frac{B}{(T+C)^2}\right] \qquad (8.10)$$

where T is in °C.
    Thus, the derivative of the vapor pressure with respect to temperature can be obtained directly from Equation (8.9) or (8.10) by direct substitution of the temperature for which the derivative is required. In addition, since dp'/dT is the slope of the vapor pressure vs temperature curve at the temperature in question, its value can be obtained directly from a p' vs T graph. Both methods are employed in the two Illustrative Examples to follow.

ILLUSTRATIVE EXAMPLE 8.4

Calculate the enthalpy of vaporization of n-butane at 270°F using the data provided below. Apply the C–C equation.

Data point	T, °F	p', atm	v_l, ft^3/lb	v_g, ft^3/lb
1	250	22.01	0.0382	0.253
2	260	24.66	0.0393	0.222
3	270	27.13	0.0408	0.192
4	280	29.79	0.0429	0.165
5	290	32.42	0.0449	0.140

The experimental value for ΔH is approximately 67.0 Btu/lb.

SOLUTION: As noted above, the derivative in the C–C equation may be evaluated using several different methods. The tangent to the curve of a graph of p' vs T is one approach. The slope of the two points "straddling" the 270°F data, i.e. 260° and 280°F, would provide another approximation. The "straddling" approach is employed in this example.

Write the Clausius–Clapeyron equation:

$$\Delta H = T(\Delta V)dp'/dT$$

Estimate the slope dp'/dT at 270°F in atm/°R:

$$dp'/dT \approx \Delta p'/\Delta T$$

"Straddling" temperatures 260° and 280°F gives

$$
\begin{aligned}
\Delta p'/\Delta T &= (29.79 - 24.66)/(280 - 260) \\
&= (29.79 - 24.66)/(740 - 720) \\
&= 0.257
\end{aligned}
$$

Calculate v in ft^3/lb at $T = 270$°F,

$$\Delta v = 0.192 - 0.0408$$
$$= 0.151\ \text{ft}^3/\text{lb}$$

Calculate the enthalpy of vaporization in Btu/lb:

$$T = 270 + 460 = 730°R$$
$$\Delta H = (730)(0.151)(0.257)(14.7)(144)/778$$
$$= 77.1\ \text{Btu/lb}$$

Note: 1 Btu = 778 ft-lb$_f$. The agreement is good considering the approach employed in estimating the derivative. ∎

ILLUSTRATIVE EXAMPLE 8.5

Refer to Illustrative Example 8.4. Use the other approaches to evaluate the p' vs T derivative in order to calculate the enthalpy of vaporization.

SOLUTION: Another method involves "straddling" the two outer data points, i.e. points (1) and (5). For this case

$$\frac{dp'}{dT} \approx \frac{\Delta p'}{\Delta T}$$
$$= \frac{32.42 - 22.01}{290 - 250}$$
$$= \frac{10.41}{40}$$
$$= 0.260 \, \text{atm}/°\text{F}$$

A second approach could be based on the average of two slopes based on the middle and two outer points. For this case

$$\left.\frac{dp'}{dT}\right|_{1-3} = \left.\frac{\Delta p'}{\Delta T}\right|_{1-3}$$
$$= \frac{27.13 - 22.01}{270 - 250}$$
$$= \frac{5.12}{20}$$
$$= 0.256$$

$$\left.\frac{dp'}{dT}\right|_{3-5} = \left.\frac{\Delta p'}{\Delta T}\right|_{3-5}$$
$$= \frac{32.42 - 27.13}{290 - 270}$$
$$= \frac{5.29}{20}$$
$$= 0.264$$

The average value is

$$\left.\frac{dp'}{dT}\right|_{av} = (0.256 + 0.264)/2$$
$$= 0.260$$

Both methods are in agreement. Since this is in reasonable agreement with the slope calculated in Illustrative Example 8.4, the enthalpy of vaporization is essentially the same.

Alternatively, Equation 8.10 may be employed in order to calculate dp'/dT directly by using the Antoine equation. Antoine equation coefficients for n-butane are found in Table 8.2:

$$A = 13.6608$$
$$B = 2154.70$$
$$C = 238.789$$

Substituting into Equation (8.10)

$$\frac{dp'}{dT} = \left(e^{A-[B/(T+C)]}\right)\left[\frac{B}{(T+C)^2}\right]$$

where T is in °C and p' is in kPa.

$$\left.\frac{dp'}{dT}\right|_{T=132.2°C} = \left(e^{13.6608-[2154.70/(132.2+238.78)]}\right)\left[\frac{2154.70}{(132.2+238.789)^2}\right]$$
$$= 40.28\,\text{kPa/K}$$
$$= 0.221\,\text{atm/°R}$$

This value of dp'/dT yields a calculated ΔH of 66.28 Btu/lb, which is in excellent agreement with the tabulated value of 67.0 Btu/lb.   ■

PREDICTIVE METHODS: NORMAL BOILING POINT

Most of the predictive methods for calculating the enthalpy of vaporization provide the value at the normal (1 atm) boiling point, T_n. The five most often used models are as follows:

1 Pitzer correlation:[3]

$$\Delta H_n = RT_c\left[7.08(1 - T_{rn})^{0.354} + 10.95\omega(1 - T_{rn})^{0.456}\right] \qquad (8.11)$$

where ΔH_n = enthalpy of vaporization at the normal boiling point, cal/(g·mol)
T_c = critical temperature, Kelvin
P_c = critical pressure, atmospheres
ω = Pitzer acentric factor
R = gas constant = 1.987 cal/gmol · K
$T_{rn} = T_n/T_c$, reduced temperature at the normal boiling point T_b
$\Delta Z = Z_v - Z_l$, the difference in the compressibility factor between the saturated vapor and saturated liquid at the normal boiling point.

2 Riedel method:[4]

$$\Delta H_n = 1.093 R T_c \left(T_{rn} \frac{\ln P_c - 1.013}{0.930 - T_{rn}} \right) \tag{8.12}$$

3 Trouton's rule (a rough estimate of the Reidel equation)

$$\frac{\Delta H_n}{R T_n} \simeq 8\text{–}13; \quad \text{a value of 10 is usually employed} \tag{8.13}$$

4 Clapeyron equation and compressibility factor:[5]

$$\Delta H_n = (R T_c \Delta Z T_{rn} \ln P_c)/(1 - T_{rn}) \tag{8.14}$$

5 Chen method:[6]

$$\Delta H_n = R T_c T_{rn} \left(\frac{3.978 T_{rn} - 3.938 + 1.555 \ln P_c}{1.07 - T_{rn}} \right) \tag{8.15}$$

The most often used of the above equations is that of Reidel. Another form of his equation (with different units) is:

$$\Delta H_n / T_n = 2.17 (\ln P_c - 1.0)/(0.930 - T_{rn}) \tag{8.16}$$

where T_n = normal boiling point, K

ΔH = enthalpy of vaporization at the normal boiling point, cal/gmol

P_c = critical pressure, atm

T_{rn} = reduced temperature at the normal boiling point, dimensionless

Interestingly, all of these equations have withstood the test of time.

ILLUSTRATIVE EXAMPLE 8.6

Estimate the enthalpy of vaporization of water at its normal boiling point. Compare the calculated value with the experimental value of 970 Btu/lb. Employ Equation (8.16).

SOLUTION: See Table 3.3 in Chapter 3 to obtain the key critical properties of water.

$$T_c = 647 \text{K}$$
$$P_c = 217.6 \text{ atm}$$

Calculate the reduced temperature at the normal boiling point:

$$T_n = 100 + 273$$
$$= 373K$$
$$T_{rn} = 373/647$$

Write the Reidel equation:

$$\Delta H_n/T_n = (2.17)(\ln P_c - 1.0)/(0.930 - T_{rn}) \tag{8.16}$$

Calculate the enthalpy of vaporization at the normal boiling point in cal/gmol:

$$\Delta H_n/T_n = (2.17)[\ln(217.6) - 1.0]/(0.930 - 0.577)$$
$$= 26.94$$
$$\Delta H_n = (26.94)(373)$$
$$= 10,049 \text{ cal/gmol}$$

Convert the enthalpy of vaporization to Btu/lb:

$$\Delta H_n = (10,049)(454)/(252)(18)$$
$$= 1006 \text{ Btu/lb}$$

Compare the calculated value with the experimental value:

$$\% \text{ difference} = (1006 - 970)/970$$
$$= 0.037$$
$$= 3.7\%$$

The correlation is excellent. ∎

ILLUSTRATIVE EXAMPLE 8.7

Calculate the enthalpy of water at its normal boiling point employing Trouton's rule.

SOLUTION: Trouton's rule states that

$$\frac{\Delta H_n}{RT_n} \simeq 8\text{--}13 \tag{8.13}$$

Therefore, use the value of 12 to calculate the enthalpy of vaporization.

$$\Delta H = (12)(1.987)(212 + 460)$$
$$= 16,023 \text{ Btu/lbmol}$$
$$= 890 \text{ Btu/lb}$$

The agreement with the calculated and literature values from Illustrative Example 8.6 is fair. ∎

PREDICTIVE METHODS: OTHER TEMPERATURES

The latent heat of vaporization at any temperature may be calculated from a known value at another temperature via

$$\frac{\Delta H_2}{\Delta H_1} = \left(\frac{1 - T_{r2}}{1 - T_{r1}}\right)^{0.38} \tag{8.17}$$

This method was proposed by Watson[7] and is the most often used for this purpose.

ILLUSTRATIVE EXAMPLE 8.8

The latent enthalpy of vaporization of water at 100°C and 101.37 kPa is 2256.9 J/g. Calculate the enthalpy of vaporization of water at 200°C. Employ Watson's equation.

SOLUTION: The appropriate reduced temperatures are

$$T_{r\,100} = \frac{373}{647}$$
$$= 0.577$$

$$T_{r\,200} = \frac{473}{647}$$
$$= 0.731$$

Apply Watson's equation

$$\Delta H_{200} = \Delta H_{100} \left(\frac{1 - T_{r\,200}}{1 - T_{r\,100}}\right)^{0.38} \tag{8.17}$$

Substituting

$$\Delta H_{200} = (2256.9)\left(\frac{1 - 0.731}{1 - 0.577}\right)^{0.38}$$
$$= (2256.9)(0.636)^{0.38}$$
$$= (2256.9)(0.842)$$
$$= 1900\,\text{J/g}$$

This is in excellent agreement with the steam tables value (following conversion) of 1939 J/g at 1555 kPa in Appendix I.A.

In lieu of key data, one of the authors[8] suggest employing Equation (8.18) to estimate the latent enthalpy of vaporization at a temperature other than the normal boiling point. The recommended equation is

$$\Delta H = \Delta H_n - \left(\frac{T - T_n}{T_c - T_n}\right)^2 \Delta H_n \qquad (8.18)$$

The only information/data required are the normal boiling point T_n, the normal latent enthalpy change ΔH_n, and the critical temperature T_c (where the latent enthalpy change reduces to zero).   ■

ILLUSTRATIVE EXAMPLE 8.9

Estimate the latent enthalpy of vaporization for water at 250°C. Assume

$$T_n = 100°C$$
$$T_c = 370°C$$
$$\Delta H_n = 2200 \, \text{kJ/kg}$$

and apply Equation (8.18).

SOLUTION: Write Equation (8.18),

$$\Delta H = \Delta H_n - \left[\frac{T - T_n}{T_c - T_n}\right]^2 \Delta H_n$$

Substituting

$$\Delta H = \left[1 - \left(\frac{250 - 100}{370 - 100}\right)^2\right] 2200$$

$$= [1 - 0.309]2200$$

$$= 1521 \, \text{kJ/kg} \qquad ■$$

INDUSTRIAL APPLICATIONS

Five industrial applications involving the latent enthalpy of vaporization are provided below. The first example involves a utility calculation of steam requirements. The second example involves the calculation of cooling water rate involving a cooling tower application. The third example is concerned with blowdown/makeup requirements for the second example. The section concludes with two process/plant applications.

ILLUSTRATIVE EXAMPLE 8.10

It is desired to evaporate 1000 lb/h of 60°F water at 1 atm at a power plant. Utility superheated steam at 40 atm and 1000°F is available, but since this steam is to be used elsewhere in the plant, it cannot drop below 20 atm and 600°F. What mass flowrate of the utility steam is required? Assume that there is no heat loss in the evaporator.

From the steam tables (see Appendix I.B)

$$P = 40\,\text{atm},\ T = 1000°\text{F}:\ H = 1572\,\text{Btu/lb}$$
$$P = 20\,\text{atm},\ T = 600°\text{F}:\ \ H = 1316\,\text{Btu/lb}$$

For saturated steam:

$$P = 1\,\text{atm}\qquad H = 1151\,\text{Btu/lb}$$

For saturated water:

$$T = 60°\text{F}\qquad H = 28.1\,\text{Btu/lb}$$

SOLUTION: A detailed flow diagram of the process is provided in Fig. 8.1. Assuming the process to be steady state and noting that there is no heat loss or shaft work, the energy balance on a rate basis is

$$\dot{Q} = \Delta\dot{H} = 0$$

This equation indicates that the sum of the enthalpy changes for the two streams must equal zero.

The change in enthalpy for the vaporization of the water stream is

$$\Delta\dot{H}_v = (1000\,\text{lb/h})(1151 - 28.1\,\text{Btu/lb})$$
$$= 1.123 \times 10^6\,\text{Btu/h}$$

The change of enthalpy for the cooling of the superheated steam may now be determined. Since the mass flowrate of one stream is unknown, its $\Delta\dot{H}$ must be expressed in terms of this mass flowrate, which is represented in Fig. 8.1 as $\dot{m}$, lb/h.

$$\Delta\dot{H}_{\text{cooling}} = \dot{m}(1572 - 1316)$$
$$= (256)\dot{m}\,\text{Btu/h}$$

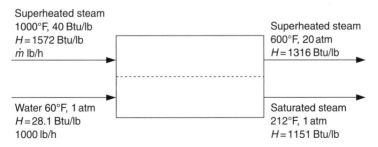

Figure 8.1   Flow diagram for Illustrative Example 8.10.

Since the overall $\Delta \dot{H}$ is zero, the enthalpy changes of the two streams must total zero. Thus,

$$\Delta \dot{H}_v + \Delta \dot{H}_{cooling} = 0$$

$$(256)\dot{m} = 1.123 \times 10^6$$

$$\dot{m} = 4387 \, lb/h \qquad\blacksquare$$

ILLUSTRATIVE EXAMPLE 8.11

Determine the total flowrate of cooling water, in gallons per minute, required for the services listed below. The cooling tower system supplies water to the units at 90°F with a return temperature of 115°F. Data is provided in Table 8.3. The heat capacity of the cooling water is 1.00 Btu/(lb · °F). The heat of vaporization at cooling tower operating conditions is 1030 Btu/lb.

SOLUTION: The total cooling load is simply the sum of the cooling loads for the five process units:

$$\dot{Q} = (12.0 + 6.0 + 23.5 + 17.0 + 31.5)(10^6)$$

$$= 90,000,000 \, Btu/h$$

The cooling water flow required, F, is the total heat duty divided by the enthalpy change per pound of water, which is the product of its heat capacity and temperature rise.

$$F = \frac{90.0 \times 10^6 \, Btu/h}{(115 - 90°F)[1.00 \, Btu/(lb \cdot °F)]}$$

$$= 3,600,000 \, lb/h$$

Since the density of water in lb/gal is 8.32,

$$F = \frac{3.6 \times 10^6 \, lb/h}{(60 \, min/h)(8.32 \, lb/gal)}$$

$$= 7200 \, gal/min$$

Table 8.3   Cooling Tower Heat Duty Data

Process unit	Heat duty (Btu/h)	Temperature (°F)
1	12,000,000	250
2	6,000,000	276–200
3	23,000,000	175–130
4	17,000,000	300
5	31,500,000	225–150

Note that the cooling water flowrate is independent of the temperatures of the process units. The sizes of the heat exchangers depend on all the temperatures, but the flowrate of cooling water does not. The heat discharge temperatures must, of course, all be higher than the cooling water temperature; otherwise the cooling water would heat, not cool, the process.   ■

ILLUSTRATIVE EXAMPLE 8.12

Refer to Illustrative Example 8.11. If 5% of the return water is sent to "blowdown," how much fresh water makeup is required?

SOLUTION: The blowdown rate, B, in this case, is 5% of the circulating water:

$$B = (0.05)(3,600,000)$$
$$= 180,000 \, \text{lb/h}$$
$$= 360 \, \text{gal/min}$$

The amount of water vaporized, V, can be calculated from the total cooling system heat load. Ultimately, all the heat is dissipated to the atmosphere by the evaporating water. Therefore, the evaporation rate is the total heat load divided by the heat of vaporization (latent heat) of the cooling water:

$$V = (90,000,000 \, \text{Btu/h})/(1030 \, \text{Btu/lb})$$
$$= 87,400 \, \text{lb/h}$$
$$= 175 \, \text{gal/min}$$

The total freshwater makeup, M, is the sum of blowdown and the amount evaporated.

$$M = 180,000 + 87,400$$
$$= 267,400 \, \text{lb/h}$$
$$= 535 \, \text{gal/min}$$

Chemical additives are often added to the circulating cooling water to reduce corrosion rates, reduce scaling and fouling, and prevent growth of bacteria and algae. Plants often contract with specialty service companies to provide these additives and supervise their use.   ■

ILLUSTRATIVE EXAMPLE 8.13

Determine how many pounds per hour of steam are required for the following process conditions if steam is provided at 500 psig. The plant has heating requirements as shown in Table 8.4 and the properties of saturated steam are given in Table 8.5.

Table 8.4   Plant Heat Duties

Process unit	Unit heat duty (UHD) (Btu/h)	Required temperature (°F)
1	10,000,000	250
2	8,000,000	450
3	12,000,000	400
4	20,000,000	300

Table 8.5   Steam Data

Pressure provided (psig)	Saturation temperature (°F)	Enthalpy of vaporization (ΔH_v) (Btu/lb)
75	320	894
500	470	751

SOLUTION: The total required flowrate of 500 psig steam, $\dot{m}_{BT}$, is given by:

$$\dot{m}_{BT} = \dot{m}_{B1} + \dot{m}_{B2} + \dot{m}_{B3} + \dot{m}_{B4}$$

For the above equation:

$\dot{m}_{B1}$ (mass flowrate of 500 psig steam through unit 1)

$\quad = \text{UHD}/\Delta H_v = 13{,}320\,\text{lb/h}$

$\dot{m}_{B2}$ (mass flowrate of 500 psig steam through unit 2)

$\quad = \text{UHD}/\Delta H_v = 10{,}655\,\text{lb/h}$

$\dot{m}_{B3}$ (mass flowrate of 500 psig steam through unit 3)

$\quad = \text{UHD}/\Delta H_v = 15{,}980\,\text{lb/h}$

$\dot{m}_{B4}$ (mass flowrate of 500 psig steam through unit 4)

$\quad = \text{UHD}/\Delta H_v = 26{,}635\,\text{lb/h}$

Thus,

$$\dot{m}_{BT} = 66{,}590\,\text{lb/h} \qquad ■$$

ILLUSTRATIVE EXAMPLE 8.14

The physical state of a material in a storage or transport container is an important factor. Define the conditions of the container temperature (T_c) and ambient temperature (T_a) that will maintain the status of a chemical with the physical characteristics provided in Table 8.6. Note that MP and BP are the melting point and boiling point of the chemical, respectively.

Table 8.6 Physical Characteristics

Problem	Physical state of material	MP/BP	Container conditions
Example	Cold or refrigerated solid	$MP > T_a$	$T_c < MP$ and T_a
a	Solid	$MP > T_a$	
b	Warm hot liquid	$BP > T_a$	
c	Cold liquid	$MP < T_a$	
d	Liquid	$BP > T_a$	
e	Hot liquid	$BP > T_a$	

SOLUTION: See Table 8.7 for a summary of the solution to this problem. ∎

Table 8.7 Solution to Illustrative Example 8.13

Problem	Physical state of material	MP/BP	Container conditions
Example	Cold or refrigerated solid	$MP > T_a$	$T_c < MP$ and T_a
a	Solid	$MP > T_a$	T_c near T_a
b	Warm hot liquid	$BP > T_a$	$T_c > MP$ and $T_a < BP$
c	Cold liquid	$MP < T_a$	$T_a > MP < T_a$ and BP
d	Liquid	$BP > T_a$	T_c near T_a
e	Hot liquid	$BP > T_a$	$BP > T_c > T_a$

REFERENCES

1. D. Green and R. Perry (editors), *"Perry's Chemical Engineers' Handbook,"* 8th edition, McGraw-Hill, New York, 2008.
2. J. SMITH, H. VAN NESS, and M. ABBOTT, *"Chemical Engineering Thermodynamics,"* 6th edition, McGraw-Hill, New York, 2001.
3. K. PITZER, *"Thermodymanics,"* 3rd edition, McGraw-Hill, New York, 1995.
4. L. REIDEL, *Chem. Ing. Tech.*, 26, 679–683, 1954.
5. C. CLAPEYRON, *J. Chem. Eng. Data*, 10, 207, 1965.
6. Y. CHEN, *Chem. Eng. Tech.*, 26, 679, 1954.
7. K. WATSON, *Ind. Eng. Chem.*, 35, 398–406, 1943.
8. L. THEODORE: personal notes, Manhattan College, 2008.

NOTE: Additional problems for each chapter are available for all readers at www. These problems may be used for additional review or homework purposes.

Chapter 9

Enthalpy of Mixing Effects

Charles Lamb [1775–1834]

The good things of life are not to be had singly, but come to us with a mixture.

—Popular Fallacies: XIII, That You Must Love Me and Love My Dog

INTRODUCTION

This third chapter of Part II addresses enthalpy effects associated with mixing processes. When two or more pure substances are mixed to form a solution, a heat effect usually results. The heat (or enthalpy) of mixing is defined as the enthalpy change that occurs when two or more pure substances in their standard states are mixed at constant temperature and pressure to form a solution. When the mixing occurs with water (H_2O), the process is normally defined as a *dilution* effect. If mixing occurs with a solvent other the H_2O, it is defined as a *solution* effect. For example, the mixing of HCl with H_2O is considered a dilution process while the mixing of HCl with H_2SO_4 is a solution process.

This chapter's contents include a brief introduction to enthalpy-concentration diagrams, applications involving the H_2SO_4–H_2O system, and applications involving the NaOH–H_2O system. Mixing effects at infinite dilution also receives treatment. The chapter concludes with an introduction to evaporator design.

ENTHALPY-CONCENTRATION DIAGRAMS[1]

The engineer is often interested in knowing how much energy in the form of heat must be transferred to or from a system if a specified amount of component 1 is dissolved in a certain amount of component 2. The heat transferred in such an operation divided by the amount of one of the components (usually on a mass or molal basis), is known as the *integral enthalpy of solution*. The differential heat of solution is the amount of heat transferred (per unit amount of solute) when a small amount of solute is dissolved

Thermodynamics for the Practicing Engineer. By L. Theodore, F. Ricci, and T. Van Vliet
Copyright © 2009 by John Wiley & Sons, Inc.

in a large amount of solvent (this definition is employed so that the concentration of the solution does not change significantly with the addition of the solute).

While some enthalpy of solution data are available in tabular form, such data are of very limited use. If the solution that one wishes to prepare is not the same concentration as the entry data in the table (and frequently only a single value is given for one solution), or if the temperature or pressure is different than that for which the table is based, one must obtain additional data by experiment or by an "engineering guess."

Enthalpy-concentration diagrams offer a convenient academic way to calculate enthalpy of mixing effects and temperature changes associated with this type of process. These diagrams, for a two component mixture, are graphs of the enthalpy of a binary solution plotted as a function of composition (mole fraction or weight fraction of one component), with the temperature as a parameter. For an ideal solution, isotherms on an enthalpy-concentration diagram would be straight lines. For a real solution, the actual isotherm is displaced vertically from the ideal-solution isotherm at a given point by the value of ΔH at that point, where ΔH is the enthalpy of mixing per unit mass of solution. With reference to the enthalpy-concentration diagram (see Fig. 9.1), ΔH is negative over the entire composition range. This means that heat is evolved whenever the pure components at a given temperature are mixed to form a solution at the same temperature. Such a system is said to be exothermic. An endothermic system is one for which the heat of solution is positive, i.e., solution at constant temperature is accompanied by the absorption of heat. Organic mixtures often fit this description. Enthalpy-concentration diagrams make it particularly useful in solving problems involving adiabatic mixing. Note that adiabatic mixing may be represented by a straight line on the diagram.

The enthalpy-concentration diagram has proved to be a very convenient way of presenting extensive data of this type for one system. Heat (enthalpy) of solution data can be obtained from such a plot by the following procedure. One of the constant-temperature (T) lines (at a given pressure) is shown in Fig. 9.1. Point A is

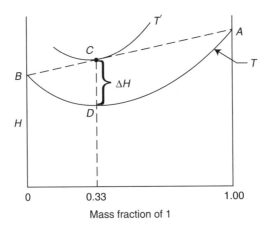

Figure 9.1   Enthalpy-concentration diagram.

the enthalpy of pure component 1. Point B is the enthalpy of pure component 2. If pure 1 and pure 2 are mixed, the resultant mixture must lie on a straight line connecting these two points because the first law, i.e., the conservation of energy principle, requires that

$$m_T H_T = m_1 H_1 + m_2 H_2 \tag{9.1}$$

for a constant-pressure process, assuming adiabatic operation. The conservation law for mass assumes that

$$m_T = m_1 + m_2 \tag{9.2}$$

The concentration (mass fraction) of component 1 in the mixture is given by $m_1/(m_1 + m_2)$. If 1 part of component 1 by mass is mixed with 2 parts of component 2 by mass, the mass fraction of A in the final mixture will be $\frac{1}{3}$. This is an illustration of the often used lever rule that has been encountered previously by many readers. On a mass diagram, the mixture point C may be considered as a fulcrum of a lever that is balanced by the weights at either end. Thus, $m_1 \overline{BC} = m_1 \overline{AC}$.

The mixture point C is the result of adiabatic mixing. If the temperature were measured, it would be found to lie on one of the higher temperature curves, e.g., T'. This would be one way of constructing additional curves for such a plot. In order to return the system to the original temperature, the amount of heat $(H_C - H_D)$ must be removed, or

$$\Delta H = -(H_C - H_D) \tag{9.3}$$

This value of ΔH would be the integral enthalpy of solution *per unit mass of solution*, since the diagram is on the basis of enthalpy per mass. The usual values of the integral heat of solution (per mass of solute) could be readily calculated from this since the mass fraction of the solute is known $(\frac{1}{3})$ for this illustration:

$$\Delta H_s = \frac{\Delta H}{1/3} = 3\Delta H = 3[-(H_C - H_D)] \tag{9.4}$$

However, with this diagram, it is not necessary to calculate heats of solution: the heat effects desired can be obtained directly from the diagram.

If two solutions are mixed, the problem can once again be solved with the aid of the enthalpy-concentration diagram and the procedures are very similar to that for pure components. Suppose that a 20 mass percent solution of component 1 (represented by point E in Fig. 9.2) is mixed with an 80 percent solution of A (represented by point F in Fig. 9.2) at constant temperature in the proportions of 1 part of E (20 percent solution) to 4 parts of F (80 percent solution). The total would be 5 parts of mixture and, by the lever rule, the mixture point would be four-fifths of the distance from E toward F, and this would be at a mass fraction of A of

$$0.2 + \tfrac{4}{5}(0.8 - 0.2) = 0.68$$

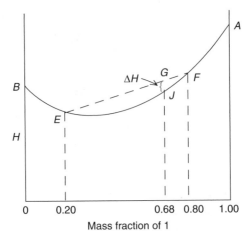

Figure 9.2   Mixing problem on enthalpy-concentration diagram.

This can also be shown by a componential balance with which many readers may be more familiar:

$$1 \text{ part } E \frac{0.2 \text{ part } A}{1 \text{ part } E} = 0.2 \text{ part } A$$

$$\frac{4 \text{ parts } F \dfrac{0.8 \text{ part } A}{1 \text{ part } F} = 3.2 \text{ parts } A}{5 \text{ parts mixture} = 3.4 \text{ parts } A}$$

Thus, the final concentration is given by:

$$\frac{3.4 \text{ parts } A}{5 \text{ parts mixture}} = 0.68 \frac{\text{part } A}{1 \text{ part mixture}}$$

The amount of heat that must be removed in this constant-pressure mixing operation in order to keep it at constant temperature is then given by $\Delta H = H_G - H_J$ per mass of final mixture.

It should also be apparent that these diagrams enable the problems of mixing to be solved when the materials are not initially at the same temperature. This procedure is demonstrated in several of the Illustrative Examples that now follow.

H$_2$SO$_4$–H$_2$O DIAGRAM

This section examines the first of two enthalpy-concentration diagrams. The sulfuric acid–water (H$_2$SO$_4$–H$_2$O) system is presented in Fig. 9.3 and receives treatment in the two Illustrative Examples that follow.

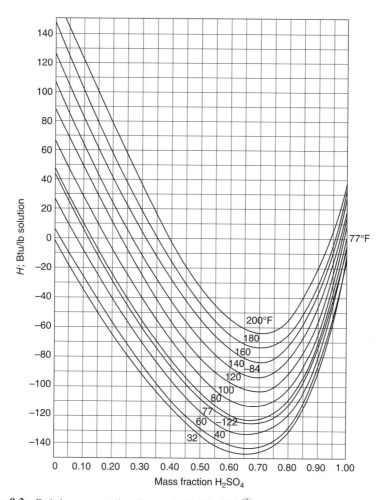

Figure 9.3   Enthalpy-concentration diagram for H$_2$SO$_4$–H$_2$O.[2]

ILLUSTRATIVE EXAMPLE 9.1

Calculate the adiabatic temperature change when 50 lb of pure H$_2$SO$_4$ at 25°C is mixed with 200 lb of a 50% by weight aqueous H$_2$SO$_4$ solution at 25°C.

SOLUTION: Calculate the mass (or weight) percent H$_2$SO$_4$ in the H$_2$SO$_4$–H$_2$O mixture:

$$H_2SO_4: \quad 50 + 100 = 150\,lb$$
$$H_2O: \quad 100\,lb$$
$$\%H_2SO_4 = (150/250)100$$
$$= 60\%$$

Referring to Fig. 9.3, construct a straight line between the 50% solution and pure H_2SO_4 at 25°C (77°F). Estimate the final temperature in °F:

$$T \approx 140°F \qquad \blacksquare$$

ILLUSTRATIVE EXAMPLE 9.2

Refer to Illustrative Example 9.1. Calculate the heat effect if the final mixture is returned to 25°C.

SOLUTION: Calculate the heat, Q, that needs to be transferred to return the mixture to 25°C:

$$H \text{ at } 140°F \approx -86 \, \text{Btu/lb}$$
$$H \text{ at } 77°F \approx -121.5 \, \text{Btu/lb}$$

Thus,

$$Q = m\Delta H$$
$$= (250)(-121.5 + 86)$$
$$= -8875 \, \text{Btu}$$

This represents the amount of heat that must be removed since Q is negative. $\qquad \blacksquare$

NaOH–H₂O DIAGRAM

This section examines the second of the two enthalpy-concentration diagrams. The sodium hydroxide–water (NaOH–H_2O) system is presented in Fig. 9.4 and receives treatment in the three Illustrative Examples that follow.

ILLUSTRATIVE EXAMPLE 9.3

Calculate the adiabatic temperature change when 65 lb of pure water at 75°C are mixed with 125 lb of 45% by mass aqueous NaOH solution at 75°C.

SOLUTION: Convert to °F

$$T = 75°C = 348K = 167°F$$

Apply an NaOH material balance to obtain the final mass fraction:

$$w_f = \frac{(65)(0) + (0.45)(125)}{125 + 65}$$
$$= 0.296$$

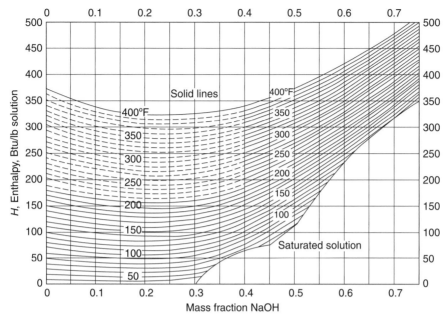

Figure 9.4   Enthalpy-concentration diagram for NaOH–H₂O system.[3]

From Fig. 9.4,

$$H_{H_2O}(167°F) = 135\,Btu/lb$$
$$H_{NaOH}(167°F) = 172\,Btu/lb$$

In addition,

$$H_{final} = 156\,Btu/lb$$

and

$$T_{final} = 208°F \qquad ■$$

ILLUSTRATIVE EXAMPLE 9.4

Refer to Illustrative Example 9.3. Calculate the adiabatic temperature change.

SOLUTION: From Fig. 9.4, the absolute temperature change, ΔT, is

$$\Delta T = 208 - 167$$
$$= 41°F \qquad ■$$

ILLUSTRATIVE EXAMPLE 9.5

Calculate the heat effect in Illustrative Example 9.3 if the final mixture is cooled to 75°C.

SOLUTION: Refer to Fig. 9.4 once again

$$H_{final}(167°F) = 118 \, \text{Btu/lb}$$

Therefore,

$$Q = 118 - 156$$
$$= -38 \, \text{Btu/lb solution}$$

and

$$Q = (-38)(65 + 125)$$
$$= -7220 \, \text{Btu (removed)} \quad \blacksquare$$

ENTHALPY OF MIXING AT INFINITE DILUTION

The reader should note that there is also an enthalpy effect when gases or solids are dissolved in liquids. This effect is normally small but can be significant in some applications, particularly acid gases in water. This type of calculation should be included in some absorber systems when temperature changes through the unit can be significant. This situation is addressed in Illustrative Example 9.7 and Problem 8 in the text's website.

When gases or solids are dissolved in liquids, the enthalpy effect is referred to as the enthalpy of solution, and is normally based on the solution of 1.0 mol of solute (gas or solid). If component 1 is the solute, set x as the mole fraction of solute in the solution. Setting ΔH as the enthalpy effect of solution, the term $\Delta H/x_1$ represents the enthalpy effect per mole of solute so that

$$\overline{\Delta H} = \Delta H/x_1 \tag{9.5}$$

where $\overline{\Delta H}$ is the enthalpy of solution based on a mole of solute. Enthalpy of solution data for gaseous HCl in H_2O is provided in Fig. 9.5.

An important property of this system is that the enthalpy at infinite dilution, ΔH_∞, represents the value of $\overline{\Delta H}$ as n approaches infinity (∞). For the $HCl(g) - H_2O(l)$ system at 25°C, $\Delta H_\infty = -75 \, \text{kJ/gmol HCl}$. In effect, this represents the quantity of energy liberated when 1 gmol of HCl at 25°C is mixed with an infinite number of gmol of H_2O at 25°C to produce a final solution at 25°C.

Data or enthalpy of mixing at infinite dilution is understandably limited. Selected values for dilution with water for several key chemicals are provided in Table 9.1.[5]

The data from Table 9.1 can be applied to the following model[6]:

$$\overline{\Delta H} = \Delta H_\infty[1 - e^{-an}] \tag{9.6}$$

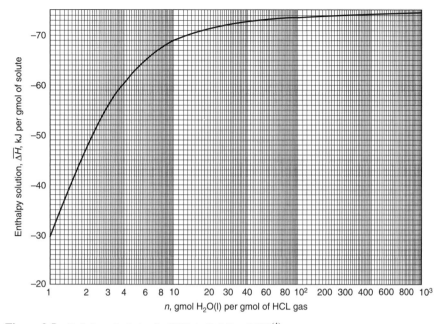

Figure 9.5   Enthalpy of solution for $HCl(g)-H_2O(l)$ at $25°C$. [4]

Table 9.1   Enthalpy of Mixing at Infinite Dilution with Water (cal/gmol)

Substance	State	ΔH_∞^0
HF	g	$-14{,}700$
HCl	g	$-17{,}888$
$HClO_4$	l	$-21{,}215$
$HClO_4 \cdot H_2O$	s	-7875
HBr	g	$-20{,}350$
HI	g	$-19{,}520$
HIO_3	c	2100
HNO_3	l	-7954
HCOOH	l	-205
CH_3COOH	l	-360
NH_3	g	-7290
LiCl	s	-8850
NaOH	s	$-10{,}637$
NaCl	s	928
KOH	s	$-13{,}769$
KCl	s	4115

where coefficient $a = 0.4375$, so that Equation (9.6), for the HCl(g)–H_2O(l) system at 25°C, takes the final form of

$$\overline{\Delta H} = \Delta H_\infty [1 - e^{-(0.4375)n}] \tag{9.7}$$

In lieu of data for other systems, it is recommended that Equation (9.7) should be used with ΔH_∞ set at the value for the system in question.

ILLUSTRATIVE EXAMPLE 9.6

Justify that Equation (9.7) does in fact satisfy conditions for $n = 0$ and $n = \infty$.

SOLUTION: The describing equation is

$$\overline{\Delta H} = \Delta H_\infty [1 - e^{-(0.4375)n}]$$

When $n = \infty$

$$\overline{\Delta H} = \Delta H_\infty [1 - 0]$$
$$= \Delta H_\infty$$

which is an agreement with the value read from Fig. 9.5.
    When $n = 0$,

$$\overline{\Delta H} = \Delta H_\infty [1 - 1]$$
$$= 0$$

which is also in agreement with the graphical data.   ∎

ILLUSTRATIVE EXAMPLE 9.7

In an attempt to quantify the effect of enthalpy of solution effects on the absorption of HCl into scrubbing water in an absorber, Ricci Consultants reviewed the literature and obtained the following rough estimates of this effect. The data provided temperature increases as a function of increasing HCl concentration (mass basis) in water:

$$0–1.5\% = 10°C$$
$$0–3.0\% = 15°C$$
$$0–5.0\% = 20°C$$

Apply Ricci's data and estimate the discharge temperature increase for the following two HCl concentration change scenarios:

Scenario 1:   0.0% inlet to 1.5% outlet

Scenario 2:   1.5% inlet to 3.0% outlet

SOLUTION: Since enthalpy is a point function, it is reasonable to assume that the temperature effects are additive. Therefore, the temperature increases are:

Scenario 1: $\Delta T = \Delta T_{1.5} - \Delta T_{0.0}$
$$= 10 - 0$$
$$= 10°C$$

Scenario 2: $\Delta T = \Delta T_{3.0} - \Delta T_{1.5}$
$$= 15 - 10$$
$$= 5°C$$

The reader should note that this is an effect that often cannot be adequately reflected in engineering applications since any increase in the temperature of the scrubbing liquid adversely affects the equilibrium, reducing the equilibrium capacity of the liquid.   ∎

EVAPORATOR DESIGN[7]

This final section provides a very brief introduction to evaporator design. There are four main types of steam-heated evaporators employed by industry. They are:

1 Short-tube evaporators

2 Long-tube evaporators

3 Agitated film evaporators

4 Coil evaporators

The presentation to follow will be limited to single-effect units employing steam as the heating medium. The steam in these evaporators condense on the inside of metal tubes. The steam is usually at or slightly above atmospheric pressure while the liquid to be evaporated is at or slightly below atmospheric pressure. By either reducing the boiling temperature of the liquid or increasing the steam temperature, or both, the temperature difference driving force between the steam and the boiling liquid increases. This, as will be shown shortly, increases the heat-transfer rate in the evaporator and reduces the area requirement.

The heat transfer rate through the heating surface of an evaporator is given by the product of three factors: the area of the heating surface, the overall temperature difference driving force, and the overall heat transfer coefficient, i.e.,

$$Q = UA(T_s - T) = UA\,\Delta T \tag{9.8}$$

where (English units) Q = rate of heat transfer, Btu/h

A = area of heating surface, ft^2

U = overall heat transfer coefficient, Btu/ft$^2 \cdot$ h $\cdot$ °F

T_s = steam condensation temperature, °F

$$T = \text{boiling temperature of solution, } °F$$

$$\Delta T = T_s - T = \text{overall temperature difference between steam and solution, } °F$$

Note: The term U in this development should not be confused with the term U that represents internal energy in Chapter 4.

The design of most evaporators do not take into account enthalpy of mixing effects. However, this effect should be considered when dealing with acid or basic solutions in calculating enthalpy changes across the evaporator. This is demonstrated in the Illustrative Examples to follow.

ILLUSTRATIVE EXAMPLE 9.8

A single-effect evaporator is to concentrate 10,000 lb/h of a 10% NaOH solution to 75%. The feed enters at 120°F and the evaporator is to operate at an absolute pressure of 14.7 psi. The 75% NaOH solution leaves at the evaporator equilibrium temperature. For what heat transfer rate (Btu/h) should the evaporator be designed?

SOLUTION: The system is shown in Fig. 9.6.

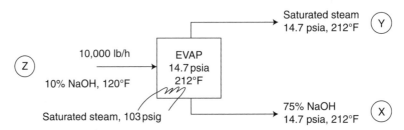

Figure 9.6   Evaporation flow diagram.

On the basis of one hour of operation, calculate the flow rate of steam Y, and the 75% NaOH–H$_2$O solution, X, leaving the evaporator.

From a NaOH balance,

$$(10,000)(0.1) = (0.75)(X)$$
$$X = 1333.3 \text{ lb}$$

From an overall material balance,

$$Z = Y + X$$
$$Y = 10,000 - 1333 = 8667 \text{ lb}$$

Estimate the enthalpy of the solution entering the unit, H_Z, in Btu/lb from Fig. 9.4:

$$H_Z \approx 81 \text{ Btu/lb solution}$$

Also estimate the enthalpy of the 75% NaOH solution, H_X, leaving the unit:

$$H_X \approx 395 \text{ Btu/lb solution}$$

Calculate the evaporator heat required, Q, in Btu/h.

$$
\begin{aligned}
Q &= \Delta H \\
&= (8667)(1150) + (1333)(395) - (10,000)(81) \\
&= 9,966,700 + 526,700 - 810,000 \\
&= 9,683,400 \text{ Btu/h}
\end{aligned}
$$

$\blacksquare$

ILLUSTRATIVE EXAMPLE 9.9

Refer to Illustrative Example 9.8. Calculate the area requirement in the evaporator if the overall heat transfer coefficient is $500 \text{ Btu/h} \cdot \text{ft}^2 \cdot °\text{F}$ and 103 psig saturated steam ($T_s = 340°\text{F}$) is employed in the steam chest.

SOLUTION: Calculate the area of the evaporator in ft^2.

$$Q = UA \, \Delta T \qquad\qquad (9.8)$$
$$A = 9,683,400/(500)(340 - 212)$$
$$= 151.3 \text{ ft}^2$$

$\blacksquare$

ILLUSTRATIVE EXAMPLE 9.10

Refer to Illustrative Example 9.8. Recalculate the area requirement if atmospheric steam is employed in the steam chest.

SOLUTION: This is an impossible condition since

$$\Delta T = T_S - T \longrightarrow 0$$

and

$$A \longrightarrow \infty$$

$\blacksquare$

REFERENCES

1. E. WILSON and H. RIES, "Principles of Chemical Engineering Thermodynamics," McGraw-Hill, New York City, 1956.
2. W. ROSS, *Chem. Eng. Prog.*, 43, 314, 1952.
3. W. McCABE, *Trans. Amer. Inst. Chem. Engrs.*, 31, 129, 1935.
4. "The NBS Tables of Chemical Thermodynamic Properties," *J. Phys. Chem. Ref. Data*, 11 (Suppl. 2), 1982.

5. National Standards Reference Data System, NSRDS–NBSZ, data unknown.
6. L. THEODORE, 2008, personal notes, Manhattan College.
7. L. THEODORE, *"Heat Transfer for the Practicing Engineer"* (in preparation).

NOTE: Additional problems for each chapter are available for all readers at www. These problems may be used for additional review or homework purposes.

Chapter **10**

Chemical Reaction Enthalpy Effects

William Congreve [1670–1729]

A sickly flame, which, if not fed, expires,
And feeding, wastes in self-consuming fires.

—*The Way of the World [1700]. Act III, Sc. 12*

INTRODUCTION

As noted in the introduction to Part II, there are many different types of enthalpy effects; these include:

Sensible (temperature)

Latent (phase)

Dilution (with water), for example, HCl with H_2O

Solution (nonaqueous), for example, HCl with a solvent other than H_2O

Reaction (chemical)

To summarize, the *sensible* enthalpy change was reviewed in Chapter 7. The *latent* enthalpy change finds application in thermodynamic calculations for determining the heat (enthalpy) of condensation or vaporization, often for water. Steam tables (or the equivalent) are usually employed for this determination. The *dilution* and *solution* enthalpy effects are often significant in some industrial absorber calculations but may safely be neglected in most thermodynamic calculations. The *enthalpy of reaction*, the subject title for this chapter, is defined as the enthalpy change of a system undergoing chemical reaction; this effect normally cannot be neglected.

The equivalence of mass and energy needs to be addressed qualitatively. This relationship is only important in nuclear reactions, the details of which are beyond the scope of this text. The energy-related effects discussed in this chapter arise because of the rearrangement of electrons outside the nucleus of the atom. In a nuclear reaction, it is the nucleus of the atom that undergoes rearrangement, releasing a significant quantity of energy; this process occurs with a miniscule loss of mass. The classic Einstein equation relates energy E to mass, as provided in Equation (10.1)

$$\Delta E = (\Delta m)c^2 \tag{10.1}$$

where $\Delta m =$ decrease in mass
         $c =$ velocity of light

Topics reviewed in this final chapter of Part II include:

1 Standard Enthalpy of Formation
2 Standard Enthalpy of Reaction
3 Effect of Temperature on Enthalpy of Reaction
4 Gross and Net Heating Values

STANDARD ENTHALPY OF FORMATION

If the reactants and products are at the same temperature and in their standard states, the enthalpy (heat) of reaction is termed the standard enthalpy (heat) of reaction. For engineering purposes, the standard state of a chemical may be taken as the *pure* chemical at 1 atm pressure. A superscript zero is often employed to identify a standard heat of reaction, for example, ΔH^0. A T subscript (ΔH_T^0) is sometimes used to indicate the temperature; standard heat of reaction data are meaningless unless the temperatures are specified. ΔH_{298}^0 data (i.e., for 298K or 25°C) for many reactions are available in the literature.[1,2]

As described in Chapter 4, the first law of thermodynamics provides that, in a steady-flow process with no mechanical work,

$$Q = \Delta H \tag{10.2}$$

Since enthalpy is a point function, it is independent of the path for any process. If heat of reaction determinations are made in a flow reactor, the energy (in the form of heat) transferred across the reactor boundary or surface is exactly equal to the heat of reaction. This is not the case for batch or nonflow systems. For this reason, *heat of reaction* is a misleading term. More recently, it has been referred to as the *enthalpy of reaction* although a more accurate term would be *enthalpy change* of reaction. The terms heat and enthalpy of reaction are used interchangeably in this chapter.

The *heat of formation* (ΔH_f) is defined as the enthalpy change occurring during a chemical reaction where 1 mol of a product is formed from its elements. The *standard heat of formation* (ΔH_f^0) is applied to formation reactions that occur at constant temperature with each element and the product in its standard state.

Consider the formation reaction for CO_2 at standard conditions at 25°C:

Enthalpy: O   O

$$C + O_2 \xrightarrow{\Delta H^0_{f298}} CO_2$$

Enthalpy: H_{CO_2} (10.3)

Once again, this reaction reads: "1 mol of carbon and 1 mol of oxygen react to form 1 mol of carbon dioxide." The enthalpies of reactants and products are printed above and below, respectively, the symbols in the reaction, and the enthalpy change for the reaction, ΔH^0_f, placed above the arrow. Note that the enthalpies of elements in their standard states (pure, 1 atm) at 25°C have arbitrarily been set to zero. The enthalpy change accompanying this reaction is the *standard heat of formation* and is given by

$$\Delta H^0_{f298} = H^0_{CO_2} - (H^0_C + H^0_{O_2}) = H^0_{CO_2} \qquad (10.4)$$

Note that this formation reaction for CO_2 is a chemical reaction; the heat of formation, in this case, is therefore equal to the heat of combustion. Since a combustion reaction is one type of chemical reaction, the development to follow will concentrate on chemical reactions in general.

STANDARD ENTHALPY OF REACTION

Chemical (stoichiometric) equations may be combined by addition or subtraction. The standard heat (enthalpy) of reaction (ΔH^0) associated with each equation may likewise be combined to give the standard heat of reaction associated with the resulting chemical equation. This is possible, once again, because enthalpy is a point function, and these changes are independent of path. In particular, formation equations and standard heats of formation may always be combined by addition and subtraction to produce any desired equation and its accompanying standard heat of reaction. This desired equation, however, cannot itself be a formation equation. Thus, the enthalpy change for a chemical reaction is the same whether it takes place in one or several steps. This is referred to as the *law of constant enthalpy summation* and is a direct consequence of the first law of thermodynamics.

Consider the general reaction

$$aA + bB + \cdots \longrightarrow cC + dD + \cdots \qquad (10.5)$$

where A, B = formulas for the reactants (r)

C, D = formulas for the products (p)

a, b, c, d = stoichiometric coefficients of the balanced reaction

To simplify the presentation that follows, Equation (10.4) is shortened to

$$aA + bB \longrightarrow cC + dD \qquad (10.6)$$

(Although this presentation plus those in later chapters of this book will deal with the hypothetical species, A, B, C, etc., application to real systems can be found in the Illustrative Examples at the end of this and those later chapters. This reaction reads: "a moles of A react with b moles of B to form c moles of C and d moles of D." The standard heat of reaction for this chemical change is given by

$$\Delta H^0 = cH_C^0 + dH_D^0 - aH_A^0 - bH_B^0 \tag{10.7}$$

where H_C^0 = enthalpy of C in its standard state

C, D, A, B = subscripts indicating chemical species C, D, A, and B, respectively

If the temperature is 25°C, the enthalpies of the elements at standard state are, by convention, equal to zero. Therefore,

$$(\Delta H_f^0)_{i298} = H_{i298}^0 \tag{10.8}$$

Substituting Equation (10.8) for each component into Equation (10.7) yields

$$\Delta H_{298}^0 = c(\Delta H_f^0)_C + d(\Delta H_f^0)_D - a(\Delta H_f^0)_A - b(\Delta H_f^0)_B \tag{10.9}$$

or

$$\Delta H_{298}^0 = \sum_p n_p (\Delta H_f^0)_p - \sum_n n_r (\Delta H_f^0)_r \tag{10.10}$$

where p = products

r = reactants

n_p, n_r = coefficients from the chemical equation

The standard heat of a reaction is therefore obtained by taking the difference between the standard heat of formation of the products and that of the reactants. If the standard heat of reaction or formation is negative (*exothermic*), as is the case with most reactions, then energy is liberated as a result of the reaction. Energy is absorbed if ΔH is positive (*endothermic*). Standard heat of formation and standard heat (enthalpy) of combustion data at 25°C are provided in Table 10.1. Both of these heat (or enthalpy) effects find extensive application in thermodynamic calculations.

The *standard heat of combustion* is defined as the enthalpy change during a chemical reaction where 1 mol of material is burned in oxygen, where all reactants and products are in their standard states. This quantity finds extensive application in calculating enthalpy changes for combustion reactions and is often given in the literature for 60°F (16°C). Although much of the literature data on standard heats of reaction are given for 25°C (76°F), there is little sensible enthalpy change between these two temperatures, and the two sets of data may be considered compatible.

Table 10.1   Standard Heats of Formation and Combustion at $25°C$ in Calories per Gram Mole[a,b]

Compound	Formula	State	ΔH^0_{f298}	$-\Delta H^0_{c298}$
Normal paraffins				
Methane	CH_4	g	$-17,889$	212,800
Ethane	C_2H_6	g	$-20,236$	372,820
Propane	C_3H_8	g	$-24,820$	530,600
n-Butane	C_4H_{10}	g	$-30,150$	687,640
n-Pentane	C_5H_{12}	g	$-35,000$	845,160
n-Hexane	C_6H_{14}	g	$-39,960$	1,002,570
Increment per C atom above C_6	—	g	-4925	157,440
Normal monoolefins (1-alkenes)				
Ethylene	C_2H_4	g	12,496	337,230
Propylene	C_3H_6	g	4879	491,990
1-Butene	C_4H_8	g	-30	649,450
1-Pentene	C_5H_{10}	g	-5000	806,850
1-Hexene	C_6H_{12}	g	-9960	964,260
Increment per C atom above C_6	—	g	-4925	157,440
Miscellaneous organic compounds				
Acetaldehyde	C_2H_4O	g	$-39,760$	
Acetic acid	$C_2H_4O_2$	l	$-116,400$	
Acetylene	C_2H_2	g	54,194	310,620
Benzene	C_6H_6	g	19,820	789,080
Benzene	C_6H_6	l	11,720	780,980
1,3-Butadiene	C_4H_6	g	26,330	607,490
Cyclohexane	C_6H_{12}	g	$-29,430$	944,790
Cyclohexane	C_6H_{12}	l	$-37,340$	936,880
Ethanol	C_2H_6O	g	$-56,240$	
Ethanol	C_2H_6O	l	$-66,356$	
Ethylbenzene	C_8H_{10}	g	7120	1,101,120
Ethylene glycol	$C_2H_6O_2$	l	108,580	
Ethylene oxide	C_2H_4O	g	$-12,190$	
Methanol	CH_4O	g	$-48,100$	
Methanol	CH_4O	l	$-57,036$	
Methylcyclohexane	C_7H_{14}	g	$-36,990$	1,099,590
Methylcyclohexane	C_7H_{14}	l	$-45,450$	1,091,130
Styrene	C_8H_8	g	35,220	1,060,900
Toluene	C_7H_8	g	11,950	943,580
Toluene	C_7H_8	l	2870	934,500
Miscellaneous inorganic compounds				
Ammonia	NH_3	g	$-11,040$	
Calcium carbide	CaC_2	s	$-15,000$	
Calcium carbonate	$CaCO_3$	s	$-288,450$	
Calcium chloride	$CaCl_2$	s	$-190,000$	
Calcium chloride	$CaCl_2 \cdot 6H_2O$	s	$-623,150$	
Calcium hydroxide	$Ca(OH)_2$	s	$-235,800$	94,052

(Continued)

TABLE 10.1   *Continued*

Compound	Formula	State	ΔH^0_{f298}	$-\Delta H^0_{c298}$
Calcium oxide	CaO	s	$-151,900$	
Carbon	C	Graphite	—	67,636
Carbon dioxide	CO_2	g	$-94,052$	
Carbon monoxide	CO	g	$-26,416$	
Hydrochloric acid	HCl	g	$-22,063$	
Hydrogen	H_2	g	—	68,317
Hydrogen sulfide	H_2S	g	-4815	
Iron oxide	FeO	s	$-64,300$	
Iron oxide	Fe_3O_4	s	$-267,000$	
Iron oxide	Fe_2O_3	s	$-196,500$	
Iron sulfide	FeS_2	s	$-42,520$	
Lithium chloride	LiCl	s	$-97,700$	
Lithium chloride	$LiCl \cdot H_2O$	s	$-170,310$	
Lithium chloride	$LiCl \cdot 2H_2O$	s	$-242,100$	
Lithium chloride	$LiCl \cdot 3H_2O$	s	$-313,500$	
Nitric acid	HNO_3	l	$-41,404$	
Nitrogen oxides	NO	g	21,600	
	NO_2	g	8041	
	N_2O	g	19,490	
	N_2O_4	g	2309	
Sodium carbonate	Na_2CO_3	s	$-270,300$	
Sodium carbonate	$Na_2CO_3 \cdot 10H_2O$	s	$-975,600$	
Sodium chloride	NaCl	s	$-98,232$	
Sodium hydroxide	NaOH	s	101,990	
Sulfur dioxide	SO_2	g	$-70,960$	
Sulfur trioxide	SO_3	g	$-94,450$	
Sulfur trioxide	SO_3	l	$-104,800$	
Sulfuric acid	H_2SO_4	l	$-193,910$	
Water	H_2O	g	$-57,798$	
Water	H_2O	l	$-68,317$	
Chlorinated organic compounds				
Methyl chloride	CH_3Cl	l	$-20,630$	
Dichloromethane	CH_2Cl_2	l	$-22,800$	
Chloroform	$CHCl_3$	l	$-24,200$	36,900
Carbon tetrachloride	CCl_4	l	$-24,000$	
Ethyl chloride	C_2H_5Cl	l	$-26,700$	
1,1-Dichloroethane	$C_2H_4Cl_2$	l	$-31,050$	
1,1,2,2-Tetrachloroethane	$C_2H_2Cl_4$	l	$-36,500$	
n-Propyl chloride	C_3H_7Cl	l	$-31,100$	
1,3-Dichloropropane	$C_3H_6Cl_2$	l	$-38,600$	
n-Butyl chloride	C_4H_9Cl	l	$-35,200$	
1-Chloropentane	$C_5H_{11}Cl$	l	$-41,800$	
1-Chloroethylene	C_2H_3Cl	l	-8400	
trans-1,2-Dichloroethylene	$C_2H_6Cl_2$	l	-1000	
Trichloroethylene	C_2HCl_3	l	-1400	510,000

(Continued)

TABLE 10.1 *Continued*

Compound	Formula	State	ΔH^0_{f298}	$-\Delta H^0_{c298}$
Tetrachloroethylene	C_2Cl_4	l	-3400	782,000
3-Chloro-1-propene	C_3H_5Cl	l	-150	1,600,000
Chlorobenzene	C_6H_5Cl	l	12,390	
p-Dichlorobenzene	$C_6H_5Cl_2$	l	5500	
Hexachlorobenzene	C_6Cl_6	l	-8100	
Benzylchloride	C_7H_7Cl	l		
1,1,1-trichloro-2,2-bis(p- chlorophenyl)ethane (DDT)	$C_{14}H_9Cl_5$	l		

[a]Selected mainly from F. D. Rossini, ed., "Selected Values of Physical and Thermodynamic Properties of Hydrocarbons and Related Compounds," *American Petroleum Institute Research Project 44*, Carnegie Institute of Technology, Pittsburgh, PA, 1953; F. D. Rossini, D. D. Wagman, W. H. Evans, S. Levine, and I. Jaffe, "Selected Values of Chemical Thermodynamic Properties," *Natl. Bur. Stand. Circ. 500*, 1952; also, from personal notes of L. Theodore and J. Reynolds, 1985.
[b]For combustion reactions, the products are $H_2O(l)$ and $CO_2(g)$.

The reader is cautioned on the use of the published heat of combustion data in the literature for chlorinated organics. It appears that the laboratory combustion tests used to generate these data were conducted in bomb calorimeters using excess oxygen. Burning carbon tetrachloride under these conditions yields carbon dioxide and chlorine oxides. Based on the published heat of combustion data, calculations suggest that chlorine monoxide or chlorine dioxide are the likely products, but there may be an equilibrium mixture of several species. The important lesson to be learned is that the published heat of combustion data for halogenated species may be useless since the products of reaction are acid gas (e.g., HCl, Cl_2) and not halogen oxides. The use of heat of formation data is therefore recommended; these can be used to determine the heat of reaction for the combustion of a particular substance that produces specific products. This avoids the pitfalls of using published data that may not apply to the reaction occurring in the combustion process.

Other tables of heat of formation, combustion, and reaction are available in the literature (particularly thermodynamics text/reference books) for a wide variety of compounds.[1,2] It is important to note that these are valueless unless the stoichiometric equation, temperature, and the state of the reactants and products are included.

ILLUSTRATIVE EXAMPLE 10.1

Calculate the standard enthalpy of reaction (25°C, 1 atm) for the combustion of methane:

$$CH_4 + O_2 \longrightarrow CO_2 + H_2O(g)$$

Standard (25°C, 1 atm) enthalpy of formation data is given below:

$$(\Delta H_f^0)_{CH_4} = -17,889 \text{ cal/gmol}$$
$$(\Delta H_f^0)_{CO_2} = -94,052 \text{ cal/gmol}$$
$$(\Delta H_f^0)_{H_2O(g)} = -57,798 \text{ cal/gmol}$$

See also Table 10.1.

SOLUTION: Balance the equation

$$CH_4 + 2O_2 \longrightarrow CO_2 + 2H_2O(g)$$

Note that the standard enthalpy of formation for oxygen is:

$$(\Delta H_f^0)_{O_2} = 0.0$$

Write the equation for the standard enthalpy operation:

$$\Delta H_{298}^0 = (\Delta H_f^0)_{CO_2} + 2(\Delta H_f^0)_{H_2O} - 2(\Delta H_f^0)_{O_2} - (\Delta H_f^0)_{CH_4}$$

Substitute and calculate the standard enthalpy of reaction at 25°C.

$$\Delta H_{298}^0 = -94,052 - (2)(57,798) - (2)(0) - (-17,889)$$
$$= -191,759 \text{ cal/gmol}$$

The tabulated value for the heat of combustion is $-212,800$ cal/gmol. However, this value assumes the water to be present in the liquid phase. The reader is left the exercise of recalculating this value with $(\Delta H_f^0)_{H_2O(l)} = -68,317$ cal/gmol. ∎

ILLUSTRATIVE EXAMPLE 10.2

Calculate the standard heat of reaction at 25°C for each of the following reactions:

1 $N_2(g) + O_2(g) \longrightarrow 2NO(g)$
2 $C_3H_8(g) \longrightarrow C_2H_4(g) + CH_4(g)$
3 $3NO_2(g) + H_2O(l) \longrightarrow 2HNO_3(l) + NO(g)$

SOLUTION: Refer to Table 10.1

1 $\Delta H_{298}^0 = 2(21,570)$
   $= 43,140$ cal/gmol N_2

2 $\Delta H_{298}^0 = 12,496 + (-17,889) - (-24,820)$
   $= 19,427$ cal/gmol C_3H_8

3 $\Delta H^0_{298} = 21{,}570 + 2(-41{,}404) - (-68{,}317) - 3(7{,}930)$
$= -16{,}711 \text{ cal/gmol } H_2O$

or

$\Delta H^0_{298} = -16{,}711 \text{ cal/3 gmol } NO_2$ ■

ILLUSTRATIVE EXAMPLE 10.3

Verify that the enthalpy of combustion (Table 10.1) may be calculated from enthalpy of formation data (also in Table 10.1). Use *n*-hexane as an example.

SOLUTION: From Table 10.1,

$$\Delta H^0_c (n\text{-hexane}) = -1{,}002{,}570 \text{ cal/gmol}$$

First, write the combustion reaction:

$$C_6H_{14} + 9.5O_2 \longrightarrow 6CO_2 + 7H_2O(l)$$

From Table 10.1, one obtains

$C_6H_{14}(g)$: $\Delta H^0_f = -36{,}960 \text{ cal/gmol}$

$CO_2(g)$: $\Delta H^0_f = -94{,}052 \text{ cal/gmol}$

$H_2O(l)$: $\Delta H^0_f = -68{,}317 \text{ cal/gmol}$

Thus,

$$\Delta H^0_c = \sum \Delta H^0_{f,p} - \sum \Delta H^0_{f,r}$$
$$= 6(-94{,}052) + 7(-68{,}317) - (-39{,}960)$$
$$= -1{,}002{,}571 \text{ cal/gmol}$$

The calculation process is verified. ■

ILLUSTRATIVE EXAMPLE 10.4

Using standard heat of combustion data, calculate the standard heat of formation of $C_{14}H_9Cl_5$. [The organic and generic names for this compound are 1,1,1-trichloro-2,2-bis(*p*-chlorophenyl)ethane and DDT, respectively. DDT is the abbreviation for dichlorodiphenyl-trichloroethene, another name sometimes used for this compound.]

SOLUTION: The standard heat of combustion for this organic is obtained directly from Table 10.1, noting that the H_2O and HCl formed are in the liquid and gaseous states, respectively:

$$\Delta H^0_c = -1600 \text{ kcal/gmol}$$

First, write a balanced stoichiometric equation for this combustion reaction:

$$C_{14}H_9Cl_5 + 15O_2 \longrightarrow 14CO_2 + 2H_2O(l) + 5HCl(g)$$

For this reaction,

$$\Delta H_c^0 = 14\Delta H_{f,CO_2}^0 + 2\Delta H_{f,H_2O(l)}^0 + 5\Delta H_{f,HCl(g)}^0 - \Delta H_{f,C_{14}H_9Cl_5}^0$$

From Table 10.1,

$$\Delta H_{f,CO_2}^0 = -94.052 \text{ kcal/gmol}$$

$$\Delta H_{f,H_2O(l)}^0 = -68.317 \text{ kcal/gmol}$$

$$\Delta H_{f,HCl(g)}^0 = -22.063 \text{ kcal/gmol}$$

Solving this equation for $\Delta H_{f,C_{14}H_9C_{15}}^0$ yields

$$\Delta H_{f,C_{14}H_9Cl_5}^0 = 36.32 \text{ kcal/gmol}$$

The reader is again reminded that the heat of formation of elements in their standard states is zero. In addition, care should be exercised in using heat of combustion data for highly chlorinated organics. ∎

EFFECT OF TEMPERATURE ON ENTHALPY OF REACTION

The heat of reaction is a function of temperature because the heat capacities of both the reactants and products vary with temperature. Smith and Van Ness[2] have described this effect mathematically in the following manner:

$$\Delta H_T^0 = \Delta H_{298}^0 + \int_{298}^{T} \Delta C_P \, dT \tag{10.11}$$

where T is the absolute temperature (K), and

$$\Delta C_P = \sum_{\text{products}} nC_P - \sum_{\text{reactants}} nC_P \tag{10.12}$$

If the heat capacity for each product and reactant is expressed by

$$C_P = \alpha + \beta T + \gamma T^2 \tag{10.13}$$

then

$$\Delta C_P = \Delta \alpha + (\Delta \beta)T + (\Delta \gamma)T^2 \tag{10.14}$$

where

$$\Delta \alpha = \sum_{products} n\alpha - \sum_{reactants} n\alpha \qquad (10.15)$$

with similar definitions for $\Delta \beta$ and $\Delta \gamma$. The following expression for the standard heat of reaction at temperature T is obtained by combining Eqs. (10.11) and (10.14),

$$\Delta H_T^0 = \Delta H_{298}^0 + \int_{298}^{T} [\Delta \alpha + (\Delta \beta)T + (\Delta \gamma)T^2]\, dT \qquad (10.16)$$

or (following integration)

$$\Delta H_T^0 = \Delta H_{298}^0 + \Delta \alpha (T - 298) + \tfrac{1}{2}\Delta \beta (T^2 - 298^2) + \tfrac{1}{3}\Delta \gamma (T^3 - 298^3) \qquad (10.17)$$

If all the constant terms in this equation are collected and lumped together into a constant designated ΔH_0, the result is

$$\Delta H_T^0 = \Delta H_0 + \Delta \alpha T + \tfrac{1}{2}(\Delta \beta)T^2 + \tfrac{1}{3}(\Delta \gamma)T^3 \qquad (10.18)$$

where ΔH_T^0 = standard heat of reaction at temperature T and

$$\Delta H_0 = \text{constant} = \Delta H_{298}^0 - 298\Delta \alpha - \tfrac{1}{2}(298)^2 \Delta \beta - \tfrac{1}{3}(298)^3 \Delta \gamma \qquad (10.19)$$

Note that the use of the equations in this section requires that the temperature T be expressed in Kelvins. If the standard heat of reaction is known at a single temperature, e.g., 25°C, ΔH_{298}^0 can be calculated. The constant ΔH_0 may then be calculated directly from Equation (10.18).

The reader should attempt to rederive the equivalent of Equations (10.17) and (10.18) if heat capacity variation with temperature is given by

$$C_P = a + bT + cT^{-2} \qquad (10.20)$$

For this case,

$$\Delta H_T^0 = \Delta H_{298}^0 - \Delta a(T - 298) + (\Delta b/2)(T^2 - 298^2) - \Delta c(T^{-1} - 298^{-1}) \qquad (10.21)$$

One application of this calculation is that associated with the adiabatic flame temperature. It represents the maximum temperature the products of combustion (flue) can achieve if the reaction is conducted adiabatically. For this condition, all the energy liberated on combustion at or near standard conditions (ΔH_c^0) appears as sensible heat in heating up the flue products, ΔH_p, that is,

$$\Delta H_c^0 + \Delta H_p = 0 \qquad (10.22)$$

where

$$\Delta H_c^0 = \text{standard heat of combustion at 298 K (25°C)} \qquad (10.23)$$

and

$$\Delta H_p = \text{sensible enthalpy change of the products as the temperature increases}$$
$$\text{from 25°C to the theoretical flame temperature} \qquad (10.24)$$

The right term of Equation (10.11), in conjunction with Equations (10.13) and (10.20), should be applied to the flue gas product in evaluating ΔH_p. Details of this calculation are presented in the Illustrative Examples to follow.

ILLUSTRATIVE EXAMPLE 10.5

Calculate the theoretical adiabatic flame temperature of C_6H_5Cl. Assume that the heat capacity variation with temperature takes the form

$$C_P = a + bT + cT^{-2}$$

SOLUTION: The theoretical adiabatic flame temperature for an organic compound has been previously defined as the maximum temperature the flue products of combustion will achieve if the reactants of combustion are at ambient conditions. This temperature is achieved with adiabatic (no heat lost to the surroundings) operation and with theoretical (stoichiometric or 0% excess) air.

The standard heat of combustion for chlorobenzene is obtained from the heats of formation data in Table 10.1. Since

$$C_6H_5Cl + 7O_2 \longrightarrow 6CO_2 + 2H_2O(g) + HCl(g)$$

$$\Delta H_c^0 = (6)(-94,052) + (2)(-57,789) + (-22,063) - 12,390$$
$$= -714,361 \text{ cal/gmol}$$

This stoichiometric reaction is now written for combustion in air. First note that there are 7.0(79/21) or 26.33 lbmol of nitrogen present in the theoretical combustion air

$$C_6H_5Cl + 7O_2 + [26.33N_2] \longrightarrow 6CO_2 + 2H_2O(g) + HCl(g) + [26.33N_2]$$

The heat capacity for the flue gas products in the form

$$C_P = a + bT + cT^{-2}$$

are read from Table 7.4. Note that only the products participate in this calculation.

$$C_{P,CO_2} = 10.57 + 2.10 \times 10^{-3}T - 2.06 \times 10^5 T^{-2}$$

$$C_{P,H_2O} = 7.30 + 2.46 \times 10^{-3}T + 0.0 \times 10^5 T^{-2}$$

$$C_{P,HCl} = 6.27 \times 1.24 \times 10^{-3}T + 0.3 \times 10^5 T^{-2}$$

$$C_{P,N_2} = 6.83 + 0.90 \times 10^{-3}T - 0.12 \times 10^5 T^{-2}$$

However,

$$\Delta C_P = 6C_{P,CO_2} + 2C_{P,H_2O} + C_{P,HCl} + 26.33C_{P,N_2}$$

Thus,

$$\Delta C_P = 6(10.57 + 2.10 \times 10^{-3}T - 2.06 \times 10^5 T^{-2})$$

$$+ 2(7.30 + 2.46 \times 10^{-3}T - 0.0 \times 10^5 T^{-2})$$

$$+ 1(6.27 + 1.24 \times 10^{-3}T + 0.30 \times 10^5 T^{-2})$$

$$+ 26.33(6.83 + 0.90 \times 10^{-3}T - 0.12 \times 10^5 T^{-2})$$

If this is expressed in the form $\Delta C_P = \Delta a + \Delta bT + \Delta c T^{-2}$, then

$$\Delta C_P = 264.12 + 0.0425T - 1.522 \times 10^6 T^{-2} \text{ cal/gmol} \cdot \text{K or Btu/lbmol} \cdot {}^\circ\text{R}$$

Equation (10.22) applies in calculating the adiabatic flame temperature. The energy liberated on combustion appears as sensible energy in heating the flue (product) gas. The sum of these two effects is zero if the operation is conducted adiabatically, i.e.,

$$\Delta H_c^0 + \Delta H_p = \Delta H = 0$$

Since $25^\circ C = 298K$, the enthalpy change associated with heating the flue products is given by

$$\Delta H_p = \int_{298}^{T_2} \Delta C_P \, dT \qquad T_2 = \text{theoretical adiabatic temperature (K)}$$

Substituting ΔC_P obtained previously and integrating yields

$$\Delta H_p = \Delta a(T_2 - 298) + \left(\frac{\Delta b}{2}\right)(T_2^2 - 298^2) - \Delta c\left(\frac{1}{T_2} - \frac{1}{298}\right)$$

$$= 264.12(T_2 - 298) + \left(\frac{0.0425}{2}\right)(T_2^2 - 298^2) + 1.522 \times 10^6 \left(\frac{1}{T_2} - \frac{1}{298}\right)$$

with

$$-\Delta H_c^0 = 714{,}361 \text{ cal/gmol}$$

Equation (10.22) may now be rewritten in the form

$$800,063 = 264.12T_2 + 0.02125T_2^2 + (1.522 \times 10^6)/T_2$$

This is a nonlinear cubic equation. It may be solved by any one of several analytical or numerical methods. The final result can also be obtained by a crude trial-and-error procedure. However, this trial-and-error process can be simplified by recognizing that the last term on the right-hand side is an order of magnitude smaller than the first two terms. Therefore, an excellent first guess can be obtained by solving the equation

$$0.02125T_2^2 + 264.12T_2 - 800,063 = 0$$

$$T_2 = \frac{-264.12 \pm \sqrt{(264.12)^2 + (4)(0.02125)(800,063)}}{(2)(0.02125)}$$

$$= 2519\text{K} = 4534°\text{R} = 4074°\text{F}$$

Final theoretical adiabatic flame temperature is approximately 4074°F.   ■

ILLUSTRATIVE EXAMPLE 10.6

Calculate the theoretical adiabatic flame temperature of chlorobenzene, C_6H_5Cl. Assume that the heat capacity variation with temperature takes the form:

$$C_P = \alpha + \beta T + \gamma T^2$$

Data:

$$C_{P,CO_2} = 6.214 + 10.396 \times 10^{-3}T - 3.545 \times 10^{-6}T^2$$
$$C_{P,H_2O} = 7.256 + 2.298 \times 10^{-3}T + 0.283 \times 10^{-6}T^2$$
$$C_{P,HCl} = 6.732 + 0.433 \times 10^{-3}T + 0.370 \times 10^{-6}T^2$$
$$C_{P,N_2} = 6.524 + 1.250 \times 10^{-3}T - 0.001 \times 10^{-6}T^2$$

SOLUTION: One can follow the same procedure provided in the previous example with the data provided.

$$C_{Pi} = \alpha + \beta T + \gamma T^2$$
$$C_{P,CO_2} = 6.214 + 10.396 \times 10^{-3}T - 3.545 \times 10^{-6}T^2$$
$$C_{P,H_2O} = 7.256 + 2.298 \times 10^{-3}T + 0.283 \times 10^{-6}T^2$$
$$C_{P,HCl} = 6.732 + 0.433 \times 10^{-3}T + 0.370 \times 10^{-6}T^2$$
$$C_{P,N_2} = 6.524 + 1.250 \times 10^{-3}T - 0.001 \times 10^{-6}T^2$$

For this case,

$$\Delta C_P = 6(C_{P,CO_2}) + 2(C_{P,H_2O}) + 1(C_{P,HCl}) + 26.33(C_{P,N_2})$$
$$= 230.305 + 1.003175 \times 10^{-1}T - 20.36033 \times 10^{-6}T^2$$

and

$$\Delta H_p = -\Delta H_c = 714{,}631$$

Therefore,

$$714{,}631 = 230.305(T - 298) + \frac{0.1003175}{2}(T^2 - 298^2)$$
$$- \frac{20.36033}{3} \times 10^{-6}(T^3 - 298^3)$$

By trial-and-error

$$T = 2511.5\text{K}$$
$$= 4061°\text{F}$$

This result is an excellent agreement with the value (4074°F) calculated in the previous example.   ∎

ILLUSTRATIVE EXAMPLE 10.7

Manhattan Generators plan to incinerate pure C_6H_5Cl in an incinerator with 100% excess air. A consulting engineer from Shannon and O'Brien Associates has assured the client that the operating temperature will exceed the permit requirement of 2100°F. Verify or dispute the consultant's claim, employing a, b, and c data.

SOLUTION: This example involves an extension of the earlier material. However, unlike the previous development, it will require an adiabatic flame temperature calculation with 100% excess air. For this condition, the stoichiometric oxygen and nitrogen (or air) requirement is increased by a factor of 2. These values and the stoichiometric equation are again given here:

$$O_2 = 14 \text{ lbmol}, \quad N_2 = 52.6 \text{ lbmol}$$

$$C_6H_5Cl + 14O_2 + [52.6N_2] \longrightarrow 6CO_2 + 2H_2O + HCl + 7O_2 + [52.6N_2]$$

The heat capacity term (ΔC_P) is now calculated using the equivalent of Equation (10.14) for a, b, and c:

$$\Delta C_P = 6C_{P,CO_2} + 2C_{P,H_2O} + C_{P,HCl} + 7C_{P,O_2} + 52.6C_{P,N_2}$$

The heat capacities are obtained from Table 7.4:

$$C_{P,CO_2} = 10.57 + 2.10 \times 10^{-3}T - 2.06 \times 10^5 T^{-2}$$

$$C_{P,H_2O} = 7.30 + 2.46 \times 10^{-3}T + 0.00 \times 10^5 T^{-2}$$

$$C_{P,HCl} = 6.27 + 1.24 \times 10^{-3}T + 0.30 \times 10^5 T^{-2}$$

$$C_{P,O_2} = 7.16 + 1.00 \times 10^{-3}T - 0.40 \times 10^5 T^{-2}$$

$$C_{P,N_2} = 6.83 + 0.90 \times 10^{-3}T - 0.12 \times 10^5 T^{-2}$$

Thus,

$$\Delta C_P = 493.67 + 0.0731T - 2.12 \times 10^6 T^{-2}$$

The sensible enthalpy change for the flue gas products is

$$\Delta H_p = \int_{298}^{T_2} \Delta C_P \, dT$$

In addition,

$$\Delta H_p = -\Delta H_c^0$$

where ΔH_c^0 was previously calculated to be $-714,361$ cal/gmol in Illustrative Example 10.5. Therefore,

$$714,361 = 493.67(T - 298) + \frac{1}{2}(0.0731)(T^2 - 298^2) + 2.12 \times 10^6 \left(\frac{1}{T} - \frac{1}{298}\right)$$

$$871,834 = 493.67T + 0.036557T^2 + \frac{2.12 \times 10^6}{T}$$

Solving for T

$$T = 1579\text{K} = 2842°\text{R} = 2382°\text{F}$$

Therefore, the operating temperature does exceed the permit requirement of $2100°$F.   ∎

ILLUSTRATIVE EXAMPLE 10.8

Calculate the operating temperature of a reactor burning pure DDT ($C_{14}H_9C_{15}$) with 150% excess air. Assume the DDT and air enter the unit at ambient conditions (25°C, 1 atm). Use a, b, and c data.

SOLUTION: This problem is very similar to Illustrative Examples 10.5 and 10.6. Here, instead of stoichiometric air, 150% excess air is used. The following values can be obtained from Table 10.1:

$$\Delta H_c = -1580.56 \text{ kcal/gmol}$$

Proceding as before,

$$C_{14}H_9O_5 + 37.5O_2 + [141.08N_2] \longrightarrow 14CO_2 + 2H_2O + 5HCl + 22.5O_2 + [141.08N_2]$$

with

$$\Delta C_P = 14C_{P,CO_2} + 2C_{P,H_2O} + 5C_{P,HCl} + 22.5C_{P,O_2} + 141.08C_{P,N_2}$$

Substituting in the heat capacities from Table 7.4:

$$\Delta C_P = 1318.60 + 0.1899T - 5.327 \times 10^6 T^{-2}$$

Since

$$\Delta H_c = \Delta H_p = \int_{298}^{T} \Delta C_P \, dT$$

$$1580.56 \times 10^3 = 1318.60(T_2 - 298) + \frac{1}{2}(0.1899)(T_2^2 - 298^2)$$

$$+ 5.327 \times 10^6 \left(\frac{1}{T_2} - \frac{1}{298} \right)$$

$$1{,}999{,}815.0 = 1318.60T_2 + 0.095T_2^2 + \frac{5.327 \times 10^6}{T_2}$$

Solving for T_2 by a trial-and-error method,

$$T_2 = 1377\text{K}$$
$$= 1104°\text{C}$$
$$= 2019°\text{F}$$

∎

ILLUSTRATIVE EXAMPLE 10.9

Determine the flame temperature, T_f, of a fuel consisting of essentially pure gaseous C_2H_6O (ethanol) operating with 10% excess air and a heat loss of 10% of the standard enthalpy of reaction. Perform the calculations using α, β, and γ data.

SOLUTION: With 10% excess air (EA), the stoichiometric equation becomes:

$$C_2H_6O + 3.3O_2 + [12.41N_2] \longrightarrow 2CO_2 + 3H_2O + 0.3O_2 + [12.41N_2]$$

For this reaction

$$\Delta H_{298}^0 = \Delta H_c^0 = 2(-94{,}054) + 3(-57{,}798) - (-56{,}240)$$
$$= -305{,}262 \text{ cal/gmol}$$

For the product (flue) gases,

$$\Delta\alpha = 2(6.214) + 3(7.256) + 0.3(6.148) + 12.41(6.524)$$
$$= 117$$
$$\Delta\beta = 0.04521$$
$$\Delta\gamma = -6.53 \times 10^{-6}$$

The key equation for this reaction remains the same, i.e.,

$$-\Delta H_c^0 = \Delta H_p = +305{,}262 \text{ cal/gmol}$$

However, with 10% heat loss,

$$\Delta H_p = (+305{,}262)(0.9) = 274{,}735.8 \text{ cal/gmol}$$

Thus,

$$274{,}735.8 = \Delta\alpha(T - 298) + \frac{\Delta\beta}{2}(T^2 - 298^2) + \frac{\Delta\gamma}{3}(T^3 - 298^3)$$

By trial-and-error,

$$T = 2025\text{K} = 3186°\text{F}$$

Interestingly, without the 10% heat loss,

$$T = 2192\text{K} = 3486°\text{F}$$ ∎

ILLUSTRATIVE EXAMPLE 10.10

Calculate the theoretical flame temperature for a mixture containing 75% by weight 1-chloro-ethylene and 25% propane. The thermochemical properties are provided in Table 10.2.

Table 10.2   Thermochemical Properties

Component	ΔH_f^0 (cal/gmol)	α	$\beta \times 10^3$	$\gamma \times 10^6$
CO_2	−94,052	6.214	10.936	−3.545
$H_2O(g)$	−57,798	7.256	2.298	0.283
HCl	−22,063	6.732	0.433	0.370
N_2	0.0	6.524	1.250	−0.001
C_2H_3Cl	−8400			
C_3H_8	−24,820			
O_2	0.0			

SOLUTION: Write balanced stoichiometric equations for the 1-chloroethylene and propane streams:

$$C_2H_3Cl + 2.5O_2 + [9.4N_2] \longrightarrow 2CO_2 + H_2O + HCl + [9.4N_2]$$

$$C_3H_8 + 5O_2 + [18.8N_2] \longrightarrow 3CO_2 + 4H_2O + [18.8N_2]$$

Calculate the standard enthalpy (heat) of combustion (reaction) for 1-chloroethylene:

$$\Delta H_c^0 = 2\Delta H_f^0(CO_2) + \Delta H_f^0(H_2O) + \Delta H_f^0(HCl) + 9.4\Delta H_f^0(N_2)$$
$$-\Delta H_f^0(C_2H_3Cl) - 2.5\Delta H_f^0(O_2) - 9.4\Delta H_f^0(N_2)$$

Substituting,

$$\Delta H_c^0 = 2(-94{,}052\,cal/gmol) + (-57{,}798\,cal/gmol)$$
$$+ (-22{,}063\,cal/gmol) - (-8{,}400\,cal/gmol)$$
$$= -259{,}565\,cal/gmol$$

Calculate the standard heat of combustion for propane:

$$\Delta H_c^0 = 3\Delta H_f^0(CO_2) + 4\Delta H_f^0(H_2O) + 18.8\Delta H_f^0(N_2)$$
$$- \Delta H_f^0(C_3H_8) - 5\Delta H_f^0(O_2) - 18.8\Delta H_f^0(N_2)$$

Substituting,

$$\Delta H_c^0 = 3(-94{,}052\,cal/gmol) + 4(-57{,}798\,cal/gmol) - (-24{,}820\,cal/gmol)$$
$$= -488{,}528\,cal/gmol$$

Determine the heat capacity values, $\Delta\alpha$, $\Delta\beta$, and $\Delta\gamma$, for the flue gas products for the chloroethylene:

$$\sum_{products} n\alpha = 2\alpha(CO_2) + \alpha(H_2O) + \alpha(HCl) + 9.4\alpha(N_2)$$

$$= 2(6.214) + (7.256) + (6.732) + 9.4(6.524)$$
$$= 87.7416$$

$$\sum_{products} n\beta = 2\beta(CO_2) + \beta(H_2O) + \beta(HCl) + 9.4\beta(N_2)$$

$$= 2(10.396 \times 10^{-3}) + (2.298 \times 10^{-3}) + (0.433 \times 10^{-3})$$
$$+ 9.4(1.250 \times 10^{-3})$$
$$= 35.273 \times 10^{-3}$$

$$\sum_{products} n\gamma = 2\gamma(CO_2) + \gamma(H_2O) + \gamma(HCl) + 9.4\gamma(N_2)$$

$$= 2(-3.545 \times 10^{-6}) + (0.283 \times 10^{-6})$$
$$+ (0.370 \times 10^{-6}) + 9.4(-0.001 \times 10^{-6})$$
$$= -6.446 \times 10^{-6}$$

Determine the heat capacity values, $\Delta\alpha$, $\Delta\beta$, and $\Delta\gamma$ for the flue gas products for the propane reaction:

$$\sum_{products} n\alpha = 3\alpha(CO_2) + 4\alpha(H_2O) + 18.8\alpha(N_2)$$

$$= 3(6.214) + 4(7.256) + 18.8(6.524)$$

$$= 170.317$$

$$\sum_{products} n\beta = 3\beta(CO_2) + 4\beta(H_2O) + 18.8\beta(N_2)$$

$$= 3(10.396 \times 10^{-3}) + 4(2.298 \times 10^{-3}) + 18.8(1.250 \times 10^{-3})$$

$$= 63.820 \times 10^{-3}$$

$$\sum_{products} n\gamma = 3\gamma(CO_2) + 4\gamma(H_2O) + 18.8\gamma(N_2)$$

$$= 3(-3.545 \times 10^{-6}) + 4(0.283 \times 10^{-6}) + 18.8(-0.001 \times 10^{-6})$$

$$= -9.5218 \times 10^{-6}$$

Calculate the mole fraction of 1-chloroethylene and propane:

$$MW \text{ of 1-chloroethylene} = 62.5; \quad 62.5 \text{ lb/lbmol}$$
$$MW \text{ of propane} = 44; \quad 44 \text{ lb/lbmol}$$

Converting from lb to lbmols on a total 100 lb basis,

$$\text{mols 1-chloroethylene} = 75 \text{ lb}/(62.5 \text{ lb/lbmol}) = 1.2$$
$$\text{mols propane} = 25 \text{ lb}/(44 \text{ lb/lbmol}) = 0.57$$
$$\text{total mols} = 1.2 + 0.57 = 1.77$$

Converting mols to mole fraction,

$$\text{mole fraction 1-chloroethylene} = 1.2/1.77 = 0.679$$
$$\text{mole fraction propane} = 0.57/1.77 = 0.321$$

Develop the expression for the sensible heat (enthalpy) of the products, ΔH_p, in terms of $\Delta\alpha$, $\Delta\beta$, $\Delta\gamma$, and T.

$$\Delta H_p^0 = \Delta\alpha(T - 298) + (\Delta\beta/2)(T^2 - 298^2) + (\Delta\gamma/3)(T^3 - 298^3)$$

Applying the appropriate mole fraction values to $\Delta\alpha$, $\Delta\beta$, and $\Delta\gamma$ gives

$$\Delta H_p^0 = [0.679(87.7416) + 0.321(170.317)](T - 298)$$
$$+ (1/2)[0.679(35.273 \times 10^{-3}) + 0.321(63.820 \times 10^{-3})](T^2 - 298^2)$$
$$+ (1/3)[0.679(-6.446 \times 10^{-6}) + 0.321(-9.5218 \times 10^{-6})](T^3 - 298^3)$$

Reducing and collecting like terms yields

$$\Delta H_p^0 = 114.25(T - 298) + 22.23 \times 10^{-3}(T^2 - 298^2)$$
$$+ (-2.4778 \times 10^{-6})(T^3 - 298^3)$$

Calculate the overall heat of combustion. The overall heat of combustion is found by again employing molar weighted averages:

$$\Delta H_c^0 = 0.679(-259,565 \text{ cal/gmol}) + 0.321(-488,528 \text{ cal/gmol})$$
$$= -333,062 \text{ cal/gmol}$$

Equate the energy released to the energy required to heat the combustion products for each reaction. As before, set

$$\Delta H_c^0 = -\Delta H_p^0$$
$$-333,062 = -[114.25(T - 298) + (22.23 \times 10^{-3})(T^2 - 298^2)$$
$$- (2.4778 \times 10^{-6})(T^3 - 298^3)]$$

Calculate the temperature by reducing like terms and rearranging,

$$0 = (-2.478 \times 10^{-6})T^3 + 0.02223T^2 + 114.25T - 369,016$$

Solving this expression, assuming the cubic term can be neglected, yields a flame temperature of 2372K. Solving the equation more rigorously using Newton's method or spreadsheet iteration yields an adiabatic flame temperature of 2406K or 3871°F. ■

Note that a comprehensive computer program is available that performs both the stoichiometric and thermochemical calculations presented in this problem set and the stoichiometry problem set. Equipment design is also included in the program. For more details on this program, contact: Dr. Lou Theodore (Consultant), Theodore Tutorials, 5 Fairview Avenue, East Williston, NY 11592, USA. Tel.: (516) 742 8939. Email: loutheodore1@verizon.net.

The program is also available in a book by J. Santoleri, J. Reynolds, and L. Theodore, see Ref. 3.

ILLUSTRATIVE EXAMPLE 10.11

The standard enthalpy of reaction per gram mole of *product* formed is

$$\Delta H_T^0 = -9140 - 7.596T + 4.243 \times 10^{-3}T^2 - 0.742 \times 10^{-6}T^3; \quad T = \text{K}$$

Specify the temperature for which the standard enthalpy of reaction is $-12,236$ cal/gmol.

SOLUTION: Set

$$\Delta H_T^0 = -12,236 \text{ cal/gmol}$$

By trial-and-error,

$$T \approx 570\text{K}$$
$$= 300°\text{C}$$

■

ILLUSTRATIVE EXAMPLE 10.12

Refer to the previous example. How much heat must be added to or removed from a flow reactor per gmole of product formed if the reaction is conducted at 250°C.

SOLUTION: Note that

$$T = 250 + 273 = 523\text{K}$$

At

$$T = 523\text{K}$$

and substituting into the equation provided in the previous example,

$$\Delta H^0_{523} = -12{,}059 \text{ cal/gmole product formed}$$

For a flow process

$$Q = \Delta H = -12{,}059 \text{ cal/gmole product formed}$$

The negative sign indicates that heat is removed.

■

ILLUSTRATIVE EXAMPLE 10.13

Refer to the previous example. What heat rate must be added to or removed from a flow reactor with a discharge flow rate of 8.0 gmole/h of the *product* specified in the standard enthalpy of reaction provided in Illustrative Example 10.11.

SOLUTION: For this case

$$\dot{n}_{\text{product}} = 8.0 \text{ gmole/h}$$

Therefore,

$$\dot{Q} = (\dot{n}_{\text{product}})\Delta H^0_{523}$$

$$= \left(\frac{8.0 \text{ gmole}}{\text{hr}}\right)\left(-12{,}059 \frac{\text{cal}}{\text{gmole}}\right)$$

$$= -96{,}470 \text{ cal/h}$$

■

ILLUSTRATIVE EXAMPLE 10.14

Refer to Illustrative Example 10.13. *Outline* how to a calculate the heat requirements if reactants enter the reactor at 250°C and products leave at 500°C.

SOLUTION: Refer to Fig. 10.1.

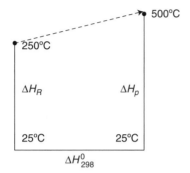

Figure 10.1   Enthalpy diagram for Illustrative Example 10.14.

Take as a basis, 1.0 gmol product upon which ΔH_T^0 is based. For this case:

$$H_{500} - H_{250} = \Delta H_R + \Delta H_{298}^0 + \Delta H_p = \Delta H$$

$$\Delta H_R \Rightarrow \text{sensible enthalpy change from } 250°C \text{ to } 25°C$$
$$\Delta H_p \Rightarrow \text{sensible enthalpy change from } 25°C \text{ to } 500°C$$
$$\Delta H_{298}^0 \Rightarrow \text{reaction at } 25°C \text{ with } 1.0 \text{ gmoles of product}$$

The calculation should proceed as illustrated earlier.   ∎

GROSS AND NET HEATING VALUES

Heat of reaction is a term rarely employed in air pollution and/or some other combustion calculations. The two terms most often used in this field are the gross (or higher) heating value and the net (or lower) heating value. The former is designated by HHV or HV_G and the latter by NHV or HV_N. The *gross heating value* represents the enthalpy change or heat released when a compound is stoichiometrically reacted (combusted) at a reference temperature with the final (flue) products also at the reference temperature and any water present in the liquid state. Most of these data are available at a reference temperature of 60°F. The *net heating value* is similar to the gross heating value except the water is in the vapor state. The difference (if any) between the two values represents the energy necessary to vaporize any water present. Thus, both the standard heat of reaction and the gross and/or net heating values employed in some industries essentially represent the same phenomenon. Gross and net heating values for a number of hydrocarbons are presented in Table 4.1 (see Chapter 4). In addition, the net heating value may be approximated by a form of Dulong's equation that includes the chlorine content:

$$\text{NHV} = 14{,}000 w_C + 45{,}000 (w_H - \tfrac{1}{8} w_O) - 760 w_{Cl} + 4500 w_S \qquad (10.25)$$

where   NHV = net heating value of the combustible mixture (Btu/lb)

m_C = mass fraction of C (carbon) in the mixture

C, H, O, Cl, S = subscripts indicating carbon, hydrogen, oxygen, chlorine, and sulfur, respectively.

Another common term employed in combustion calculations is the *available heat*, usually designated as HA_T. The available heat at any temperature T is the gross heating value minus the amount of heat ($\sum \Delta H$) required to take the product(s) of combustion (flue gas) from the reference temperature to that temperature T. Thus,

$$HA_T = \text{HHV} - \sum \Delta H \qquad (10.26)$$

If all the heat liberated by the reaction goes into heating up the products of combustion (the flue gas), the temperature achieved is defined as the *flame temperature* (FT). If the combustion process is conducted adiabatically, with no heat transfer to the surroundings, the final temperature achieved by the flue gas is defined as the *adiabatic flame temperature* (AFT). If the combustion process is conducted with theoretical or stoichiometric air (0% excess), the resulting temperature is defined as the *theoretical adiabatic flame temperature* (TAFT). It is rare that a thermal device operates at the adiabatic flame temperature. This is never the case in a typical combustion process due to losses through the burner, refractories, and insulation. The actual flame temperature is usually a few hundred degrees below the calculated or theoretical adiabatic furnace temperature. Again, most combustion processes whether batch or continuous are conducted with air in excess of that required for theoretical combustion.

Heat losses (conduction and convection) through the enclosure (tube walls) due to construction materials and external shell conditions (air-cooled, water-cooled, or natural radiation and convection to the surrounding atmosphere), as well as heat losses by radiation to downstream equipment (boiler tubes, spray quench chamber, etc.), will affect the final temperature conditions of the reaction (combustion) products.[3]

To calculate fuel requirements, operating temperature, and (excess) air requirements for a combustion operation, one must apply the conservation laws for mass and energy in conjunction with thermochemical principles. This is an extremely involved, rigorous calculation. An enthalpy balance is applied around the combustor following comprehensive overall and componental material balances. The enthalpy balance must account for all temperature changes of both the feed and fuel (reactants) as well as the flue gas (products). Latent (phase) and combustion (reaction) enthalpy effects must also be included in the analysis. Fortunately, simple algorithms are available to perform these detailed thermodynamic calculations; these can be found in the next paragraph and the Illustrative Examples that follow.

As noted earlier, an important parameter in combustion/incineration calculations is the operating temperature. The operating temperature in an incinerator is a function of many variables. For most incinerators, the operating temperature is calculated by determining the flame temperature under adiabatic or near-adiabatic conditions. From a calculational point-of-view, the flame temperature has a strong dependence on the excess air requirement and the heating value of the combined combustible mixture. The Theodore–Reynolds equation[3,4] shown below can be used to estimate the average temperature in the thermal device in lieu of using a rigorous model that may require extensive experimental data and physical/chemical properties,

$$T = 60 + \frac{\text{NHV}}{(0.325)[1 + (1 + \text{EA})(7.5 \times 10^{-4})(\text{NHV})]} \qquad (10.27)$$

where T = temperature, °F

NHV = net heating value of the inlet mixture, Btu/lb

EA = excess air on a fractional basis

The value of EA may be estimated by

$$\text{EA} = \frac{0.95Y}{(21 - Y)} \qquad (10.28)$$

where Y is the dry mol % O_2 in the combustion (incinerated) gas.

Additional details on these two equations (including their derivations) are available in the literature.[3,4]

ILLUSTRATIVE EXAMPLE 10.15

The composition (mole fraction) of a gas mixture is given in Table 10.3. Using combustion properties provided in Table 10.4, estimate the gross heating value of the gas mixture in Btu/scf.

SOLUTION: Combine Tables 10.3 and 10.4 as shown in Table 10.5.

Table 10.3   Composition of Natural Gas

Component	x_i
N_2	0.0515
CH_4	0.8111
C_2H_6	0.0967
C_3H_8	0.0351
C_4H_{10}	0.0056
	1.0000

Table 10.4   Natural Gas Gross Heating Values

Component	HV_G (Btu/scf)
N_2	0
CH_4	1013
C_2H_6	1792
C_3H_8	2590
C_4H_{10}	3370

Table 10.5   Natural Gas Composition and Gross Heating Values

Component	HV_G (Btu/scf)	x_i
N_2	0	0.0515
CH_4	1013	0.8111
C_2H_6	1792	0.0967
C_3H_8	2590	0.0351
C_4H_{10}	3370	0.0056

Calculate the gross heating value of the gas mixture, HV_G in Btu/scf:

$$HHV = HV_G = \sum (HHV_i) x_i \qquad (10.29)$$

$$
\begin{aligned}
HV_G &= (0.0515)(0) + (0.8111)(1013) + (0.0967)(1792) \\
&\quad + (0.0351)(2590) + (0.0056)(3370) \\
&= 1105 \, \text{Btu/scf}
\end{aligned}
$$

■

ILLUSTRATIVE EXAMPLE 10.16

Estimate the theoretical flame temperature of a hazardous waste mixture containing 25% cellulose, 35% motor oil, 15% water (vapor), and 25% inerts, by mass. Assume 5% radiant heat losses. The flue gas contains 11.8% CO_2, 13 ppm CO, and 10.4% O_2 (dry basis) by volume.

NHV of cellulose = 14,000 Btu/lb

NHV of motor oil = 25,000 Btu/lb

NHV of water = 0 Btu/lb

NHV of inerts (effective) = −1000 Btu/lb

SOLUTION: The NHV for the mixture is obtained by multiplying the component mass fractions by their respective NHVs and taking the sum of the products. Thus,

$$HV_N = NHV = \sum_{i=1}^{n} (NHV)_i x_i \qquad (10.30)$$

Substituting,

$$NHV = 0.25(14{,}000\,Btu/lb) + 0.35(25{,}000\,Btu/lb)$$
$$+ 0.15(0.0\,Btu/lb) + 0.25(-1000\,Btu/lb)$$
$$= 12{,}000\,Btu/lb$$

The excess air employed is obtained from Equation (10.28):

$$EA = \frac{0.95Y}{21 - Y}$$
$$= \frac{0.95(10.4)}{21 - 10.4}$$
$$= 0.932$$

The flame temperature is estimated using the Theodore–Reynolds equation provided in Equation (10.27).

$$T = 60 + \frac{NHV}{(0.325)[1 + (1 + EA)(7.5 \times 10^{-4})(NHV)]}$$
$$= 60 + \frac{12{,}000}{(0.325)[1 + (1 + 0.932)(7.5 \times 10^{-4})(12{,}000)]}$$
$$= 2068°F$$

The reader should also note that this equation is sensitive to the value assigned to the coefficient which represents the average heat capacity of the flue gas, in this case, 0.325. This is a function of both the temperature (T) and EA fraction and also depends on the flue products since the heat capacities of air and CO_2 are about half that of H_2O. In addition, the 7.5×10^{-4} term may vary slightly with the composition of the mixture incinerated. The overall relationship between operating temperature and composition is therefore rather complex, and its prediction is not necessarily as straightforward as shown here. ∎

ILLUSTRATIVE EXAMPLE 10.17

A radioactive mixture is to be burned in an incinerator at an operating temperature of 1900°F. Calculate the minimum net heating value (NHV) of the mixture in Btu/lb if 0 and 100% excess air is employed. Use the following equation, a slight modification of Equation (10.27), to perform the calculations:

$$NHV = \frac{(0.3)(T - 60)}{[1 - (1 + EA)(7.5 \times 10^{-4})(0.3)(T - 60)]} \tag{10.31}$$

SOLUTION: For this problem,

$$T = 1900°F$$

For 0% excess air:

$$\text{NHV} = \frac{(0.3)(1900 - 60)}{[1 - (1)(7.5 \times 10^{-4})(0.3)(1900 - 60)]}$$

$$= 942 \, \text{Btu/lb}$$

Similarly,

$$\text{NHV (100\% excess air)} = 3209 \, \text{Btu/lb}$$

As expected, the NHV for 100% excess is significantly higher than that for 0% excess air. ∎

ILLUSTRATIVE EXAMPLE 10.18

Calculate the net heating value (NHV) of methane, chloroform, benzene(g), chlorobenzene, and hydrogen sulfide. This assumes that the water product is in the vapor state. Compare these values with those calculated using Dulong's equation. Also calculate the relative percent difference between the "true" NHVs as determined by thermodynamic calculations and the "estimated" values calculated using Dulong's equation.

SOLUTION: Dulong's equation can be written as follows:

$$\text{NHV} \approx 14{,}000 w_C + 45{,}000(w_H - \tfrac{1}{8} w_O) - 760(w_{Cl}) + 4500(w_S)$$

where m_i is the mass fraction of component i.

The first step in the solution of this problem is to write the balanced oxidation reaction (combustion) equation for each compound, and from these balanced equations, calculate the standard heat of combustion.

$$CH_4: \; CH_4 + 2O_2 \; \longrightarrow \; CO_2 + 2H_2O(g)$$
$$CHCl_3: \; CHCl_3 + \tfrac{7}{6}O_2 \; \longrightarrow \; CO_2 + \tfrac{1}{3}H_2O + \tfrac{1}{3}HCl + \tfrac{4}{3}Cl_2$$
$$C_6H_6: \; C_6H_6 + \tfrac{15}{2}O_2 \; \longrightarrow \; 6CO_2 + 3H_2O(g)$$
$$C_6H_5Cl: \; C_6H_5Cl + 7O_2 \; \longrightarrow \; 6CO_2 + 2H_2O(g) + HCl$$
$$H_2S: \; H_2S + \tfrac{3}{2}O_2 \; \longrightarrow \; SO_2 + H_2O(g)$$

For the methane reaction, the heat of combustion is calculated from heats of formation, as follows:

$$\Delta H_c^0 = (-94{,}052 \, \text{cal/gmol} \, CO_2) + 2(-57{,}798 \, \text{cal/gmol} \, H_2O(g))$$
$$- (-17{,}889 \, \text{cal/gmol} \, CH_4)$$
$$= -191{,}759 \, \text{cal/gmol}$$

Heat of combustion values for the other compounds are calculated in a similar manner or can be found in the literature.[5] The results of these calculations are summarized in Table 10.6 using the conversion factor $1.8 \, \text{cal/g} = 1.0 \, \text{Btu/lb}$.

The results using Dulong's equation are shown in Table 10.7.

Table 10.6   Net Heating Value

Compound	NHV (cal/gmol)	MW (g/gmol)	NHV (cal/g)	NHV (Btu/lb)
CH_4	191,759	16	11,985	21,593
$CHCl_3$	96,472	119.5	807	1453
C_6H_6	717,886	78	9204	16,583
C_6H_5Cl	714,361	112.5	6350	11,441
H_2S	123,943	34	3645	6567

Table 10.7   Net Heating Value Result

Compound	Mass %C	Mass %H	Mass %O	Mass %Cl	Mass %S	NHV (Btu/lb)
CH_4	0.75	0.25	0	0	0	21,750
$CHCl_3$	0.10	0.0084	0	0.891	0	1101
C_6H_6	0.92	0.08	0	0	0	16,480
C_6H_5Cl	0.64	0.04	0	0.32	0	10,520
H_2S	0	0.06	0	0	0.94	6930

Table 10.8   Net Heating Value Comparison

Compound	NHV_{thermo} (Btu/lb)	NHV_{Dulong} (Btu/lb)	% Difference
CH_4	21,593	21,750	0.73
$CHCl_3$	1453	1101	6.7
C_6H_6	16,583	16,480	0.62
C_6H_5Cl	11,441	10,520	8.05
H_2S	6567	6930	5.53

Based on these calculations, the difference between the thermodynamically based NHV values and those estimated using Dulong's equation are given in Table 10.8.   ■

REFERENCES

1. R. C. WEAST, ed., "*CRC Handbook of Chemistry and Physics*," 80th edition, CRC Press, Boca Raton, FL, 1999.
2. J. M. SMITH and H. C. VAN NESS, "*Introduction to Chemical Engineering Thermodynamics*," 4th edition, McGraw-Hill, New York, 1992.
3. J. SANTOLERI, J. REYNOLDS, and L. THEODORE, "*Introduction to Hazardous Waste Incineration*," 2nd edition, John Wiley & Sons, Hoboken, NJ, 2000.
4. L. THEODORE and J. REYNOLDS, personal notes, Manhattan College, 1986.
5. D. GREEN and R. PERRY, "*Perry's Chemical Engineers' Handbook*," 8th edition, McGraw-Hill, New York, 2008.

NOTE: Additional problems for each chapter are available for all readers at www. These problems may be used for additional review or homework purposes.

Part III

Equilibrium Thermodynamics

Léon Blum [1872–1950]

No government can remain stable in an unstable society and an unstable world.

—*A l'Échelle Humaine (1945). Page 54*

This third Part serves to introduce the general subject of equilibrium thermodynamics with the state of equilibrium defined in the literature in terms of time invariance. The magic word in this Part is *equilibrium*; all the material presented in this Part apply to/ for *equilibrium* conditions.

Two key subject areas receive treatment: phase equilibrium and chemical reaction equilibrium. The first two chapters deal with the former topic while the last two chapters deal with the latter topic. The chapter subject titles are: Phase Equilibrium Principles; Vapor–Liquid Equilibrium Calculations; Chemical Reaction Equilibrium Principles; Chemical Reaction Equilibrium Applications.

Thermodynamics for the Practicing Engineer. By L. Theodore, F. Ricci, and T. Van Vliet
Copyright © 2009 by John Wiley & Sons, Inc.

Chapter 11

Phase Equilibrium Principles

Joseph Stalin [1879–1953]

We are witnessing a temporary stabilization of capitalism and the stabilization of the Soviet regime. A temporary equilibrium has been established between the two stabilizations. This compromise is the basic feature of the present situation.

—Speech to Party Officials, May 9, 1925

INTRODUCTION

Relationships governing the equilibrium distribution of a substance between two phases, particularly gas and liquid phases, are the principal subject matter of phase-equilibrium thermodynamics. These relationships form the basis of calculational procedures that are employed in the design and the prediction of performance of several mass transfer processes.[1]

The term *phase*, for a pure substance, indicates a *state of matter*, i.e., solid, liquid, or gas. For mixtures, however, a more stringent connotation must be used since a totally liquid or totally solid system may contain more than one phase (e.g., a mixture of oil and water). A phase is characterized by uniformity or *homogeneity*; the same composition and properties must exist throughout the phase region. At most temperatures and pressures, a pure substance normally exists as a single phase. At certain temperatures and pressures, two or perhaps even three phases can coexist in equilibrium. This is shown on the phase diagram for water (see Fig. 11.1). Regarding the interpretation of this diagram, the following points should be noted:

1 The line between the gas and liquid phase regions is the *boiling point* line and represents equilibrium between the gas and liquid.

2 The boiling point of a liquid is the temperature at which its vapor pressure is equal to the *external* pressure. The temperature at which the vapor pressure is equal to 1 atm is the *normal* boiling point.

3 The line between the solid and gas phase regions is the *sublimation point* line and represents equilibrium between the solid and gas.

Thermodynamics for the Practicing Engineer. By L. Theodore, F. Ricci, and T. Van Vliet
Copyright © 2009 by John Wiley & Sons, Inc.

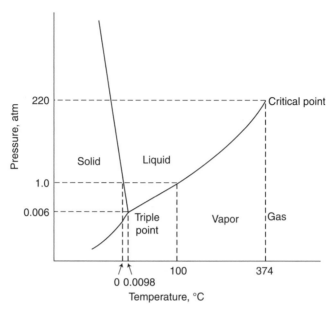

Figure 11.1   Phase diagram for water.

4 The line between the solid and liquid phase regions is the *melting point* or *freezing point* line and represents equilibrium between the liquid and solid.

5 The point at which all three equilibrium lines meet (i.e., the one pressure and temperature where solid, liquid, and gas phases can all coexist) is the *triple point*.

6 The liquid–gas equilibrium line is bounded on one end by the triple point and the other end by the *critical* point. The *critical temperature* (the temperature coordinate of the critical point) is defined as the temperature above which a gas or vapor cannot be liquefied by the application of pressure alone.

The term *vapor*, strictly speaking, is used only for a *condensable* gas (i.e., a gas below its *critical* temperature), and should not be applied to a non-condensable gas. It should also be pointed out that the phase diagram for water (Fig. 11.1) differs from that of other substances in one respect—the freezing point line is *negatively* sloped; for other materials, the slope of this line is *positive*. This is a consequence of the fact that liquid water is denser than ice, and isothermal (constant temperature) compression of the liquid can result in a transformation to the solid.

Thermodynamic calculations, however, rarely involve single (pure) components. Phase equilibria for multicomponent systems are considerably more complex, mainly because of the addition of *composition* variables (e.g. in a *ternary* or three-component system, the mole fractions of two of the components are pertinent variables along with temperature and pressure). In a single-component system, dynamic equilibrium between two phases is achieved when the rate of molecular transfer from one phase to the second equals that in the opposite direction. In multicomponent systems, the equilibrium requirement is more stringent—the rate of transfer of *each component* must be the same in both directions.

In thermodynamic applications, the most important equilibrium phase relationship, as noted above, is that between liquid and vapor. Raoult's and Henry's laws theoretically describe liquid–vapor behavior and under certain conditions are applicable in practice. Raoult's law and Henry's law are the two equations most often used in the study of phase equilibrium and, specifically, within the boundaries of vapor–liquid equilibrium (VLE). Both Raoult's and Henry's law are able to aid in the understanding of the equilibrium properties of liquid mixtures. These two laws are reviewed later in this chapter.

Phase equilibrium examines the physical properties of various classes of mixture and analyzes how different components affect each other within those mixtures. There are three classes of mixtures:

1 vapor–liquid

2 vapor–solid

3 liquid–solid

PSYCHOMETRIC CHART

A vapor–liquid phase equilibrium example involving raw data is the psychrometric or humidity chart[2] (see Fig. 11.2). A humidity chart is used to determine the properties of moist air and to calculate moisture content in air. The ordinate of the chart is the absolute humidity $\mathcal{H}$, which is defined as the mass of water vapor per mass of bone-dry air. (Some charts base the ordinate on moles instead of mass.)

Based on this definition, Equation (11.1) gives $\mathcal{H}$ in terms of moles and also in terms of partial pressure:

$$\mathcal{H} = \frac{18 n_{H_2O}}{29(n_T - n_{H_2O})} = \frac{18 p_{H_2O}}{29(P - p_{H_2O})} \qquad (11.1)$$

where n_{H_2O} = number of moles of water vapor
  n_T = total number of moles in gas
  p_{H_2O} = partial pressure of water vapor
  P = total system pressure

Curves showing the *relative humidity* (ratio of the mass of the water vapor in the air to the maximum mass of water vapor the air could hold at that temperature, that is, if the air were saturated) of humid air also appear on the charts. The curve for 100% relative humidity is also referred to as the *saturation curve*. The abscissa of the humidity chart is air temperature, also known as the *dry-bulb* temperature (T_{DB}). The *wet-bulb* temperature (T_{WB}) is another measure of humidity; it is the temperature at which a thermometer with a wet wick wrapped around the bulb stabilizes. As water evaporates from the wick to the ambient air, the bulb is cooled; the rate of cooling depends on how humid the air is. No evaporation occurs if the air is saturated with water;

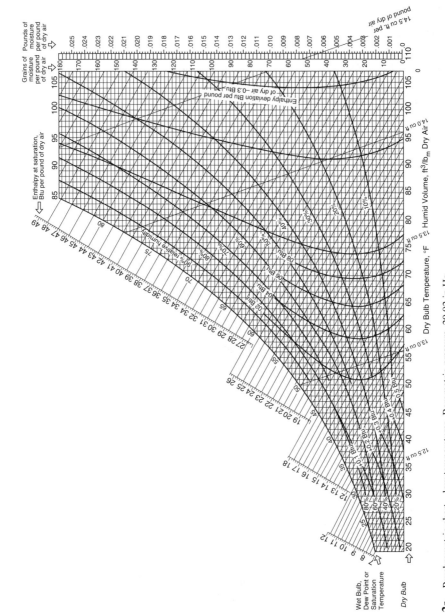

Figure 11.2a Psychrometric chart—low temperatures. Barometric pressure, 29.92 in. Hg.

204

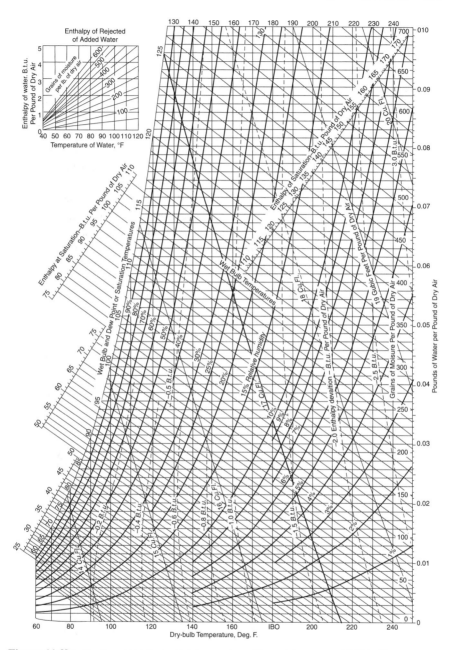

Figure 11.2b   Psychrometric chart—high temperatures. Barometric pressure, 29.92 in Hg.

hence, T_{WB} and T_{DB} are the same. The lower the humidity, the greater the difference between these two temperatures. On the psychrometric chart, constant wet-bulb temperature lines are straight with negative slopes. The value of T_{WB} corresponds to the value of the abscissa at the point of intersection of this line with the saturation curve.

Given the dry bulb and wet bulb temperatures, the relative humidity (along with any other quantity on the chart) may be determined by finding the point of intersection between the dry bulb abscissca and wet bulb ordinate. The point of intersection describes all humidity properties of the system.

The *humid volume* is the volume of wet air per mass of dry air and is linearly related to the humidity. (In Fig. 11.2, this quantity is used as an alternative ordinate. Note the straight parallel lines labeled with units of cubic feet.) The quotient of the humid volume and the absolute humidity gives the volume of the moist air per pound of H_2O. The *humid enthalpy* (also called *humid heat*) is the enthalpy of the moist air on a bone-dry air basis. The term *enthalpy* is a measure of the energy content of the mixture as defined earlier. The enthalpy for *saturated* air can be read from the chart by extending the approximate wet-bulb temperature line upwards to the diagonal scale labeled *enthalpy at saturation*.

The reader should also be familiar with the following definitions, which are pertinent to vapor–liquid systems (some specifically to air–water systems).

1 *Absolute saturation*: mass of vapor/mass of bone-dry gas.

2 *Absolute humidity* ($\mathcal{H}_a$): same definition as absolute saturation but applied to the air–water system, i.e.

$$\mathcal{H}_a = \text{mass of water vapor/mass of bone-dry air.}$$

3 *Relative saturation*:

$$\left(\frac{\text{partial pressure of vapor}}{\text{vapor pressure}}\right) 100\%$$

4 *Relative humidity* (H_r): same definition as relative saturation, but applied to the air–water system, i.e.

$$\left(\frac{\text{partial pressure of water vapor}}{\text{vapor pressure of water}}\right) 100\%$$

or

$$\mathcal{H}_r = (p_w/p'_w)100\% \tag{11.2}$$

Note that since the vapor pressure is the maximum partial pressure that a gas can have at a given temperature, the relative humidity (or relative saturation) is a measure of how much vapor the gas possesses relative to the maximum (saturation) amount it can hold.

Another term that measures how concentrated the vapor is in the gas phase is the *dew point*. Since vapor pressure decreases with decreasing temperature, a gas possessing a specific amount of vapor can be brought to the saturation point by lowering the temperature. The temperature at which the gas reaches saturation in this fashion is called the *dew point*. As an example, suppose the air in a room at 70°F is at 50% relative humidity. From steam tables, the vapor pressure of water at 70°F is 0.3631 psi, which means that the air at 50% humidity holds water vapor with a partial pressure of (0.50)(0.3631) or 0.1812 psi. If the temperature is dropped to the point where 0.1812 psi equals the water vapor pressure (around 52°F), the air becomes saturated with water and any further drop in temperature will cause condensation. The dew point of this air mixture is then 52°F. Obviously, if the air were already saturated at 70°F (i.e., 100% relative humidity), the dew point would also be 70°F.

The following are some helpful points on the use of psychrometric charts:

1 Heating or cooling at temperatures above the *dew point* (temperature at which the vapor begins to condense) corresponds to a horizontal movement on the chart. As long as no condensation occurs, the absolute humidity stays constant.

2 If the air is cooled, the system follows the appropriate horizontal line to the left until it reaches the saturation curve and follows it thereafter.

3 In problems involving use of the humidity chart, it is convenient to choose the *mass of dry air* as a basis, since the chart uses this basis.

ILLUSTRATIVE EXAMPLE 11.1

Refer to the psychrometric chart in Fig. 11.2 to answer the following. List key properties for humid air at a dry-bulb temperature of 160°F and a wet-bulb temperature of 100°F.

SOLUTION: If the air were to be cooled until the moisture just begins to condense, the dew point would be reached. This is represented by a horizontal line at constant humidity intersecting the saturation curve at a dew point of 87.5°F.

The relative humidity is approximately 14% (interpolating between the 15 and 10% relative humidity lines). The absolute humidity is the horizontal line extended to the right intersecting the ordinate at a humidity of 0.0285 lb H_2O/lb dry air. *Note*: dry air $\equiv$ BDA.

The humid volume is approximately 16.3 ft^3 moist air/lb dry air (interpolating between 16 and 17 ft^3 moist air volume).

The enthalpy for saturated air at a T_{WB} of 100°F is 71.8 Btu/lb dry air. For the unsaturated air, the enthalpy deviation is -1.0 Btu/lb dry air; therefore, the actual enthalpy for the moist air at a T_{WB} of 100°F and a T_{DB} of 160°F is 70.8 Btu/lb dry air.   ■

ILLUSTRATIVE EXAMPLE 11.2

Refer once again to the psychrometric chart in Fig. 11.2 to answer the following. A stream of moist air is cooled and humidified adiabatically from a T_{DB} of 100°F and T_{WB} of 70°F to a T_{DB} of 80°F at 1 atm. How much moisture is added per pound of dry air?

SOLUTION: Adiabatic cooling follows the wet-bulb temperature line upwards (toward the saturation curve). The difference in the final and initial humidities is the required additional moisture

$$\text{Initial } \mathcal{H} = 0.0090$$
$$\text{Final } \mathcal{H} = 0.0133$$

Therefore

$$\Delta\mathcal{H} = 0.0133 - 0.0090$$
$$= 4.3 \times 10^{-3} \text{ lb H}_2\text{O/lb dry air} \quad \blacksquare$$

ILLUSTRATIVE EXAMPLE 11.3

A gas is discharged at 120°F from an HCl absorber. If 9000 lb/h (MW = 30) of gas enters the absorber essentially dry (negligible water) at 560°F, calculate the moisture content, the mass flow rate, and the volumetric flow rate of the discharge gas. The discharge gas from the absorber may safely be assumed to be saturated with water vapor.

SOLUTION: From Fig. 11.2, the discharge humidity of the gas is approximately

$$\mathcal{H}_{\text{out}} = 0.0814 \text{ lb H}_2\text{O/lb bone-dry air}$$

This represents the moisture content of the gas at outlet conditions in lb H$_2$O/lb dry air. If the gas is assumed to have the properties of air, the discharge water vapor rate is

$$\dot{m}_{\text{H}_2\text{O}} = (0.0814)(9000)$$
$$= 733 \text{ lb/h}$$

The total flow rate leaving the absorber is

$$\dot{m}_{\text{total}} = 733 + 9000$$
$$= 9733 \text{ lb/h}$$

The volumetric (or molar) flow rate can only be calculated if the molecular weight of the gas is known. The average molecular weight of the discharge gas must first be calculated from the mole fraction of the gas (g) and water vapor (wv):

$$y_g = \frac{9000/30}{(9000/30) + (733/18)} = 0.88$$

$$y_{wv} = \frac{733/18}{(733/18) + (9000/30)} = 0.12$$

$$\overline{\text{MW}} = (0.88)(30) + (0.12)(18) = 28.6 \text{ lb}$$

The ideal gas law is employed to calculate the volumetric flow rate, q_a:

$$Pq_a = \frac{\dot{m}}{MW} RT$$

$$q_a = \frac{9733\,(0.73)(460 + 140)}{28.6 \qquad\qquad 1.0}$$

$$= 1.49 \times 10^5 \text{ ft}^3/\text{h} \qquad\qquad\qquad ■$$

RAOULT'S LAW

Raoult's law states that at *equilibrium* the partial pressure of each component (p_i) in a solution is proportional to its vapor pressure as a pure liquid (p_i'). The "proportionality constant" is the mole fraction (x_i) of that component in the liquid mixture being studied. Therefore, for component i in a mixture, Raoult's law can be expressed as:

$$p_i = x_i p_i' \qquad\qquad\qquad (11.3)$$

where p_i is the partial pressure of component i in the vapor, p_i' is the vapor pressure of pure i at the same temperature and x_i is the mole fraction of component i in the liquid. This expression may be applied to all components so that the total pressure P is given by the sum of all the partial pressures from Equation (11.3). If the gas phase is ideal, Dalton's law applies and

$$p_i = y_i P \qquad\qquad\qquad (11.4)$$

Therefore, Equation (11.3) then can be written as follows:

$$y_i = x_i (p_i'/P) \qquad\qquad\qquad (11.5)$$

where y_i is the mole fraction of component i in the vapor and P is the total system pressure. Thus, the mole fraction of water vapor in a gas that is saturated, i.e., in equilibrium contact with pure water ($x = 1.0$), is simply given by the ratio of the vapor pressure of water at the system temperature divided by the system pressure. These equations primarily find application in distillation, and to a lesser extent, absorption and stripping calculations.[1]

The origin of Raoult's law can be described in molecular terms by considering the rates at which molecules leave and return to the liquid. Raoult's law shows how the presence of a second component, say B, reduces the rate at which component A molecules leave the surface of the liquid, but it does not inhibit the rate at which they return.[3] In order to finally obtain Raoult's law, one must draw on additional experimental information about the relation between the vapor pressures and the composition of the liquid. Raoult himself obtained this data from experiments on mixtures of closely related liquids in order to develop his law.

Most mixtures obey Raoult's law to *some* extent, however small it may be. The mixtures that closely obey Raoult's law are those whose components are structurally

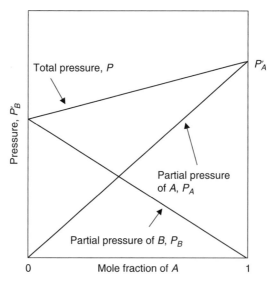

Figure 11.3 Graphical representation of the vapor pressure of an ideal binary solution; that is, one that obeys Raoult's law for the entire composition range.

similar and, therefore, this law is most useful when dealing with these solutions. Mixtures that obey Raoult's law for the entire composition range are called ideal solutions. A graphical representation of this behavior can be seen in Fig. 11.3.

Many solutions do deviate significantly from Raoult's law. However, the law is obeyed even in these cases for the component in excess (which, in this case, is usually referred to as the solvent) as it approaches purity. Therefore, if the solution is dilute, the properties of the solvent can be approximated using Raoult's law. Raoult's law, however, does not have universal applications. There are several aspects of this law that give it limitations. Firstly, it assumes that the vapor is an ideal gas, which is not necessarily the case. Secondly, Raoult's law only applies to ideal solutions, and, in reality, there are no ideal solutions. Also, a third problem to be aware of is that Raoult's law only really works for solutes that do not change their nature when they dissolve (i.e. they do not ionize or associate).

ILLUSTRATIVE EXAMPLE 11.4

If the vapor pressure of acetone at 0°C is 71 mm Hg, calculate the maximum concentration (in mole fraction units) of acetone in air at 1 atm total pressure.

SOLUTION: The maximum concentration of a component in a noncondensable gas is given by its vapor pressure p' divided by the total pressure, P. Any increase in concentration will result in the condensation of the component in question. The maximum mole fraction of acetone in air at

0°C and 1 atm is therefore:

$$y_{max} = \frac{p'}{P}$$

$$= \frac{71}{760} = 0.0934$$

■

It is important to note that vapor mixtures do not condense at one temperature as a pure vapor does. The temperature at which a vapor begins to condense as the temperature is lowered was defined earlier as the *dew point*. It is determined by calculating the temperature at which a given vapor mixture is saturated. The *bubble point* of a liquid mixture is defined as the temperature at which it begins to vaporize as the temperature is increased. The bubble point may also be viewed as the temperature at which the last vapor condenses, while the dew point is the temperature at which the last liquid vaporizes. These examples are based on a given and constant pressure, and are thus referred to as the dew point and bubble point temperatures. Calculations based on holding the temperature constant lead to the dew point and bubble point pressures.

Dew point and bubble point calculations enable vapor–liquid equilibrium (VLE) relationships to be obtained for a binary mixture; these are often provided as a *P-x, y* diagram (with the temperature constant) or a *T-x, y* diagram (with the pressure constant), or both. VLE data can be generated assuming Raoult's law applies, and there are two aforementioned types of diagrams of interest: *P-x, y* and *T-x, y*. Additional details and procedures for obtaining these graphs are provided in the next chapter; algorithms are included. The next chapter also provides procedures for preparing phase diagrams for real systems.

The next two Illustrative Examples serve as an introduction to VLE calculations assuming Raoult's law applies. As already noted, the actual generation of VLE diagrams receives treatment in the next chapter.

ILLUSTRATIVE EXAMPLE 11.5

A liquid stream contains 5 gmol % ethane (*A*) and 95 gmol % *n*-hexane (*B*) at 25°C. The vapor pressure of ethane at 25°C is 4150 kPa and the vapor pressure of *n*-hexane at 25°C is 16.1 kPa. If the pressure is such that this is a saturated liquid, what is the pressure and what will be the composition of the first vapor to form?

SOLUTION: For a two component (*A–B*) system, Equations (11.3) and (11.4) reduce to

$$P = x_A \, p'_A + x_B \, p'_B$$

Substituting,

$$P = (0.05)(4150) + 0.95(16.1)$$
$$= 224.6\,\text{kPa} = 32.6\,\text{psia}$$

From Equation (11.5),

$$y_A = x_A p_A'/P$$

Substituting

$$y_A = (0.05)(4150)/224.6$$
$$= 0.924$$
$$y_B = 1 - 0.924$$
$$= 0.076$$

This is an example of a bubble point pressure calculation (since the temperature is fixed) for a liquid mixture. ∎

ILLUSTRATIVE EXAMPLE 11.6

The vapor pressures of ethyl acetate and carbon tetrachloride at 27°C are:

carbon tetrachloride $(A) = 111\,\text{mm Hg}$

ethyl acetate $(B) = 92$ mm Hg

Calculate the composition of the liquid and vapor phases in equilibrium at this temperature and a total pressure of 100 mm Hg.

SOLUTION: Substitute into Equation (11.3) for the total pressure and solve for x_A.

$$P = p_A' x_A + p_B'(1 - x_A)$$
$$100 = 111 x_A + 92(1 - x_A)$$
$$x_A = 0.421$$

The vapor phase composition is given by

$$y_A = p_A' x_A/P$$
$$y_A = (111)(0.421)/100 \tag{11.5}$$
$$y_A = 0.467$$

∎

As already mentioned, the application of Raoult's law will receive extensive treatment in the following chapter. Predictions from this law will be compared to those that include non-ideal effects.

HENRY'S LAW

Unfortunately, relatively few mixtures follow Raoult's law. Henry's law is another empirical relationship used for representing data on many systems. Henry's law states that at *equilibrium* the partial pressure of the solute is proportional to its mole fraction, but the proportionality constant is not the vapor pressure of the pure substance (as it is in Raoult's law). Instead, the proportionality constant is some empirical constant, denoted H_B. Another use of Henry's law is when one of the system's components is above its critical temperature (e.g., air–water) since the component in question no longer has a measurable "vapor pressure". Therefore, for some component B in a solution, Henry's law[4] can be written:

$$p_B = H_B x_B \qquad (11.6)$$

The value of the Henry's law constant, H_B, is found to be temperature dependent. The value generally increases with increasing temperature. As a consequence, the solubility of gases generally decreases with increasing temperature. In other words, the dissolved gases in a liquid can be driven off by heating the liquid. Mixtures that obey Henry's law are known as ideal-dilute solutions. The above equation has also been written, for component A in this instance, as

$$y_A = m_A x_A \qquad (11.7)$$

where m_A is once again an empirical constant.

   To compare the relative uses of Raoult's and Henry's law, take note of the proceeding explanation. Raoult's law is applicable (under limited validity) to a vapor-liquid system where each component is below its critical temperature (e.g., benzene-toluene at 100°F). However, in a vapor–liquid system where one of the components is a non-condensable, above its critical temperature (e.g., air–water at 100°F), then Henry's law may be used to perform calculations on the non-condensable ("air") concentrations in the liquid ("water") phase. Raoult's law should still apply to the water in such a case.

   The results obtained from Henry's law for the mole fraction of dissolved gas is valid for the liquid layer just beneath the interface, but not necessarily the entire liquid. The latter will be the case only when thermodynamic equilibrium is established throughout the entire liquid body. There, the use of Henry's law is limited to dilute-gas-liquid solutions (that is, liquids with a small amount of gas dissolved in them). The linear relationship of Henry's law does not apply in the case when the gas is highly soluble in the liquid (or when the solute and solvent are structurally similar).

   Henry's law has been found to hold experimentally for all dilute solutions in which the molecular species is the same in the solution as in the gas. One of the most conspicuous and apparent exceptions to this is the class of electrolytic solutions,

Table 11.1   Henry's Law Constants for Gases in Water at Approximately 25°C

Gas	H, $(\text{bar})^{-1}$
Acetylene	1350
Air	73,000
Carbon dioxide	1700
Ethane	31,000
Ethylene	11,500
Methane	42,000

or solutions in which the solute has ionized or dissociated. As in the case of Raoult's law, Henry's law in this particular case does not hold.

Several constants for Henry's law are provided in Table 11.1.

ILLUSTRATIVE EXAMPLE 11.7

Explain why there is a temperature increase when a gas is "dissolved" in a liquid.

SOLUTION: Refer to Chapter 9. Since gases usually liberate heat when they dissolve in liquids, thermodynamics reveals, as noted earlier, that an increase of temperature will result in a decrease in solubility. This is why gases may be readily removed from solution by heating.

Another important factor influencing the solubility of a gas is pressure. As is to be expected from kinetic considerations, compression of the gas will tend to increase its solubility.   ■

ILLUSTRATIVE EXAMPLE 11.8

Given Henry's law constant for and the partial pressure of H_2S, determine the maximum mole fraction of H_2S that can be dissolved in solution. Data are provided below:

Partial pressure of $H_2S = 0.01$ atm

Total pressure $= 1.0$ atm

Temperature $= 60°F$

Henry's law constant, $H_{H_2S} = 483$ atm/mole fraction (1 atm, 60°F)

SOLUTION: Write the equation describing Henry's law:

$$p_{H_2S} = H_{H_2S} x_{H_2S}$$
$$p_{H_2S} = y_{H_2S} P$$

Calculate the maximum mole fraction of H_2S that can be dissolved in solution:

$$x_{H_2S} = p_{H_2S}/H_{H_2S}$$
$$= 0.01/483$$
$$= 2.07 \times 10^{-5}$$

■

To illustrate the application of Henry's law to an absorption process,[1] consider the air–water–ammonia system at T and P pictured in Fig. 11.4a–f. In this system,

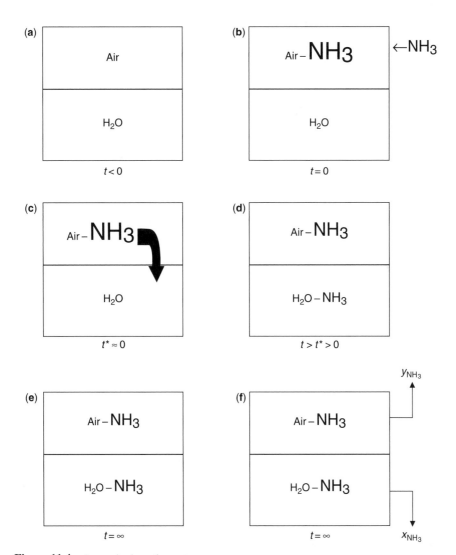

Figure 11.4   Ammonia absorption system.

NH_3 is added to the air at time $t = 0$ (b). The NH_3 slowly proceeds to distribute itself (c, d) until "equilibrium" is reached (e). The mole fractions in both phases are measured (f). This data is represented as point (1) in Fig. 11.5.

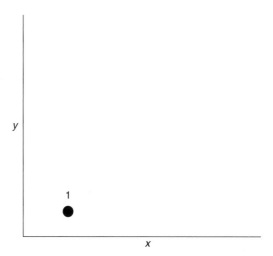

Figure 11.5 NH_3–air–H_2O x-y equilibrium plot.

If the process in Fig. 11.4 is repeated several times with additional quantities of NH_3, additional equilibrium points will be generated. These points are now plotted in Fig. 11.6.

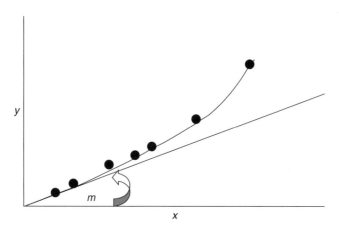

Figure 11.6 x-y equilibrium diagram.

Although the plot in Fig. 11.6 curves upwards, the data approaches the straight line of slope m at low values of x. It is this region where it is assumed that Henry's law applies. One of the coauthors[5] of this text believes that as $x \rightarrow 0$, $m \rightarrow 0$. This conclusion, and its ramifications, has received further attention in the literature.[5]

ILLUSTRATIVE EXAMPLE 11.9

Convert the ammonia–water equilibrium data given in Table 11.2 to an x–y plot at 30°C and 1 atm. Evaluate Henry's law constant for this system. Over what range of liquid mole fraction will Henry's law predict the equilibrium ammonia vapor content to within 5% of the experimental data?

Table 11.2   Equilibrium Data for Ammonia–Water System

x, mole fraction NH_3 in liquid	p_{NH_3}, mm Hg
0	0
0.0126	11.5
0.0167	15.3
0.0208	19.3
0.0258	24.4
0.0309	29.6
0.0405	40.1
0.0503	51.0
0.0737	79.7
0.0960	110.0
0.1370	179.0
0.1750	260.0
0.2100	352.0
0.2410	454.0
0.2970	719.0

SOLUTION: The partial pressure of ammonia is converted to mole fraction in vapor as shown in Table 11.3.

Table 11.3   Ammonia–Water Equilibrium (x–y) Data

x_{NH_3}	$y_{NH_3} = p_{NH_3}/P$
0	0
0.0126	0.015
0.0167	0.0210
0.0208	0.0254
0.0258	0.0321
0.0309	0.03894
0.0405	0.05276
0.0503	0.06710
0.0737	0.10486
0.0960	0.145
0.1370	0.236
0.1750	0.342
0.2100	0.463
0.2410	0.596
0.2970	0.945

These results are plotted in Fig. 11.7.

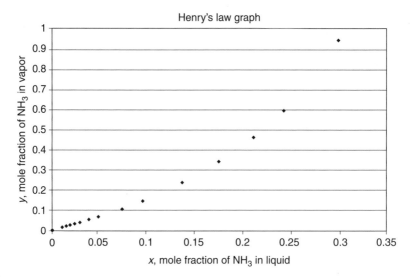

Figure 11.7 Ammonia–water equilibrium at 30°C and 1 atm.

Henry's law constant from the graph is approximately 1.485 at $x = 0.095$. Since

$$y_{actual} = 0.148$$
$$y_{calculated} = 1.485(0.095) = 0.141$$
$$\text{Percent agreement} = \left(\frac{0.141}{0.148}\right)100 = 95.27\%$$

Thus, from $x = 0$ to $x = 0.095$, Henry's law equation, $y = 1.485x$, predicts the equilibrium vapor content to within 5% of the experimental data. ∎

RAOULT'S LAW VS HENRY'S LAW[6]

A basic difference between Raoult's and Henry's laws is that Raoult's law applies to the solvent, while Henry's law applies to the solute. However, in ideal-*dilute* solutions, the solute obeys Henry's law whereas the solvent obeys Raoult's law. In Fig. 11.8, one can see the differences between Raoult's and Henry's law in graphical form, in addition to the behavior of a real solution. It is important to realize that as the solvent (or the component in excess) approaches purity, or rather, as a solution approaches ideality, Raoult's law and Henry's law as applied to the solution are identical.

There are several other significant differences between these two laws that must be recognized. Raoult's law is much more theory-based, since it only applies to ideal situations. Henry's law, on the other hand, is more empirically based, making it a more general and practical law.

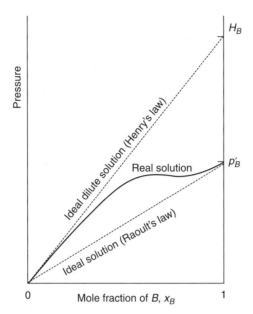

Figure 11.8 Graphical representation of the differences between Raoult's law and Henry's law; component B represents the solute.

In addition to providing a link between the mole fraction of the solute and its partial pressure, Henry's law constants may also be used to calculate gas solubilities. This plays an important role in biological functions, such as the transportation of gases in the bloodstream. A knowledge of Henry's law constants for gases in fats and lipids is also important in discussing respiration.[3] For example, consider scuba-diving. In this recreational activity, air is supplied at a higher pressure so as to allow the pressure within the diver's chest to match the pressure exerted by the surrounding water. This water pressure of the ocean increases at approximately 1 atm per 10 meters of depth. However, air inhaled at higher pressures makes nitrogen more soluble in fats and lipids rather than water. This causes nitrogen to enter the central nervous system, bone marrow, and fat reserves of the body. The nitrogen then bubbles out of its lipid solution if the diver rises to the surface too quickly and causes a condition called decompression sickness (also known as *the bends*). This condition can be fatal as the nitrogen gas bubbles can block arteries and cause unconsciousness as they rise into the diver's head.

An interesting application of Henry's law arises in the treatment of carbon monoxide poisoning. In a hyperbaric oxygen chamber, oxygen is raised to an elevated partial pressure. When an individual with carbon monoxide poisoning steps inside this chamber, there is a steep pressure gradient between the partial pressure of the arterial blood's oxygen and the partial pressure of the oxygen in freshly-inhaled air. In this way, oxygen floods quickly into the arterial blood and allows a rapid re-supply of oxygen to the bloodstream.

Another application of interest arises in gas absorption operations.[1] The equilibrium of interest is that between a nonvolatile absorbing liquid (solvent) and a solute gas. The solute is ordinarily removed from its mixture in a relatively large amount of a carrier gas that does not dissolve in the absorbing liquid. Therefore, it is often possible, and frequently the case when considering the removal of a gaseous component by absorption, to assume that only the component in question is transferred between phases. Both the solubility of the non-diffusing (inert) gas in the liquid and the presence of vapor from the liquid gas are usually neglected. The important variables to be considered then are the pressure, temperature, and the concentrations of the component in the liquid and the gas phase. The temperature and pressure may be fixed and the concentration(s) of the component(s) in the various phases are defined from phase equilibrium relationships. An equation that may be employed to relate the equilibrium concentrations in these two phases is the aforementioned Henry's law.

As noted above, Henry's law and Raoult's law have several similarities and differences, but both have restrictions. The knowledge of both of these laws is invaluable not only academically, but for practical purposes as well. While Henry's law has a wide range of practical applications, it "stems from" Raoult's law. This fact makes the knowledge of *both* these laws and the concepts behind them very useful. Finally, it should be noted that unlike Raoult's law, which provides information on how two components are distributed between the vapor and liquid phase (e.g., acetone–water), Henry's law provides information on how a component (e.g., acetone) will be distributed between a gas and a liquid phase (e.g., air and water).

ILLUSTRATIVE EXAMPLE 11.10

A stream of dry air and carbon tetrachloride (CCl_4) from a dry cleaning operation has a flow rate of 2000 acfm and contains 13% CCl_4 by volume. One proposed control method to remove the CCl_4 is to design a condenser (heat exchanger) where the mixture enters the condenser at 80°F and atmospheric pressure. Estimate the surface area required to condense 95% of the CCl_4 if the average temperature of the cooling surface is -70°F and the overall heat transfer coefficient U is 4.0 Btu/ft^2· h · °F. Assume the average heat capacity and enthalpy of vaporization for CCl_4 to be 0.125 Btu/lb °F liquid and 93.7 Btu/lb (80°F), respectively. Vapor pressure (saturation) data for CCl_4 are provided below:

$$p' = 7.74 \text{ mm Hg} \quad \text{at } 0°F$$
$$p' = 5.64 \text{ mm Hg} \quad \text{at } -18°F$$
$$p' = 2.33 \text{ mm Hg} \quad \text{at } -40°F$$

Assume the gas/liquid mixture leaves the condenser at equilibrium (saturation) conditions.

SOLUTION: Assuming 100 lbmols of entering gas as a basis, determine the lbmols of CCl_4 and air:

$$n_{CCl_4} = (0.13)(100)$$
$$= 13 \text{ lbmol}$$
$$n_{air} = (0.87)(100)$$
$$= 87 \text{ lbmol}$$

Calculate the mole fraction of the CCl_4 in the gas discharged from the condenser.

For 95% recovery,

$$n_{CCl_4} = (0.05)(13)$$
$$= 0.65 \, \text{lbmol}$$
$$y_{CCl_4} = (0.65)/(87 + 0.65)$$
$$= 0.00742$$

Calculate the partial pressure of the CCl_4 in the exiting gas in mm Hg:

$$p_{CCl_4} = (0.00742)(760)$$
$$= 5.64 \, \text{mm Hg}$$

Proceed to estimate the condensation temperature in the heat exchanger. At saturation conditions, $T \approx -18°F$ from the vapor pressure data provided.

Calculate the mass flow rate of air and CCl_4 entering the condenser in lb/h:

$$\dot{m}_{air} = (2000)(0.87)(60)[(14.7)/(460 + 80)(10.73)](29)$$
$$= 7681 \, \text{lb/h}$$
$$\dot{m}_{CCl_4} = (2000)(0.13)(60)[(14.7)/(460 + 80)(10.73)](154)$$
$$= 6095 \, \text{lb/h}$$

Calculate the heat transferred from the air in Btu/h:

$$Q_{air} = (7681)(0.24)(-18 - 80)$$
$$= -180{,}660 \, \text{Btu/h}$$

Calculate the heat transferred from CCl_4 in Btu/h (including the heat of condensation):

$$Q_{CCl_4} = (6095)(0.125)(-18 - 80) + (6095)(0.95)(-93.7)$$
$$= -74{,}663 - 542{,}546$$
$$= -617{,}200 \, \text{Btu/h}$$

Calculate the total heat load on the condenser in Btu/h:

$$Q = -180{,}660 - 617{,}200$$
$$= -797{,}900 \, \text{Btu/h}$$

Estimate the log mean temperature difference driving force in °F (see Chapter 6 for additional details on LMTD):

$$\text{LMTD} = \{[80 - (-70)] - [-18 - (-70)]\}/\ln(150/52)$$
$$= 91.4°F$$

Finally, calculate the required heat transfer area in ft^2. See Chapter 6 for additional details.

$$A = Q/U\Delta T_{LM}$$
$$= 797,900/(4)(91.4)$$
$$= 2182 \text{ ft}^2$$

∎

VAPOR–SOLID EQUILIBRIUM

This section is concerned with a discussion of vapor–solid equilibria. The relation between the amount of substance adsorbed by an adsorbent (solid) and the equilibrium partial pressure or concentration at constant temperature is called the adsorption isotherm. The adsorption isotherm is the most important and by far the most often used of the various equilibria data that can be measured. Also note that the notation adopted by the adsorption industry is employed in the presentation to follow.

Most available data on adsorption systems are determined at equilibrium conditions. *Adsorption equilibrium* is the set of conditions at which the number of molecules arriving on the surface of the adsorbent equals the number of molecules that are leaving. The adsorbent bed is then said to be "saturated with vapors" and can remove no more vapors from a gaseous stream. Equilibrium determines the maximum amount of vapor that may be adsorbed on the solid at a given set of operating conditions. Although a number of variables affect adsorption, the two most important ones in determining equilibrium for a given system are temperature and pressure. Three types of equilibrium graphs and/or data are used to describe adsorption systems: isotherm at constant temperature, isobar at constant pressure, and isostere at constant amount of vapors adsorbed.

The most common and useful adsorption equilibrium data is the adsorption *isotherm*. The isotherm is a plot of the adsorbent capacity vs the partial pressure of the adsorbate at a constant temperature. Adsorbent capacity is usually given in weight percent expressed as grams of adsorbate per 100 g of adsorbent. Figure 11.9 shows a typical example of an adsorption isotherm for carbon tetrachloride on activated carbon. Graphs of this type are used to estimate the size of adsorption systems.[1] Attempts have been made to develop generalized equations that can predict adsorption equilibrium from physical data. This is very difficult because adsorption isotherms take many shapes depending on the forces involved. Isotherms may be concave upward, concave downward, or "S"-shaped. To date, most of the theories agree with data only for specific adsorbate-systems and are valid over limited concentration ranges.

Two additional adsorption equilibrium relationships are the isostere and the isobar. The isostere is a plot of the $\ln p$ vs $1/T$ at a constant amount of vapor adsorbed. Adsorption isostere lines are usually straight for most adsorbate–adsorbent systems. The isostere is important in that the slope of the isostere (approximately) corresponds to the heat of adsorption. The isobar is a plot of the amount of vapors adsorbed vs temperature at a constant partial pressure. However, in the design of most engineering

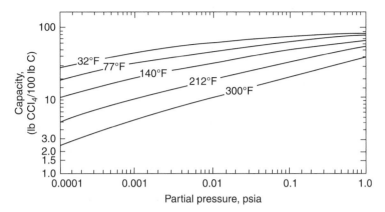

Figure 11.9   Adsorption isotherms for carbon tetrachloride on activated carbon.

systems, the adsorption isotherm is by far the most commonly used equilibrium relationship.

Several models have been proposed to describe this vapor–solid equilibrium phenomena. To represent the variation of the amount of adsorption per unit area or unit mass with partial pressure, Freundlich proposed the equation:

$$Y = kp^{1/n} \tag{11.8}$$

where Y is the weight or volume of gas (or vapor) adsorbed per unit area or unit mass of adsorbent and p is the equilibrium partial pressure. The k and n terms are the empirical constants dependent on the nature of solid and adsorbate, and on the temperature. Equation (11.8) may be rewritten as follows. Taking logarithms of both side,

$$\log Y = \log k + \left(\frac{1}{n}\right)\log p \tag{11.9}$$

If $\log Y$ is now plotted against $\log p$, a straight line should result with slope equal to $1/n$ and an ordinate intercept equal to $\log k$. Although the requirements of the equation are met satisfactorily at lower pressures, the experimental points curve away from the straight line at higher pressures, indicating that this equation does not have general applicability in reproducing adsorption of gases (or vapors) by solids.

A much better equation for isotherms was deduced by Langmuir from theoretical considerations. The final form is given as

$$Y = \frac{ap}{1 + bp} \tag{11.10}$$

which can be rewritten as

$$\frac{p}{Y} = \frac{1}{a} + \left(\frac{b}{a}\right)p \tag{11.11}$$

Since a and b are constants, a plot of p/Y vs p should yield a straight line with slope equal to b/a and an ordinate intercept equal to $1/a$.

By assuming that adsorption on a solid surfaces takes place with the formation of secondary, tertiary, and finally multilayers upon the primary monomolecular layer, the researchers Brunauer, Emmett, and Teller[7] derived the relation

$$\frac{P}{v(p'-p)} = \frac{1}{v_m c} + \left(\frac{c-1}{v_m c}\right)\frac{p}{p'} \tag{11.12}$$

where v is the volume, reduced standard to conditions of gas (or vapor) adsorbed at system pressure P and T; p is the equilibrium partial pressure of adsorbate; T is the temperature; p$'$ is the vapor pressure of the adsorbate at temperature T; v_m is the volume of gas (or vapor), reduced to standard conditions, adsorbed when the surface is covered with a unimolecular layer; and, c is the constant at any given temperature. The constant c is approximately given by

$$c = e^{(E_1 - E_L)/RT} \tag{11.13}$$

where E_1 is the enthalpy of adsorption of the first adsorbed layer and E_L is the enthalpy of liquefaction of the gas (or vapor).

The most useful theory from an engineering design viewpoint in trying to predict adsorption isotherms is the Polanyi potential theory. The Polanyi theory states that the adsorption potential is a function of the reversible isothermal work done by the system. Polyani[8] and Dubinin and co-workers[9-11] showed that the adsorption isotherms of various vapors can be represented by the following equations:

$$\ln(W) = \ln(W_0) - k(E/\beta)^2 \tag{11.14}$$

where

$$E = (RT)\ln(p'/p) \tag{11.15}$$

and W is the volume of condensed adsorbate per gram of carbon (m^3/g); W_0 is the active pore volume of carbon (m^3/g); k is the constant related to pore structure $(cal/gmol)^{-2}$; β is the affinity coefficient that permits comparison of the adsorption potential of the (test) adsorbate to a reference adsorbate; R is the gas constant, 1.987 cal/gmol $\cdot$ K; T is the absolute temperature (K); p is the equilibrium partial pressure of adsorbate; and p' is the vapor pressure of adsorbate at T.

The above equations were suggested by Dubinin for the case when the pores of the adsorbent are comparable in size to the adsorbed molecules. These can be used to determine the equilibrium adsorption isotherm of any given vapor from the adsorption isotherm of a reference vapor, provided that the value of the affinity coefficient, β, of this vapor is available.

The adsorption of nonpolar and weakly polar vapors is dominated by dispersion forces. For this situation, Dubinin proposed that β may be determined from the ratio of the molar volume, V, of a test solvent to that of a reference solvent used to obtain the values of W_0 and k for the given carbon.

ILLUSTRATIVE EXAMPLE 11.11

The carbon dioxide adsorption data on Columbia (Columbia is a registered trade-mark of Union Carbide Corporation) activated carbon are presented in Table 11.4 for a temperature of 50°C. Determine the constants of the Freundlich equation.

SOLUTION: Refer to Equations (11.8) and (11.9). Table 11.5 can be generated for the Freundlich equation using the data in the problem statement. A plot of log Y vs log p yields the equation (see the Freundlich equation plot in the LHS of Fig. 11.10)

$$Y = 30p^{0.7}$$
■

Table 11.4 Carbon Dioxide Equilibrium Data

Equilibrium capacity, cm^3/g	Partial pressure CO$_2$, atm
30	1
51	2
67	3
81	4
93	5
104	6

Table 11.5 Freundlich Calculations

Equilibrium capacity, cm^3/g	Partial pressure CO$_2$, atm	log Y	log p
30	1	1.477	0.000
51	2	1.708	0.301
67	3	1.826	0.477
81	4	1.909	0.602
93	5	1.969	0.699
104	6	2.017	0.778

ILLUSTRATIVE EXAMPLE 11.12

Refer to Illustrative Example 11.11. Determine the constants of the Langmuir equations.

SOLUTION: For the Langmuir equation,

$$\frac{p}{Y} = \frac{1}{a} + \left(\frac{b}{a}\right)p \tag{11.11}$$

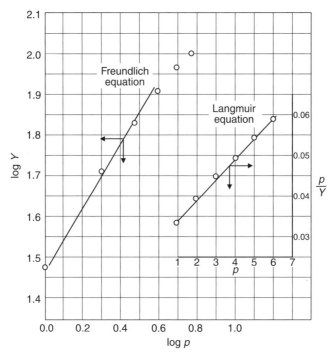

Figure 11.10   Plots of log Y vs log p and p/Y vs p.

Key calculations are given in Table 11.6. A plot of p/Y vs p yields the equation (see RHS insert of Fig. 11.10)

$$\frac{p}{Y} = \frac{3.57p}{(1 + 0.186)p}$$

Table 11.6   Langmuir Calculations

p/Y	P
0.063	1
0.039	2
0.045	3
0.049	4
0.054	5
0.058	6

ILLUSTRATIVE EXAMPLE 11.13

Use the graph in Fig. 11.11 to solve the following problem. Estimate the pounds of CO_2 that can be adsorbed by 100 lb of Davison 4 Å molecular sieve from a discharge gas mixture at 77°F and 40 psia containing 10,000 ppmv (ppm by volume) CO_2.

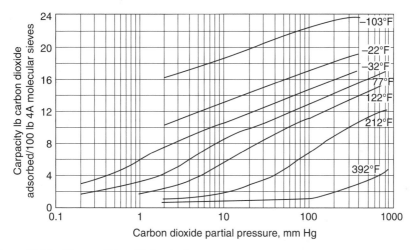

Figure 11.11   Vapor–solid equilibrium isotherms.

SOLUTION: Calculate the mole fraction of CO_2 in the discharge gas mixture:

$$y_{CO_2} = \text{ppm}/10^6$$
$$= 10,000/10^6$$
$$= 0.01$$

Also determine the partial pressure of CO_2 in psia and mm Hg:

$$p_{CO_2} = y_{CO_2}P$$
$$= (0.01)(40)$$
$$= 0.4 \,\text{psia}$$
$$= (0.4)(760/14.7)$$
$$= 20.7 \,\text{mm Hg}$$

Estimate the adsorbent capacity at 77°F from Fig. 11.11:

$$Y = 9.8 \,\text{lb } CO_2/100 \,\text{lb sieve} \qquad ■$$

LIQUID–SOLID EQUILIBRIUM

The simplest system in liquid–solid equilibria is one in which the components are completely miscible in the liquid state and the solid phase consists of a pure component. These systems find application in water purification processes. The equilibrium equations and relationships that have been presented in the Vapor–Solid Equilibrium section generally apply to these systems as well. Details are available in the literature.[12]

ILLUSTRATIVE EXAMPLE 11.14

Adsorption processes are often used as a follow-up to chemical wastewater treatment to remove organic reaction products that cause taste, odor, color, and toxicity problems. The equilibrium relationship between adsorbents (solid materials that adsorb organic matter, e.g., activated carbon) and adsorbates (substances that are bound to the adsorbates, e.g., benzene) are often simply expressed as

$$q = Kc^n \qquad (11.16)$$

where q = amount of organic matter adsorbed per amount of adsorbent
   c = concentration of organic matter in water
   n = experimentally determined constant
   K = equilibrium distribution constant

Based upon the data given in Table 11.7, which were obtained from a laboratory sorption experiment, a college student who worked for a company as an intern was asked to evaluate the K and n values for a certain type of activated carbon to be used to remove undesirable by-products formed during chemical treatment. What are the values of K and n that the intern should have generated?

Table 11.7   Adsorption Data Collected in Laboratory Experiments

c (mg/L)	q (μg/g)
50	118
100	316
200	894
300	1640
400	2530
650	5240

SOLUTION: The equilibrium relationship given in the problem statement can be linearized by taking the log (logarithm) of both sides of the equation. This yields the following equation:

$$\log(q) = \log(K) + n\log(c)$$

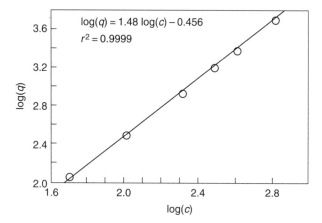

Figure 11.12 Log–log plot of adsorption data.

A plot of log(q) vs log(c) yields a straight line if this relationship can be used to represent the experimental data. The slope of this line is equal to n, while the intercept is log(K). The experimental data analyzed using the equation above are plotted in Fig. 11.12, showing that the data fit this linearized isotherm quite well.

The equation generated from a regression analysis indicates that

$$n = 1.48$$
$$\log(K) = -0.456 \quad \text{or} \quad K = 0.35$$

provided the units of c and q are mg/L and µg/g, respectively. The describing equation is therefore

$$q = 0.35c^{1.48}$$

∎

REFERENCES

1. L. THEODORE, "*Mass Transfer Operations for the Practicing Engineer*," John Wiley & Sons, Hoboken, NJ, 2009.
2. L. THEODORE, personal files, Manhattan College.
3. P. ATKINS and J. DE PAULA, "*Atkins' Physical Chemistry*," 8th edition, W. H. Freeman and Co., San Francisco, 2006.
4. W. HENRY, Publication unknown, 1803.
5. L. THEODORE, "*Air Pollution Control Equipment*," John Wiley & Sons, Hoboken, NJ, 2009.
6. A. MOHAN, Adapted from homework assignment submitted to L. Theodore, 2007.
7. S. BRUNAUER, P. H. EMMETT, and E. TELLER, *J. Am. Chem. Soc.*, 60, 309, 1938.
8. K. POLANYI, *Trans. Faraday Soc.*, 28, 316, 1932.
9. J. DUBININ, *Chem. Rev.*, 60, 235, 1960.
10. A. KADLEC and J. DUBININ, *Coll. Int. Sci.*, 31, 479, 1969.

11. J. DUBININ, "*Chemistry and Physics of Carbon*," Vol. 2, P. L. Walker, ed., Marcel Dekker, New York, 51, 1966.

12. D. GREEN and R. PERRY, "*Perry's Chemical Engineers' Handbook*," 8th edition, McGraw-Hill, New York, 2008.

NOTE: Additional problems for each chapter are available for all readers at www. These problems may be used for additional review or homework purpose.

Chapter **12**

Vapor–Liquid Equilibrium Calculations

The Bible: Matthew

And before him shall be gathered all nations: and he shall separate them one from another, as a shepherd divideth his sheep from the goats.

—Matthew. XXV, 32

INTRODUCTION

It is important to understand how and why a vapor is converted into a liquid and when a liquid is converted into a vapor. Four key definitions involving this conversion process follow. The bubble point temperature (BPT) is the temperature at which the first bubble of vapor forms when the liquid is heated slowly at a constant pressure (see Fig. 12.1a). The bubble point pressure (BPP) is the pressure at which the first bubble of vapor forms when the pressure above a liquid is reduced at a constant temperature (see Fig. 12.1b). The dew point temperature (DPT) is the temperature at which the first drop of liquid forms when the vapor is cooled at a constant pressure (see Fig. 12.1c), while the dew point pressure (DPP) is the pressure at which the first drop of liquid appears when the pressure of a vapor is increased at a constant temperature (see Fig. 12.1d).

If the system is ideal, Raoult's law can be used to perform these calculations. If the system is not ideal, other equations are available to account for the deviation from ideality. This information can be better understood by providing either a *P-x, y* (pressure–mole fraction liquid, mole fraction vapor) or *T-x, y* (temperature) diagram from the binary system under study. A *T-x, y* diagram is a graph of equilibrium temperatures vs the liquid and vapor mole fraction of one of the species of the binary system. The diagrams are made up of two curves, a "liquid curve" and a "vapor curve," with a two phase region existing between these two curves. It is the generation and interpretation of these diagrams that the remainder of the chapter primarily addresses.

Thermodynamics for the Practicing Engineer. By L. Theodore, F. Ricci, and T. Van Vliet
Copyright © 2009 by John Wiley & Sons, Inc.

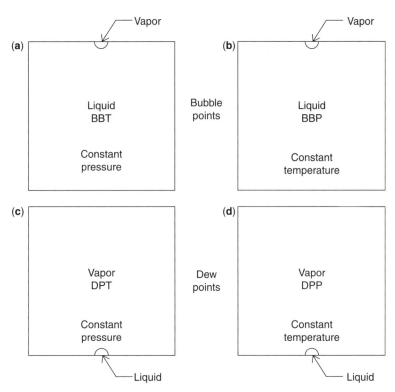

Figure 12.1 Bubble and dew points.

As noted in the previous chapter, vapor–liquid equilibrium (VLE) relationships for a binary mixture are often provided as the aforementioned P-x, y diagram or a T-x, y diagram, or both, and it is these two types of diagrams that are of interest to the practicing engineer. VLE data can be generated assuming Raoult's law applies. VLE diagrams can also be developed for two-component systems where the liquid phase is not assumed to be ideal. Once again, the same two types of diagrams are of interest, P-x, y and T-x, y. Procedures for generating these figures are provided later in this chapter and each presentation contains an applicable algorithm. These tasks were accomplished through a Microsoft Excel spread sheet program that generated vapor liquid equilibrium diagrams of two given chemical species at a specified temperature or pressure.

This second chapter on phase equilibrium deals exclusively with vapor–liquid equilibrium (VLE) applications. The section that immediately follows is concerned with the DePriester charts—a valuable source of VLE data for many hydrocarbons that approach ideal behavior. However, it should be noted that the DePriester chart data are based on experimental data. The fact that these compounds approach ideal behavior allows the data to be presented in a simple form, i.e., as a function solely of temperature and pressure alone.

The remainder of the chapter keys on the generation of VLE diagrams for both ideal and non-ideal behavior. Although Raoult's law (see previous chapter) is included in the early analysis, non-ideal deviations are accounted for by two theoretical models that have been verified by rather extensive experimental data. The two models are the:

1 NRTL model
2 Wilson's method

The presentation of the non-ideal behavior of real two-component systems through the NRTL and Wilson equations is provided in two sections, with each section dealing with a different system and a different set of conditions. These are:

1 *T-x, y*: ethanol and toluene (NRTL)
2 *P-x, y*: methanol and water (NRTL)
3 *T-x, y*: ethanol and toluene (Wilson)
4 *P-x, y*: ethanol and toluene (Wilson)

The material contains the VLE calculations for the system and conditions specified plus the VLE calculation predicted by Raoult's law. The reader is thus able to compare the predictions of both the ideal and non-ideal results.

Finally, the reader should note that all the phase equilibrium calculations in this chapter are based on the assumption that the vapor and liquid are in *equilibrium*. The presentation is also limited to two-component systems. An approach to multiphase (two or more coexisting liquid phases) VLE is available in the literature.[1]

ILLUSTRATIVE EXAMPLE 12.1

Using the vapor–liquid equilibrium data for the ethanol–water system provided in Table 12.1, generate a *T-x, y* diagram. What are the liquid and vapor mole fractions of water if the liquid mixture is 30 mol% ethanol?

SOLUTION: The *T-x, y* diagram is plotted in Fig. 12.2. The top curve (*y* vs *T*) represents saturated vapor and the bottom curve (*x* vs *T*) is saturated liquid. From the diagram, when $x_{EtOH} = 0.3$, $y_{EtOH} = 0.57$ (see tie line at about 179°C); therefore, the ethanol vapor composition is 57%. The water vapor mole fraction is given by

$$y_{water} = 1 - 0.57 = 0.43$$ ■

ILLUSTRATIVE EXAMPLE 12.2

Refer to the previous example. What are the liquid and vapor mole fractions of water if the liquid mixture contains 30% (mole basis) ethanol.

SOLUTION: Based on the problem statement

$$x_{water} = 0.70$$ ■

Table 12.1   Vapor–Liquid Equilibrium at 1 atm (mole fractions); Ethanol–Water System

T (°C)	x_{EtOH}	y_{EtOH}
212	0.000	0.000
192	0.072	0.390
186	0.124	0.470
181	0.238	0.545
180	0.261	0.557
177	0.397	0.612
176	0.520	0.661
174	0.676	0.738
173	0.750	0.812
172	0.862	0.925
171	1.000	1.000

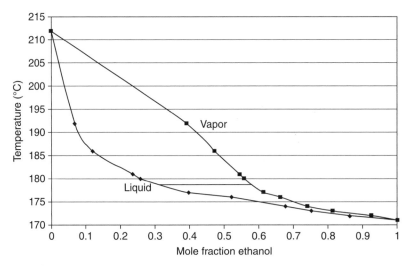

Figure 12.2   *T-x, y* diagram for the ethanol–water system.

THE DEPRIESTER CHARTS

Mixtures of condensable vapors and noncondensable gases must be handled in some engineering applications. A common example is water vapor and air; a mixture of organic vapors and air is another such example that often appears in thermodynamic problems. Condensers can be used to control organic emissions to the atmosphere by lowering the temperature of the gaseous stream (see Illustrative Example 11.10), although an increase in pressure will produce the same result. The calculation for this can be accomplished using the phase equilibrium constant (K_i). This constant has been referred to in industry as a componential split factor since it provides the ratio of the mole fractions of a component in two equilibrium phases. The defining equation is

$$K_i = y_i/x_i \tag{12.1}$$

where K_i = phase equilibrium constant for component i (dimensionless).

This equilibrium constant is a function of the temperature, pressure, the other components in the system, and the mole fractions of these components. However, as a first approximation, K_i is generally treated as a function only of the temperature and pressure. For ideal gas conditions, K_i may be represented by

$$K_i = p_i'/P \tag{12.2}$$

where p_i' is the vapor pressure.

Many of the phase equilibrium calculations involve hydrocarbons, particularly in petrochemical applications. Fortunately, most hydrocarbons approach ideal behavior over a fairly wide range of temperature and pressure. Values of K_i for a large number of hydrocarbons are provided in the two DePriester nomographs (see Figs. 12.3 and 12.4). These two nomographs or charts were originally provided by DePriester in 1953.[2] Both have withstood the test of time and continue to be used in some phase equilibrium calculations.

It is important to note that the DePriester charts are based on the assumption of ideal gas behavior. This assumption is generally not valid for most mixtures containing inorganic components. No simple and reliable method is available for estimating K for both inorganics and organics in an inorganic–organic mixture. However, calculations involving bubble and dew points can be accomplished using the aforementioned phase equilibrium constant, K_i.

Bubble point and dew point calculations are centered around Equation (12.1),

$$y_i = K_i x_i$$

In a bubble point calculation, x_is are normally known and the y_is can be calculated from

$$\Sigma y_i = \Sigma(K_i x_i) = 1.0 \tag{12.3}$$

A trial-and-error calculation on either the temperature or pressure is dictated in order to determine the bubble point temperature (BPT) or pressure (BPP), respectively. In a dew point calculation (DPT or DPP), the procedure is reversed and

$$\Sigma x_i = \Sigma(y_i/K_i) = 1.0 \tag{12.4}$$

ILLUSTRATIVE EXAMPLE 12.3

Calculate the bubble point pressure and composition of the vapor in equilibrium with a liquid containing 5% CH_4, 10% C_2H_6, 30% C_3H_8, 25% i-C_4H_{10}, and 30% n-C_4H_{10} (mole %) that is at

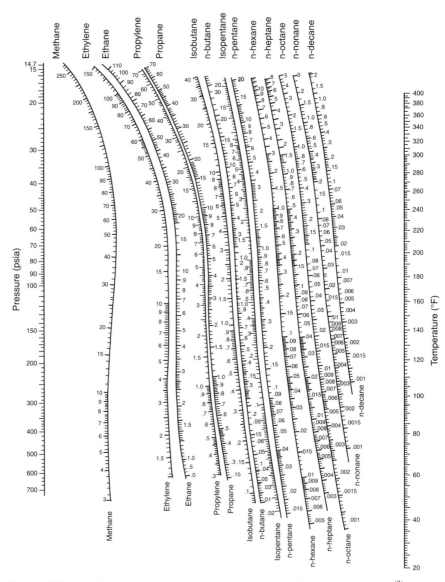

Figure 12.3   Equilibrium constants in light-hydrocarbon systems. High-temperature range.[2]

100°F. Also calculate the dew point pressure and composition of the liquid in equilibrium with a vapor of the same composition, i.e., containing 5% CH_4, 10% C_2H_6, 30% C_3H_8, 25% iso-C_4H_{10}, and 30% n-C_4H_{10} (mole %) that is also at 100°F.

SOLUTION: Write the describing equation for the bubble point

$$\Sigma y_i = \Sigma(K_i x_i) = 1.0 \tag{12.3}$$

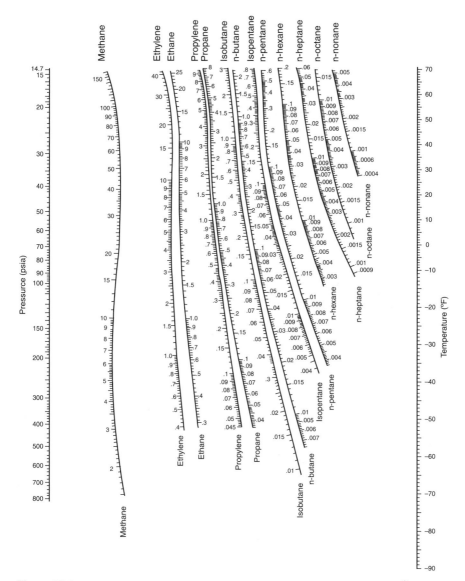

Figure 12.4   Equilibrium constants in light-hydrocarbon systems. Low-temperature range.[2]

Calculate the BPP in psia from the results provided in Table 12.2. Linear extrapolation indicates that the BPP is approximately 304 psia.

Write the describing equation for the dew point:

$$\Sigma x_i = \Sigma(y_i/K_i) = 1.0 \tag{12.4}$$

Calculate the DPP in psia from the results provided in Table 12.3. Linear interpolation indicates a DPP of approximately 97 psia.   ■

Table 12.2 Bubble Point Pressure Calculation

Component	x	K (150 psia)	$K_i x_i$	K (300 psia)	$K_i x_i$
C_1	0.05	17.0	0.850	8.8	0.44
C_2	0.10	3.65	0.365	2.1	0.21
C_3	0.30	1.30	0.390	0.72	0.22
iC_4	0.25	0.57	0.143	0.34	0.08
nC_4	0.30	0.41	0.123	0.24	0.07
			$\Sigma = 1.871$		$\Sigma = 1.02$

Table 12.3 Dew Point Pressure Calculation

Component	y	K (90 psia)	y_i/K_i	K (100 psia)	y_i/K_i
C_1	0.05	27	0.00186	25	0.0020
C_2	0.10	5.7	0.0175	5.2	0.0190
C_3	0.30	2.0	0.150	1.80	0.1667
iC_4	0.25	0.88	0.284	0.79	0.3160
nC_4	0.30	0.63	0.476	0.57	0.5260
			$\Sigma = 0.9296$		$\Sigma = 1.030$

ILLUSTRATIVE EXAMPLE 12.4

If the mixture in Illustrative Example 12.3 is 400 psia and 100°F, is the state of the mixture all liquid, all vapor, or a combination of both? If the above mixture is at 250 psia and 100°F, is the state of the mixture all liquid, all vapor, or a combination of both?

SOLUTION: If the system pressure is above the BBP, the state is all liquid. If the system pressure is below the DPP, the state is all vapor. If the system pressure is between the BBP and DPP, the state is a mixture of vapor and liquid. Thus, the state of the mixture at 100°F and 400 psia is all liquid. The state of the mixture at 100°F and 250 psia consists of both liquid and vapor.   ■

ILLUSTRATIVE EXAMPLE 12.5

Calculate the dew-point and bubble-point temperatures of a 1.0 atm mixture consisting of:

heptane:	0.10 (mole fraction)
octane:	0.55
nonane:	0.30
decane:	0.05

Employ the DePriester charts in performing the calculations.

SOLUTION: The dew point temperature calculation employs Equation (12.4) and the results are provided in Table 12.4. Another calculation should be performed at 285°F. However, by linearly interpolating the above results, $T = 283°F$.

Table 12.4   Dew Point Temperature Calculation

Component	y	$K_{250°F}$	y/K	$K_{300°F}$	y/K
C_7	0.10	1.8	0.0556	3.4	0.0294
C_8	0.55	0.93	0.5914	1.8	0.3056
C_9	0.30	0.39	0.7693	0.9	0.3333
C_{10}	0.05	0.18	0.2778	0.46	0.1089
			$\Sigma = 1.6941$		$\Sigma = 0.7771$

The bubble point temperature calculation employs Equation (12.3). Results can be found in Table 12.5. Linearly interpolating gives $T = 264°F$.

Table 12.5   Bubble Point Temperature Calculation

Component	K_{200}	Kx	K_{270}	Kx
C_7	2.05	0.2051	2.4	0.242
C_8	1.05	0.5775	1.2	0.661
C_9	0.47	0.1411	0.55	0.164
C_{10}	0.22	0.0110	0.26	0.013
		$\Sigma = 0.9347$		$\Sigma = 1.079$

The bubble and dew point material presented here can be extended to calculating the equilibrium vapor and liquid molar split from a hydrocarbon mixture. If F, L, and V represent the feed, liquid, and vapor, respectively, in molar units, then

$$F = L + V \tag{12.5}$$

For a basis of one mole of overall mixture,

$$F = 1$$

and

$$V = 1 - L \tag{12.6}$$

A mass balance for component i is then

$$z_i = x_i L + y_i(1 - L) \tag{12.7}$$

where z_i is the moles of component i in the feed, and x and y are the moles (and mole fractions) of liquid and vapor, respectively.

Introducing the phase equilibrium relationship [see Equation (12.1)] presented earlier,

$$K_i = y_i/x_i$$

results in the componential split equation

$$\Sigma x_i = \Sigma\{z_i/[L + K_i(1 - L)]\} = 1.0 \tag{12.8}$$

Alternatively, the above equation can be written as

$$\Sigma y_i = \Sigma\{K_i z_i/[L + K_i(1 - L)]\} = 1.0 \tag{12.9}$$

ILLUSTRATIVE EXAMPLE 12.6

A mixture of propane and n-butane contains 20 mole % n-butane. If the pressure is maintained constant at 280 psia, calculate:

1 the bubble point temperature;

2 the dew point temperature;

3 the temperature when 40 mole % of the mixture is in the vapor phase.

SOLUTION: Write the describing equation for the BPT:

$$\Sigma y_i = \Sigma(K_i x_i) = 1.0 \tag{12.3}$$

Calculate the BPT in °F. Try BPT $= 140°$ F:

$$K_P \approx 1.15$$
$$K_B \approx 0.41$$

$$
\begin{aligned}
\Sigma y_i &= \Sigma(K_i x_i) \\
&= (0.2)(0.41) + (0.8)(1.15) \\
&= 1.002
\end{aligned}
$$

Close enough, so that

$$BPT = 140°F \text{ (or marginally lower)}$$

Write the describing equation for the DPT:

$$\Sigma x_i = \Sigma(y_i/K_i) = 1.0 \tag{12.4}$$

Calculate the DPT in °F. Try DPT $= 160°F$:

$$K_P \approx 1.37$$
$$K_B \approx 0.525$$

$$\Sigma x_i = (0.2/0.525) + (0.8/1.37)$$
$$= 0.965 \text{ (assumed DPT too high)}$$

Try DPT $= 150°F$:

$$K_P \approx 1.28$$
$$K_B \approx 0.48$$

$$\Sigma x_i = (0.2/0.48) + (0.8/1.28)$$
$$= 1.042 \text{ (assumed DPT too low)}$$

Linear interpolation yields

$$DPT = 154°F$$

Write the componential split equation:

$$\Sigma x_i = \Sigma \{z_i/[L + K_i(1 - L)]\} \tag{12.8}$$

Calculate the system temperature for a 60/40 liquid/vapor molar split ($L = 0.6$). Try $T = 145°F$:

$$K_P \approx 1.20$$
$$K_B \approx 0.45$$

$$\Sigma x_i = 0.2/[0.6 + (0.45)(0.4)] + 0.8/[0.6 + (1.20)(0.4)]$$
$$= 0.256 + 0.741$$
$$= 0.997$$

Therefore, $T = 145°F$ (or perhaps marginally lower).   ■

ILLUSTRATIVE EXAMPLE 12.7

Refer to Illustrative Example 12.6. Calculate the fraction of vapor present (mole basis) if the temperature is 115°F.

SOLUTION: Since T is below the BPT, it is all liquid.   ■

ILLUSTRATIVE EXAMPLE 12.8

A liquid of composition 25 mole % ethane, 15 mole % n-butane and a third unknown component enters a partial reboiler from a distillation column at a pressure of 120 psia and 40°F. If the bottoms (discharged from reboiler at bottom of column) to overhead (discharged from top of column) molar split is $4/1$ at these conditions, determine the chemical name of the third remaining component in the liquid mixture.

SOLUTION: Write the componential split equation:

$$\Sigma x_i = \Sigma\{z_i/[L + K_i(1 - L)]\} = 1.0 \qquad (12.8)$$

Set L and V. The bottoms to boilup ratio is $4/1$. Therefore,

$$L = 0.80$$
$$V = 1 - L$$
$$= 0.20$$

Obtain K for ethane (E) and n-butane (B) at 120 psia and 40°F:

$$K_E = 2.60$$
$$K_B = 0.18$$

Calculate x_E and x_B by employing Equation (12.8):

$$x_i = z_i/[0.80 + K_i(0.20)]$$

Substituting

$$x_E = 0.19$$
$$x_B = 0.18$$

Set Y as the unknown component (see Table 12.6), and then calculate x_Y.

Table 12.6   Illustrative Example 12.8 Calculations

Component	z_i	K_i	x_i
Ethane	0.25	2.60	0.19
n-Butane	0.15	0.18	0.18
Y	0.60	—	x_Y
			$\Sigma = 1.0$

$$x_Y = 1.0 - (0.19 + 0.18)$$
$$= 1.0 - 0.19 - 0.18$$
$$= 0.63$$

Calculate K_Y by applying Equation (12.4) to component Y.

$$0.63 = 0.6/[0.8 + K_Y(0.2)]$$
$$K_Y = 0.76$$

Obtain the chemical name of the unknown component. Refer to Fig. 12.3 or 12.4. A K_Y (120 psia and 40°F) with a value of 0.76 appears to be *propane*.   ■

RAOULT'S LAW DIAGRAMS

This section presents two VLE diagrams which are based on Raoult's law, with methanol–water as the system of choice. One is a T-x, y diagram at a constant pressure of 1 atm; the second is a P-x, y diagram at a constant temperature of 40°C. A Microsoft Excel spread sheet program generated the VLE diagrams of this system at either the specified temperature or pressure. Although details of the calculation plus an algorithm are included with both cases (see Illustrative Examples 12.9 and 12.10), it is recommended that the reader review the Raoult's Law section of Chapter 11.

ILLUSTRATIVE EXAMPLE 12.9

Generate a T-x, y diagram for the methanol (m)–water (w) system at 1 atm (101.325 kPa) assuming Raoult's law to apply. Antoine equation coefficients, drawn from Table 9.1, are provided in Table 12.7.[3]

Table 12.7   Antoine Equation Constants

	Methanol	Water
A	16.5938	16.262
B	3644.3	3799.89
C	239.76	226.35

Note that this form of the Antoine equation has the natural log of pressure in kPa and temperature in °C.

SOLUTION: The general algorithm for generating a T-x, y diagram is provided in Fig. 12.5. The details of the solution to this Illustrative Example follows.

Obtain the vapor pressure of each pure component (methanol and water) at 1 atm. Solve Antoine's equation for T (see Chapter 8),

$$\ln p' = A - [B/(T + C)] \tag{8.8}$$

For component i,

$$T = \frac{B_i}{A_i - \ln p_i'} - C_i$$

Note that the following coefficients used employ the natural log form of the Antoine equation, the pressure is in kPa and the temperature in °C. The saturation temperatures are

$$T_{\text{sat},m} = \frac{B_m}{A_m - \ln p_m} - C_m = \frac{3644.30}{16.5938 - \ln 101.325} - 239.76$$
$$= 64.55°C$$

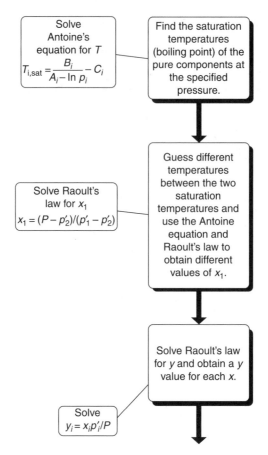

Figure 12.5   Generating a *T-x, y* diagram; Raoult's law.

and

$$T_{\text{sat},w} = \frac{B_w}{A_w - \ln p_w} - C_w = \frac{3799.89}{16.262 - \ln 101.325} - 226.35$$

$$= 100.0°C \text{ (as expected)}$$

From Raoult's law,

$$x_i p_i' = y_i P \tag{11.5}$$

and

$$\Sigma y_i = 1 \tag{12.3}$$

For the two components,

$$P = x_m p_m' + x_w p_w'$$

and since this is a binary system,

$$P = x_m p'_m + (1 - x_m)p'_w$$

Solving the previous equation for x_m,

$$x_m = \frac{P - p'_w}{p'_m - p'_w} = \frac{P - \exp\left[A_w - \dfrac{B_w}{T + C_w}\right]}{\exp\left[A_m - \dfrac{B_m}{T + C_m}\right] - \exp\left[A_w - \dfrac{B_w}{T + C_w}\right]}$$

Determine x_m using various values of T in the 64.6°C–100°C range. These two temperatures represent the saturation temperature of methanol and water, respectively, at 1 atm. For example, at $T = 70$°C,

$$
\begin{aligned}
x_m &= \frac{P - \exp\left[A_w - \left(\dfrac{B_w}{T + C_w}\right)\right]}{\exp\left[A_m - \left(\dfrac{B_m}{T + C_m}\right)\right] - \exp\left[A_w - \left(\dfrac{B_w}{T + C_w}\right)\right]} \\[2ex]
&= \frac{101.325 - \exp\left[16.262 - \dfrac{3799.89}{70 + 226.35}\right]}{\exp\left[16.5938 - \left(\dfrac{3644.30}{70 + 239.76}\right)\right] - \exp\left[16.262 - \left(\dfrac{3799.89}{70 + 226.35}\right)\right]} \\[2ex]
&= 0.747
\end{aligned}
$$

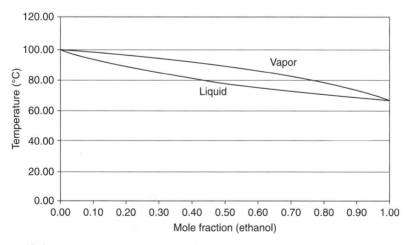

Figure 12.6   *T-x, y* diagram for the methanol–water system. Raoult's law $P = 1$ atm.

Solve Raoult's law for y_m,

$$y_m = \frac{x_m p'_m}{P} = \frac{(0.747)(125.07)}{101.325} = 0.922$$

Proceed to determine y_m for x_m at a different T.

To generate a T-x, y diagram, plot the x_m and y_m data as the ordinate and temperature as the abscissa. See Fig. 12.6. Generating an x_m–y_m plot is left as an exercise for the reader. Note that it is standard to use the light component when generating phase equilibrium diagrams.   ∎

ILLUSTRATIVE EXAMPLE 12.10

Generate a P-x, y and x, y diagram for the methanol–water system at 40°C. The general algorithm for generating a P-x, y diagram is given in Fig. 12.7.

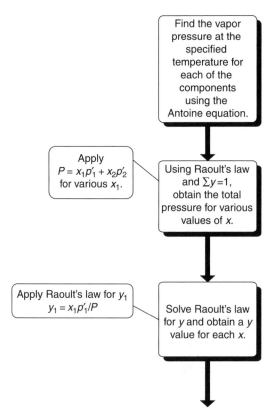

Figure 12.7   Generating a P-x, y diagram; Raoult's law.

SOLUTION: First obtain the vapor pressure of each component at the specified temperature. From the Antoine equation

$$p'_i = \exp\left[A_i - \left(\frac{B_i}{T + C_i}\right)\right]$$

one obtains

$$p'_m = \exp\left[A_m - \left(\frac{B_m}{T + C_m}\right)\right] = \exp\left[16.5938 - \left(\frac{3644.3}{40 + 239.76}\right)\right] = 35.420\,\text{kPa}$$

$$p'_w = \exp\left[A_w - \left(\frac{B_w}{T + C_w}\right)\right] = \exp\left[16.262 - \left(\frac{3799.89}{40 + 226.35}\right)\right] = 7.356\,\text{kPa}$$

Note once again that this form of the Antoine equation uses the natural log with the pressure in kPa and temperature in °C.

From Raoult's law,

$$x_i p'_i = y_i P \qquad\qquad (11.5)$$

with

$$\Sigma y_i = 1 \qquad\qquad (12.3)$$

Then

$$P = x_m p'_m + x_w p'_w$$

Table 12.8 *P-x* Data for Methanol–Water System

x_m	P, kPa
0.0	7.356
0.1	10.16
0.2	12.97
0.3	15.77
0.4	18.58
0.5	21.39
0.6	24.19
0.7	27.00
0.8	29.81
0.9	32.61
1.0	35.42

and since this is a binary system,

$$P = x_m p'_m + (1 - x_m) p'_w$$

The corresponding pressure can be calculated for various values of x_m. These are provided in Table 12.8.

Solve Raoult's law for y_m.

$$y_m = \frac{x_m p'_m}{P} \tag{11.5}$$

Determining y_m for each x_m leads to the results given in Table 12.9.

Table 12.9 x–y Data for Methanol–Water System

x_m	y_m
0.0	0.000
0.1	0.349
0.2	0.546
0.3	0.674
0.4	0.762
0.5	0.828
0.6	0.878
0.7	0.918
0.8	0.951
0.9	0.977
1.0	1.000

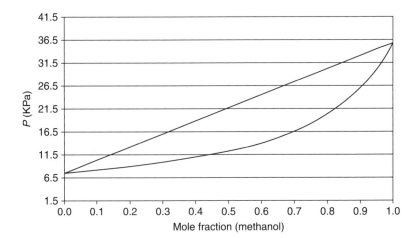

Figure 12.8   *P-x, y* diagram for the methanol–water at 40°C Raoult's law.

To generate an P-x, y diagram, plot the x_m and y_m data as the ordinate and pressure as the abscissa (Fig. 12.8).

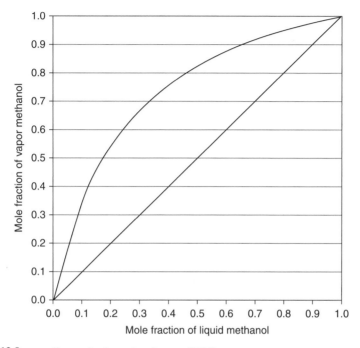

Figure 12.9 x–y diagram for the methanol–water (40°C).

To generate an x–y diagram, simply plot the x_m as the ordinate and y_m as the abscissa (Fig. 12.9). Again, the convention is to plot only the more volatile compound on phase equilibria diagrams. Also, for x–y diagrams, it is standard to plot data on a square coordinate system. ■

VAPOR–LIQUID EQUILIBRIUM IN NONIDEAL SOLUTIONS

In the case where liquid solutions cannot be considered ideal, Raoult's law will give highly inaccurate results. For these nonideal liquid solutions, various alternatives are available. These are considered in most standard thermodynamics texts and will be treated to some extent below. One important case will be mentioned because it is frequently encountered in distillation, namely, where the liquid phase is not an ideal solution but the pressure is low enough so that the vapor phase behaves as an ideal gas. In this case, the deviations from ideality are localized in the liquid and treatment is possible by quantitatively considering deviations from Raoult's law. These deviations are taken into account by incorporating a correction factor, γ, into Raoult's law. The purpose of γ, defined as the activity coefficient, is to account for the departure of the liquid phase from ideal solution behavior. It is introduced into the Raoult's law equation (for component i) as follows:

$$y_i P = p_i = \gamma_i x_i p_i' \tag{12.10}$$

The activity coefficient is a function of liquid phase composition and temperature. Phase-equilibria problems of the above type are often effectively reduced to evaluating γ.

Vapor–liquid equilibrium calculations performed with Equation (12.10) are slightly more complex than those made with Raoult's law. The key equation becomes

$$P = \Sigma p_i = \Sigma(y_i P) = \Sigma = \gamma_i x_i p_i' \tag{12.11}$$

When applied to a two-component $(A-B)$ system, Equation (12.11) becomes

$$P = y_A P + y_B P = \gamma_A x_A p_A' + \gamma_B x_B p_B' \tag{12.12}$$

so that

$$y_A P = p_A = \gamma_A x_A p_A' \tag{12.13}$$

$$y_B P = p_B = \gamma_B x_B p_B' \tag{12.14}$$

Methods to determine the activity coefficient(s) now follow.

Theoretical developments in the molecular thermodynamics of liquid-solution behavior are often based on the concept of *local composition*. Within a liquid solution, local compositions, different from the overall mixture composition, are presumed to account for the short-range order and nonrandom molecular orientations that result from differences in molecular size and intermolecular forces. The concept was introduced by G. M. Wilson in 1964 with the publication of a model of solution behavior, since known as the Wilson equation.[4] The success of this equation in the correlations of vapor–liquid equilibrium data prompted the devolvement of several alternative local-composition models. Perhaps the most notable of those is the NRTL (non-random-two liquid) equation of Renon and Prausnitz.[5]

The Wilson equation contains just two parameters for a binary system (E_{AB} with E_{BA}), with the activity coefficients are written as:

$$\ln \gamma_A = -\ln(x_A + x_B E_{AB}) + x_B \left(\frac{E_{AB}}{x_A + x_B E_{AB}} - \frac{E_{BA}}{x_B + x_A E_{BA}} \right) \tag{12.15}$$

$$\ln \gamma_B = -\ln(x_B + x_A E_{BA}) - x_A \left(\frac{E_{AB}}{x_A + x_B E_{AB}} - \frac{E_{BA}}{x_B + x_A E_{BA}} \right) \tag{12.16}$$

For infinite dilution, these equations become:

$$\ln \gamma_A^\infty = -\ln(E_{AB}) + 1 - E_{BA} \tag{12.17}$$

$$\ln \gamma_B^\infty = -\ln(E_{BA}) + 1 - E_{AB} \tag{12.18}$$

Note that the Wilson parameters, E_{AB} and E_{BA}, must always be positive numbers. The temperature dependence of the Wilson parameter is given by:

$$E_{ij} = \frac{V_j}{V_i} \exp\frac{-a_{ij}}{RT} \quad i \neq j \tag{12.19}$$

where V_j and V_i are the molar volumes at temperature T of pure liquids j and i, respectively, and a_{ij} is a constant independent of composition and temperature. Thus, the Wilson equation has built into it an approximate temperature dependence for the parameters. Wilson model coefficients are provided in Table 12.10 for a number of binary systems.[4,5]

Table 12.10   Wilson Equation Parameters[4,5]

System	V_A, V_B (cm^3/gmol)	Wilson equation (cal/gmol)	
		a_{AB}	a_{BA}
Acetone (A)	74.05	291.27	1448.01
Water (B)	18.07		
Methanol (A)	40.73	107.38	469.55
Water (B)	18.07		
1-Propanol (A)	75.14	775.48	1351.90
Water (B)	18.07		
Water (A)	18.07	1696.98	−219.39
1,4-Dioxane (B)	85.71		
Methanol (A)	40.73	504.31	196.75
Acetonitrile (B)	66.30		
Acetone (A)	74.05	−161.88	583.11
Methanol (B)	40.73		
Methyl acetate (A)	79.84	−31.19	813.18
Methanol (B)	40.73		
Methanol (A)	40.73	17,423.42	183.04
Benzene (B)	89.41		
Ethanol (A)	58.68	1556.45	210.52
Toluene (B)	106.85		

The NRTL equation, containing three parameters for a binary system, is:

$$\ln \gamma_A = x_B^2 \left[\tau_{BA} \left(\frac{G_{BA}}{x_A + x_B G_{BA}} \right)^2 + \frac{G_{AB}\tau_{AB}}{(x_B + x_A G_{AB})^2} \right] \tag{12.20}$$

$$\ln \gamma_B = x_A^2 \left[\tau_{AB} \left(\frac{G_{AB}}{x_B + x_A G_{AB}} \right)^2 + \frac{G_{BA}\tau_{BA}}{(x_A + x_B G_{BA})^2} \right] \tag{12.21}$$

The parameters G and τ can be obtained using the following equations:

$$G_{AB} = \exp(-\alpha\tau_{AB}) \tag{12.22}$$

$$G_{BA} = \exp(-\alpha\tau_{BA}) \tag{12.23}$$

Furthermore,

$$\tau_{AB} = \frac{b_{AB}}{RT} \tag{12.24}$$

$$\tau_{BA} = \frac{b_{BA}}{RT} \tag{12.25}$$

Table 12.11   NRTL Equation Parameters[5–7]

| System | NRTL equation (cal/gmol) | | α |
	b_{AB}	b_{BA}	
Acetone	631.05	1197.41	0.5343
Water			
Methanol	−253.88	845.21	0.2994
Water			
1-Propanol	500.40	1636.57	0.5081
Water			
Water	715.96	548.90	0.2920
1,4-Dioxane			
Methanol	343.70	314.59	0.2981
Acetonitrile			
Acetone	184.70	222.64	0.3084
Methanol			
Methyl acetate	381.46	346.54	0.2965
Methanol			
Methanol	730.09	1175.41	0.4743
Benzene			
Ethanol	713.57	1147.86	0.5292
Toluene			

where α, b_{AB}, b_{BA} are parameters specific to a particular pair of species, independent of composition and temperature. The corresponding infinite-dilution values of the activity coefficients are given by:

$$\ln \gamma_A^\infty = \tau_{BA} + \tau_{AB} \exp(-\alpha\tau_{AB}) \tag{12.26}$$

$$\ln \gamma_B^\infty = \tau_{AB} + \tau_{BA} \exp(-\alpha\tau_{BA}) \tag{12.27}$$

Tabulated values of the parameters for the NRTL model can be found in Table 12.11.[5–7]

The application of both the Wilson equation and the NRTL method receive treatment in the next two sections. The NRTL diagrams are treated first because of the interest in this model.

NRTL DIAGRAMS

VLE diagrams can be developed employing the NRTL method for two-component systems where the liquid phase is not assumed to be ideal. Once again, two types of diagrams are of interest: T-x, y and P-x, y. Procedures for generating these figures are provided in Illustrative Example 12.11 for a T-x, y diagram and in Illustrative Example 12.12 for a P-x, y diagram. Each solution contains a common applicable algorithm.

ILLUSTRATIVE EXAMPLE 12.11

Generate a T-x, y diagram for the ethanol–toluene system at 1 atm. Employ the NRTL method. The Antoine equation constants can be found in Table 12.12.

Table 12.12 Antoine Equation Constants

	Ethanol	Toluene
A	8.1122	6.95805
B	1592.864	1346.773
C	226.184	219.693

Note that this form of the Antoine equation has the log of the pressure in kPa and temperature in °C.

SOLUTION: The general algorithm for generating a T-x, y diagram, including non-ideal solution effects, is presented in Fig. 12.10. The following is a detailed solution (plus sample calculations) to this Illustrative Example.

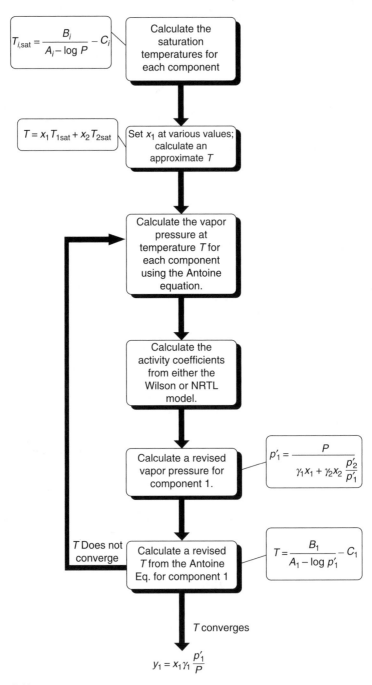

$$T_{i,\text{sat}} = \frac{B_i}{A_i - \log P} - C_i$$

Calculate the saturation temperatures for each component

$$T = x_1 T_{1\text{sat}} + x_2 T_{2\text{sat}}$$

Set x_1 at various values; calculate an approximate T

Calculate the vapor pressure at temperature T for each component using the Antoine equation.

Calculate the activity coefficients from either the Wilson or NRTL model.

Calculate a revised vapor pressure for component 1.

$$p'_1 = \frac{P}{\gamma_1 x_1 + \gamma_2 x_2 \dfrac{p'_2}{p'_1}}$$

T Does not converge

Calculate a revised T from the Antoine Eq. for component 1

$$T = \frac{B_1}{A_1 - \log p'_1} - C_1$$

T converges

$$y_1 = x_1 \gamma_1 \frac{p'_1}{P}$$

Figure 12.10   Generating a T-x, y diagram; modified Raoult's law.

1 First calculate the pure component saturation temperatures using the Antoine equation:

$$T_{\text{sat},e} = [B_e/(A_e - \log p'_e)] - C_e = [1592.864/(8.1122 - \log 760)] - 226.184 = 78.3°C$$

$$T_{\text{sat},t} = [B_t/(A_t - \log p'_t)] - C_t = [1346.773/(6.95805 - \log 760)] - 219.693$$
$$= 110.62°C$$

2 Set $x_e = 0.5$ (and therefore $x_t = 0.5$) and calculate an approximate temperature, T:

$$T = x_e T_{\text{sat},e} + x_t T_{\text{sat},t} = (0.5)(78.3) + (0.5)(110.62)$$
$$= 94.46°C$$

3 Calculate the vapor pressure at temperature T for each component:

$$\log p' = A - \left(\frac{B}{T + C}\right)$$

so that

$$p'_e = 10^{8.1122 - [1592.864/(94.46 + 226.184)]} = 1394.75 \text{ mm Hg}$$

$$p'_t = 10^{6.95805 - [1346.773/(94.46 + 219.693)]} = 468.87 \text{ mm Hg}$$

4 Obtain the parameters from the NRTL method (see Table 12.11).

$$\alpha = 0.5292$$
$$b_{et} = 713.57$$
$$b_{te} = 1147.86$$

Calculate the activity coefficients for both components.

$$\tau_{et} = \frac{b_{et}}{RT}$$
$$= \frac{713.57}{1.987(94.46 + 273)}$$
$$= 0.977 \qquad\qquad (12.24)$$

$$\tau_{te} = \frac{b_{te}}{RT}$$
$$= \frac{1147.86}{1.987(94.46 + 273)}$$
$$= 1.572 \qquad\qquad (12.25)$$

$$G_{et} = \exp[-\alpha \tau_{et}]$$ (12.22)
$$= \exp[-(0.5292)(0.977)]$$
$$= 0.596$$

$$G_{te} = \exp[-\alpha \tau_{te}]$$ (12.23)
$$= \exp[-(0.5292)(1.572)]$$
$$= 0.435$$

$$\gamma_e = \exp\left[x_t^2\left(\tau_{te}\left(\frac{G_{te}}{x_e + x_t G_{te}}\right)^2 + \frac{G_{et}\tau_{et}}{(x_t + x_e G_{et})^2}\right)\right]$$ (12.20)

$$= \exp\left[(0.5)^2\left((1.572)\left(\frac{0.435}{0.5 + (0.5)(0.435)}\right)^2 + \frac{(0.596)(0.977)}{[0.5 + (0.5)(0.596)]^2}\right)\right]$$

$$= 1.41$$

$$\gamma_t = \exp\left[x_e^2\left(\tau_{et}\left(\frac{G_{et}}{x_t + x_e G_{et}}\right)^2 + \frac{G_{te}\tau_{te}}{(x_e + x_t G_{te})^2}\right)\right]$$ (12.21)

$$= \exp\left[(0.5)^2\left(0.977\left(\frac{0.596}{0.5 + (0.5)(0.596)}\right)^2 + \frac{(0.435)(1.572)}{[0.5 + (0.5)(0.435)]^2}\right)\right]$$

$$= 1.60$$

5 Calculate a revised vapor pressure for component 1 (ethanol) using the activity coefficients determined above and a rearranged form of equation (12.11).

$$p'_e = \frac{P}{\gamma_e x_e + \gamma_t x_t \dfrac{p'_t}{p'_e}} = \frac{760}{1.41(0.5) + 1.60(0.5)\left(\dfrac{468.78}{1394.75}\right)}$$

$$= 780.38$$

6 Using this revised vapor pressure, calculate a revised temperature using Antoine's equation:

$$T = \frac{B_e}{A_e - \log p'_e} - C_e$$

$$= \frac{1592.864}{8.1122 - \log 780.38} - 226.184$$

$$= 78.97°C$$

7 Compare this temperature with the temperature estimate from step 2. If these values do not converge, then use the temperature from step 6 and return to step 3. If they do converge, calculate a vapor mole fraction using the following equation:

$$y_e = x_e \gamma_e \frac{p'_e}{P}$$ (12.10)

8 Return to step 2 and use a different value for x_e. Continue this until an entire T-x, y diagram is formed.

Table 12.13 contains the results of the above calculation for various values of x_e.

Table 12.13   NRTL Temperature Calculation for
Ethanol–Toluene

x_e	T °C (NRTL)
0.0	110.62
0.1	90.82
0.2	84.41
0.3	81.25
0.4	79.4
0.5	78.2
0.6	77.4
0.7	76.88
0.8	76.65
0.9	76.89
1.0	78.298

A T-x, y diagram for ethanol and toluene, employing the NRTL method can be found in Fig. 12.11. To complete the algorithm, apply Raoult's law with the activity coefficient to determine the y_e corresponding to the assumed value of x_e. These results are given in Table 12.14. Note that an azeotrope ($x_e = y_e$) is formed at approximately $x_e = 0.8$.

To generate an x–y diagram, simply plot the x_e as the ordinate and y_e as the abscissa. ∎

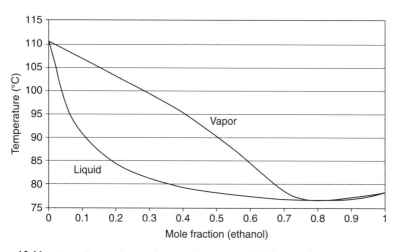

Figure 12.11 T-x, y diagram for the ethanol–toluene; 1 atm (NRTL method).

Table 12.14 x–y Calculations for Ethanol–Toluene

x_e	y_e (NRTL)
0.0	0
0.1	0.26
0.2	0.43
0.3	0.56
0.4	0.66
0.5	0.74
0.6	0.81
0.7	0.87
0.8	0.92
0.9	0.96
1.0	1.0

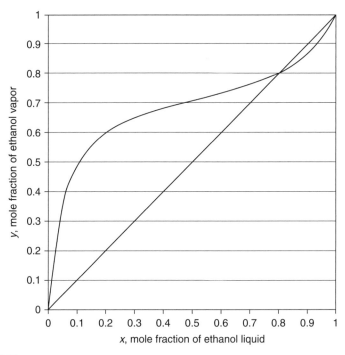

Figure 12.12 x–y diagram for the ethanol–toluene system (NRTL method).

ILLUSTRATIVE EXAMPLE 12.12

Generate a P-x, y diagram using the NRTL method for a methanol (m)–water (w) system at 40°C.

SOLUTION: The solution for this example is somewhat similar to Illustrative Example 12.11. First obtain the vapor pressure of each component at the specified temperature. Since (see Chapter 8)

$$p'_i = e^{A_i - [B_i/(T+C_i)]}$$

one obtains (using the same Antoine coefficients as before—see Table 12.7):

$$p'_m = e^{A_m - [B_m/(T+C_m)]} = e^{16.5938 - [3644.3/(40+239.76)]} = 35.420\,\text{kPa}$$

$$p'_w = e^{A_w - [B_w/(T+C_w)]} = e^{16.262 - [3799.89/(40+226.35)]} = 7.356\,\text{kPa}$$

Using the NRTL model, calculate γ_m and γ_w employing the following equations presented before:

$$\ln \gamma_m = x_w^2 \left[\tau_{wm} \left(\frac{G_{wm}}{x_m + x_w G_{wm}} \right)^2 + \frac{G_{mw}\tau_{mw}}{(x_w + x_m G_{mw})^2} \right] \quad (12.20)$$

$$\ln \gamma_w = x_m^2 \left[\tau_{mw} \left(\frac{G_{mw}}{x_w + x_m G_{mw}} \right)^2 + \frac{G_{wm}\tau_{wm}}{(x_m + x_w G_{wm})^2} \right] \quad (12.21)$$

The parameters G and τ can be obtained using the following equations:

$$G_{mw} = \exp(-\alpha\tau_{mw}) \quad (12.22)$$

$$G_{wm} = \exp(-\alpha\tau_{wm}) \quad (12.23)$$

Furthermore,

$$\tau_{mw} = \frac{b_{mw}}{RT} \quad (12.24)$$

$$\tau_{wm} = \frac{b_{wm}}{RT} \quad (12.25)$$

From Table 12.11,

$$\alpha = 0.2994$$
$$b_{mw} = -253.88\,\text{cal/gmol}$$
$$b_{wm} = 845.21\,\text{cal/gmol}$$

The results shown in Table 12.15 are obtained by substituting into the equations above and varying the methanol mole fraction.

Table 12.15 Results for Illustrative Example 12.12 (NRTL method)

x_m	P, kPa
0.0	7.356
0.1	15.192
0.2	21.149
0.3	25.527
0.4	28.631
0.5	30.765
0.6	32.227
0.7	33.291
0.8	34.188
0.9	35.016
1.0	35.42

The corresponding y_m is calculated from the following equation:

$$y_m = \frac{x_m \gamma_m P'_m}{P} \qquad (12.10)$$

The y_m values as shown in Table 12.16 are obtained from this equation. To generate a P-x, y diagram, plot x_m and y_m data as the ordinate and pressure as the abscissa (see Fig. 12.13). An expanded plot would indicate an azeotrope in the $x = 0.9$–1.0 range. ∎

Table 12.16 Calculated y_m values

x_m	y_m
0.0	0
0.1	0.561
0.2	0.713
0.3	0.783
0.4	0.823
0.5	0.848
0.6	0.866
0.7	0.882
0.8	0.901
0.9	0.931
1.0	1

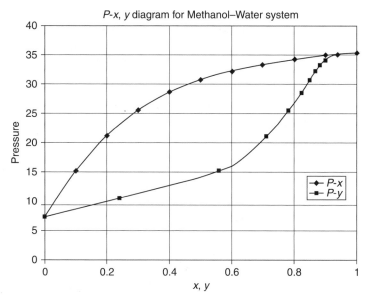

Figure 12.13 *P-x, y* diagram generated using NRTL method.

ILLUSTRATIVE EXAMPLE 12.13

Compare the *P-x, y* diagram generated earlier using Raoult's law for a methanol–water system at 40°C with the plot generated using the NRTL method and compare the answers.

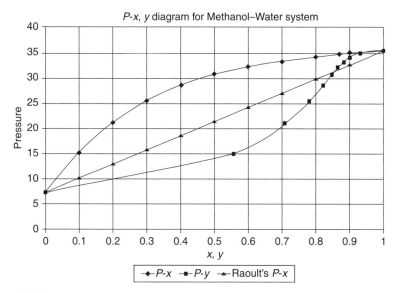

Figure 12.14 Comparison of *P-x, y* diagram generated from two different VLE methods.

SOLUTION: Combining the curves generated from both methods into one figure (see Fig. 12.14), it can be observed that the plot generated using Raoult's law gives lower values of pressure at the same x_m values that the NRTL method gives for higher values. Also the bubble point curve from Raoult's law is (as expected) a straight line compared to the curve generated by the NRTL method, which is concave down. ∎

WILSON DIAGRAMS

VLE diagrams can also be developed by applying Wilson's method for two-component systems where the liquid phase is not assumed to be ideal. As with Raoult's law and the NRTL method, there are two diagrams of interest: T-x,y and P-x,y. Procedures for generating these figures are provided in the next two Illustrative Examples. Once again, each contains an applicable algorithm.

ILLUSTRATIVE EXAMPLE 12.14

Generate T-x, y and x-y diagrams using Wilson's method for the ethanol–toluene system at 1 atm. Refer to Table 12.12 again for Antoine equation coefficients.[8]

SOLUTION: The general algorithm for generating a T-x, y diagram, including non-ideal solution effects, is presented in Fig. 12.10. Presented below is a detailed solution (plus sample calculations) to this Illustrative Example.

1 First calculate the pure component saturation temperatures:

$$T_{\text{sat},e} = \frac{B_e}{A_e - \log p'_e} - C_e = \frac{1592.864}{8.1122 - \log 760} - 226.184 = 78.3°C$$

$$T_{\text{sat},t} = \frac{B_t}{A_t - \log p'_t} - C_t = \frac{1346.773}{6.95805 - \log 760} - 219.693 = 110.62°C$$

2 Set $x_e = 0.5$ (and therefore $x_t = 0.5$) and calculate an approximate temperature, T.

$$T = x_e T_{\text{sat},e} + x_t T_{\text{sat},t} = (0.5)(78.3) + (0.5)(110.62) = 94.46°C$$

3 Employ the Antoine equation and calculate the vapor pressure at temperature T for each component.

$$p'_e = 10^{A_e - [B_e/(T+C_e)]} = 10^{8.1122 - [1592.864/(94.46 + 226.184)]} = 1394.75\,\text{mm Hg}$$

$$p'_t = 10^{A_t - [B_t/(T+C_t)]} = 10^{6.95805 - [1346.773/(94.46 + 219.693)]} = 468.87\,\text{mm Hg}$$

4 Using the parameters for the Wilson equation, calculate the activity coefficients for both components. The Wilson equation constants are obtained from Table 12.10.

$$V_A = 58.68 \text{ cal/gmol}$$
$$V_B = 106.85 \text{ cal/gmol}$$
$$a_{AB} = 1556.45 \text{ cal/gmol}$$
$$a_{BA} = 210.52 \text{ cal/gmol}$$

Using V_A and V_B, plus a_{AB} and a_{BA}, determine the temperature-dependent value of E_{AB} and E_{BA} using Equation (12.19):

$$E_{AB} = \frac{V_B}{V_A} \exp \frac{-a_{AB}}{RT}$$
$$= \frac{106.85}{58.68} \exp \frac{-1556.45}{(1.987)(94.46 + 273)}$$
$$= 0.216$$

$$E_{BA} = \frac{V_A}{V_B} \exp \frac{-a_{BA}}{RT}$$
$$= \frac{58.68}{106.85} \exp \frac{-210.52}{(1.987)(94.46 + 273)}$$
$$= 0.411$$

Using Equations (12.15) and (12.16), determine the value of the activity coefficients, γ_A and γ_B:

$$\ln \gamma_A = -\ln(x_A + x_B E_{AB}) + x_B \left(\frac{E_{AB}}{x_A + x_B E_{AB}} - \frac{E_{BA}}{x_B + x_A E_{BA}} \right)$$
$$= -\ln[0.5 + 0.5(0.216)] + 0.5 \left(\frac{0.216}{0.5 + 0.5(0.216)} - \frac{0.411}{0.5 + 0.5(0.411)} \right)$$
$$= 0.384$$
$$\gamma_A = 1.468$$

$$\ln \gamma_B = -\ln(x_B + x_A E_{BA}) + x_A \left(\frac{E_{AB}}{x_A + x_B E_{AB}} - \frac{E_{BA}}{x_B + x_A E_{BA}} \right)$$
$$= -\ln[0.5 + 0.5(0.411)] + 0.5 \left(\frac{0.216}{0.5 + 0.5(0.216)} - \frac{0.411}{0.5 + 0.5(0.411)} \right)$$
$$= 0.235$$
$$\gamma_B = 1.265$$

5 Calculate a revised vapor pressure for component 1 (ethanol) using the activity coefficients determined above. Rearrange Equation (12.12) and solve.

$$p'_e = \frac{P}{\gamma_e x_e + \gamma_t x_t (p'_t/p'_e)}$$

$$= \frac{760}{1.468(0.5) + 1.265(0.5)(468.78/1394.75)}$$

$$= 802.89 \text{ mm Hg}$$

6 Using the vapor pressure obtained in step 5, calculate a revised temperature using Antoine's equation.

$$T = \frac{B_e}{A_e - \log p'_e} - C_e$$

$$= \frac{1592.864}{8.1122 - \log 802.89} - 226.184$$

$$= 79.69°C$$

7 Compare this temperature with the temperature estimate from step 2. If these values do not converge, use the temperature from step 6 and return to step 3. If they do converge, then calculate a vapor mole fraction using the following equation:

$$y_e = x_e \gamma_e \frac{p'_e}{P} \qquad (12.13)$$

8 Return to step 2 and use a different value for x_e. Continue this until an entire T-x, y diagram is formed.

Table 12.17 T-x Data for the Ethanol–Toluene System

x_e	$T°C$ (Wilson's)
0.0	110.62
0.1	84.62
0.2	80.46
0.3	78.93
0.4	78.11
0.5	77.54
0.6	77.09
0.7	76.75
0.8	76.60
0.9	76.88
1.0	78.299

T-x data can be found in Table 12.17. The equation

$$y_e = \frac{x_e \gamma_e p'_e}{P}$$

is employed to obtain y_e. A *T-x, y* diagram for ethanol and toluene employing Wilson's method can be found in Fig. 12.15. Note that an azeotrope is formed at $x = y = 0.8$.

Generate an *x–y* diagram from the results obtained above. Refer to Table 12.18 for the *x–y* data. To generate an *x–y* diagram, simply plot the x_e as the ordinate and y_e as the abscissa. This is provided in Fig. 12.16.

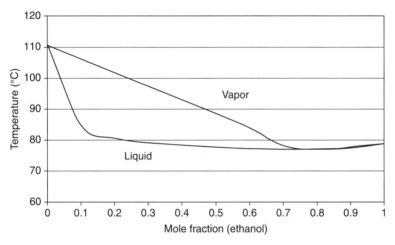

Figure 12.15   *T-x, y* diagram for ethanol–toluene; 1 atm (Wilson's method).

Table 12.18   *x–y* Data for the Ethanol–Toluene System (Wilson's)

x_e	y_e (Wilson's)
0.0	0
0.1	0.26
0.2	0.43
0.3	0.56
0.4	0.66
0.5	0.74
0.6	0.81
0.7	0.87
0.8	0.92
0.9	0.96
1.0	1

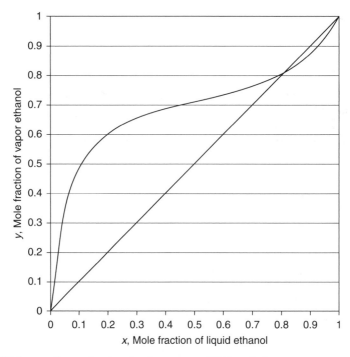

Figure 12.16 x–y diagram for ethanol–toluene system. NRTL method.   ■

ILLUSTRATIVE EXAMPLE 12.15

Generate a P–x, y diagram for the ethanol–toluene system at 35°C. Use the Wilson method.

SOLUTION: In a very real sense, this example is similar to the previous example and Illustrative Example 12.12 that employed the NRTL method. This is left as an exercise for the reader.   ■

RELATIVE VOLATILITY

For a two-component (A, B) system, the relative volatility α_{AB} is defined as

$$\alpha_{AB} = \frac{y_A/x_A}{y_B/x_B} \tag{12.28}$$

with component A the more volatile of the two. For a binary mixture obeying Raoult's law

$$y_A/x_A = p'_A/P$$
$$y_B/x_B = p'_B/P$$

so that

$$\alpha_{AB} = \frac{p'_A}{p'_B} \qquad (12.29)$$

The relative volatility provides a convenient index of the relative ease (or difficulty) of separating A from B by distillation since separation is achieved because of differences in vapor pressure. It should be noted that distillation is rarely employed if α is less than 1.1.

For binary mixtures following Henry's law,

$$\alpha_{AB} = H_A/H_B \qquad (12.30)$$

Thus, the relative volatility is constant with composition for substances following either Raoult's or Henry's laws. For many other mixtures, the relative volatility varies. For example, the relative volatility of methanol–water systems varies because this system deviates widely from Raoult's law.

ILLUSTRATIVE EXAMPLE 12.16

Calculate the relative volatility, α_{AB}, at $0°C$ for the acetone-B system. Assume the vapor pressure of B at $0°C$ is 35 mm Hg.

SOLUTION: As noted in an earlier example

$$p'_A = 70.01 \text{ mm Hg}$$

By definition,

$$\alpha_{AB} = \frac{p'_A}{p'_B} \qquad (12.29)$$

Substituting

$$\alpha_{AB} = \frac{70.01}{35}$$

$$= 2.0 \qquad \blacksquare$$

ILLUSTRATIVE EXAMPLE 12.17

Outline how to generate an x–y equilibrium diagram for a two component (A–B) system if the relative volatility, α_{AB}, equals 1.5.

SOLUTION: Replace component B by component A in Equation (12.28)

$$\alpha_{AB} = \frac{y_A(1 - x_A)}{(1 - y_A)(x_A)}$$

Table 12.19 $x-y$ Results for the $A-B$ System; Illustrative Example 12.15

x_A	y_A
0.2	0.43
0.4	0.67
0.6	0.82
0.8	0.925

and note that

$$\frac{y_A(1 - x_A)}{(1 - y_A)(x_A)} = 1.5$$

Pairs of $x-y$ points may now be generated by selecting values of x_A between 0 and 1.0, and solving for y_A. The results are listed in Table 12.19.   ■

REFERENCES

1. M. Leach, "An Approach to Multiphase Vapor–Liquid Equilibria," *Chem. Eng.*, New York, May 27, 1977.
2. C. DePriester, *Chem. Eng. Progr. Symp. Ser.*, 49(7), 42, 1953.
3. J. Kibuthu: class assignment submitted to L. Theodore, Manhattan College, 2004.
4. G. Wilson, *J. Am. Chem. Soc.*, 86, 27–130, 1964.
5. H. Renon and J. Prausnitz, *AIChE J.*, 14, 135–144, 1968.
6. J. Smith, H. Van Ness, and M. Abbott, "Chemical Engineering Thermodynamics," Sixth edition, McGraw-Hill, New York, 2001.
7. R. Gmehling et al., *"Vapor–Liquid Equilibrium Data Collection,"* Chemistry Data Series, Vol. I, Parts 1a, 1b, 2c and 2e, DEA & EMA, Frankfurt/Main, 1981–88.
8. N. May: class assignment submitted to L. Theodore, Manhattan College, 2004.

NOTE: Additional problems for each chapter are available for all readers at www. These problems may be used for additional review or homework purposes.

Chapter 13

Chemical Reaction Equilibrium Principles

Plato [427–347 B.C.]

A fit of laughter which has been indulged to excess almost always produces a violent reaction.

—The Republic, Book III, 388–E

INTRODUCTION

As noted at the beginning of this part of the book, equilibrium is a state in which there is no macroscopic change with respect to time. In chemical reaction equilibrium (CRE), where attention is focused upon a particular quantity of material, this indicates no change in the properties of the material with time. For a chemical reaction, the material can be either a reactant or product, or both.

A rigorous, detailed presentation of this topic is beyond the scope of this text. However, a superficial treatment is presented in the hope that it may at least provide a qualitative introduction to chemical reaction equilibrium. This topic does find application in thermodynamic calculations and will hopefully explain, in part, the role of reaction equilibria in these calculations. The four major equilibrium calculation considerations are:

1 the standard free energy of reactions, ΔG^0;

2 the chemical reaction equilibrium constant, K;

3 the effect of temperature on ΔG^0 and K;

4 the extent of an equilibrium reaction.

This chapter addresses topics (1)–(3); the next chapter treats (4). The first of these two chapters on chemical reaction equilibrium serves to introduce this subject. Topics covered include:

1 Standard Free Energy of Formation, ΔG_f^0

Thermodynamics for the Practicing Engineer. By L. Theodore, F. Ricci, and T. Van Vliet

In a very real sense, the purpose of this chapter to prepare the reader for the applications to be presented in the next chapter. The emphasis here is on the development of CRE principles that will ultimately lead to a better and more comprehensive understanding of the subject.

STANDARD FREE ENERGY OF FORMATION, ΔG_f^0

Chemical reaction equilibrium calculations are structured around another thermodynamic term referred to as *free energy*. This so-called *free energy* (G) is a thermodynamic property that cannot be easily defined without some basic grounding in thermodynamics. No attempt will be made to define it here in terms of fundamental principles, and the interested reader is again directed to the literature for further development of this topic.[1,2] In the opinion of the authors, the fact that standard free energy data are available in the literature and can be used to calculate chemical equilibrium constants warrants the inclusion of this property in the discussion to follow. Free energy has the same units as enthalpy and internal energy and may be used on a mole or total mass basis. The uppercase G signifies a mole basis.

Tabulated values of the standard free energy of formation, denoted as ΔG_f^0 are provided in Table 13.1. The major application of ΔG_f^0 is in determining the standard free energy of a reaction, denoted as ΔG^0. This receives treatment in the next section.

ILLUSTRATIVE EXAMPLE 13.1

What is the enthalpy of formation of benzene (g)?

SOLUTION: Ordinarily, one would refer to Table 10.1 to obtain the answer. But, since no temperature is specified, one cannot answer this question.   ■

ILLUSTRATIVE EXAMPLE 13.2

What is the standard enthalpy of formation of benzene (g)?

SOLUTION: Refer to Table 10.1.

$$\Delta H_f^0 = +19{,}820 \text{ cal/gmol}$$ ■

Table 13.1   Standard Free Energy of Formation at 25°C in Calories per Gram Mole[a]

Compound	Formula	State	ΔG_{f298}^0
Normal paraffins			
Methane	CH_4	g	$-12{,}140$
Ethane	C_2H_6	g	-7860
Propane	C_3H_8	g	-5614
n-Butane	C_4H_{10}	g	-4100
n-Pentane	C_5H_{12}	g	-2000
n-Hexane	C_6H_{14}	g	-70
n-Heptane	C_7H_{16}	g	1920
n-Octane	C_8H_{18}	g	3920
Increment per C atom above C_8		g	2010
Normal monoolefins (1-alkenes)			
Ethylene	C_2H_4	g	$16{,}282$
Propylene	C_3H_6	g	$14{,}990$
1-Butene	C_4H_8	g	$17{,}090$
1-Pentene	C_5H_{10}	g	$18{,}960$
1-Hexene	C_6H_{12}	g	$20{,}940$
Increment per C atom above C_6		g	2010
Miscellaneous organic compounds			
Acetaldehyde	C_2H_4O	g	$-31{,}960$
Acetic acid	$C_2H_4O_2$	l	$-93{,}800$
Acetylene	C_2H_2	g	$50{,}000$
Benzene	C_6H_6	g	$30{,}989$
Benzene	C_6H_6	l	$29{,}756$
1,3-Butadiene	C_4H_6	g	$36{,}010$
Cyclohexane	C_6H_{12}	g	7590
Cyclohexane	C_6H_{12}	l	6370
Ethanol	C_2H_6O	g	$-40{,}130$
Ethanol	C_2H_6O	l	$-41{,}650$
Ethylbenzene	C_8H_{10}	g	$31{,}208$
Ethylene glycol	$C_2H_6O_2$	l	$-77{,}120$
Ethylene oxide	C_2H_4O	g	-2790
Methanol	CH_4O	g	$-38{,}810$
Methanol	CH_4O	l	$-39{,}850$
Methylcyclohexane	C_6H_{14}	g	6520
Methylcyclohexane	C_6H_{14}	l	4860
Styrene	C_8H_8	g	$51{,}100$

(*Continued*)

Table 13.1 *Continued*

Compound	Formula	State	ΔG^0_{f298}
Toluene	C_7H_8	g	29,228
Toluene	C_7H_8	l	27,282
Miscellaneous inorganic compounds			
Ammonia	NH_3	g	-3976
Ammonia	NH_3	aq	-6370
Calcium carbide	CaC_2	s	$-16,200$
Calcium carbonate	$CaCO_3$	s	$-269,780$
Calcium chloride	$CaCl_2$	s	$-179,300$
Calcium chloride	$CaCl_2$	aq	$-194,880$
Calcium hydroxide	$Ca(OH)_2$	s	$-214,330$
Calcium hydroxide	$Ca(OH)_2$	aq	$-207,370$
Calcium oxide	CaO	s	$-144,400$
Carbon dioxide	CO_2	g	$-94,258$
Carbon monoxide	CO	g	$-32,781$
Hydrochloric acid	HCl	g	$-22,778$
Hydrogen sulfide	H_2S	g	-7892
Iron oxide	Fe_3O_4	s	$-242,400$
Iron oxide	Fe_2O_3	s	$-177,100$
Iron sulfide	FeS_2	s	$-39,840$
Nitric acid	HNO_3	l	$-19,100$
Nitric acid	HNO_3	aq	$-26,410$
Nitrogen oxides	NO	g	20,690
	NO_2	g	12,265
	N_2O	g	24,933
	N_2O_4	g	23,395
Sodium carbonate	Na_2CO_3	s	$-250,400$
Sodium chloride	$NaCl$	s	$-91,785$
Sodium chloride	$NaCl$	aq	$-93,939$
Sodium hydroxide	$NaOH$	s	$-90,600$
Sodium hydroxide	$NaOH$	aq	$-100,184$
Sulfur dioxide	SO_2	g	$-71,790$
Sulfur trioxide	SO_3	g	$-88,520$
Sulfuric acid	H_2SO_4	aq	$-177,340$
Water	H_2O	g	$-54,635$
Water	H_2O	l	$-56,690$
Chlorinated organics			
Methyl chloride	CH_3Cl	l	$-15,030$
Dichloromethane	CH_2Cl_2	l	$-16,460$

(Continued)

Table 13.1   *Continued*

Compound	Formula	State	ΔG^0_{f298}
Chloroform	$CHCl_3$	1	$-16{,}380$
Carbon tetrachloride	CCl_4	1	$-13{,}920$
Ethyl chloride	C_2H_5Cl	1	$-14{,}340$
1,1-Dichloroethane	$C_2H_4Cl_2$	1	$-17{,}470$
1,1,2,2-Tetrachloroethane	$C_2H_2Cl_4$	1	$-20{,}480$
n-Propyl chloride	C_3H_7Cl	1	$-12{,}110$
1,3-Dichloropropane	$C_3H_6Cl_2$	1	$-19{,}740$
n-Butyl chloride	C_4H_9Cl	1	-9270
1-Chloropentane	$C_5H_{11}Cl$	1	-8940
1-Chloroethylene	C_2H_3Cl	1	$-12{,}310$
trans-1,2-Dichloroethylene	$C_2H_6Cl_2$	1	-6350
Trichloroethylene	C_2HCl_3	1	-4750
Tetrachloroethylene	C_2Cl_4	1	-4900
3-Chloro-1-propene	C_3H_5Cl	1	$10{,}420$
Chlorobenzene	C_6H_5Cl	1	$23{,}700$
p-Dichlorobenzene	$C_6H_5Cl_2$	1	$18{,}440$
Hexachlorobenzene	C_6Cl_6	1	$10{,}560$

[a]Selected mainly from F. D. Rossini, ed., "Selected Values of Physical and Thermodynamic Properties of Hydrocarbons and Related Compounds," *American Petroleum Institute Research Project*, 44, Carnegie Institute of Technology, Pittsburgh, PA, 1953; F. D. Rossini, D. D. Wagman, W. H. Evans, S. Levine, and I. Jaffee, "Selected Values of Chemical Thermodynamic Properties," *Natl. Bur. Stand. Circ.* 500, 1952; also, from personal notes, L. Theodore and J. Reynolds, 1986.

ILLUSTRATIVE EXAMPLE 13.3

What is the standard free energy of formation of benzene (g)?

SOLUTION: Refer to Table 13.1,

$$\Delta G^0_f = +30{,}989 \, \text{cal/gmol}$$ ■

STANDARD FREE ENERGY OF REACTION, ΔG^0

Consider the following formation reaction:

$$\underset{G_{CO_2}}{\overset{G_C \quad G_{O_2}}{C + O_2 \longrightarrow CO_2}}$$ (13.1)

The free energy change for this reaction is

$$\Delta G = G_{CO_2} - G_C - G_{O_2} \tag{13.2}$$

If applied to a standard state, Equation (13.2) becomes

$$\Delta G^0 = G^0_{CO_2} - G^0_C - G^0_{O_2} \tag{13.3}$$

By definition, for elements in their standard state,

$$G^0_C = G^0_{O_2} = 0 \tag{13.4}$$

so that

$$\Delta G^0_{f,CO_2} = G^0_{CO_2} \tag{13.5}$$

Thus, one can conclude that the standard free energy of formation for a compound is equal to the standard free energy of that compound. This development is now applied to the authors' favorite equilibrium reaction

$$aA + bB = cC + dD \tag{13.6}$$

where $A, B, C, D =$ chemical formulas of the reactant and product species

$a, b, c, d =$ stoichiometric coefficients

and the equal sign is a reminder that the reacting system is at equilibrium. For this reaction (as with enthalpy),

$$\Delta G^0 = cG^0_C + dG^0_D - aG^0_A - bG^0_B \tag{13.7}$$

where ΔG^0 now represents the free energy change for this reaction when reactants and products are in their standard states. Note that ΔG^0 may be calculated and obtained in a manner similar to ΔH^0 (refer to Chapter 10).

Equation (13.7) allows the standard free energy of reaction to be calculated for any reaction provided standard free energy of formation data is available and the standard state is specified. As with the standard enthalpy of formation, the standard state is almost always 25°C (298K) and 1 atm. Thus,

$$\Delta G^0_{298} = c(\Delta G^0_f)_C + d(\Delta G^0_f)_D - a(\Delta G^0_f)_A - b(\Delta G^0_f)_B \tag{13.8}$$

The reader should note that rather than try to list the free energy change for all known reactions, it has been found much more efficient to list the free energy of formation (see Table 13.1) for all known compounds for which the data are available. In this way, the free energy change for any reaction, known or unknown, can be calculated, providing that the free energies of formation are known for all the participating reactants and products.

ILLUSTRATIVE EXAMPLE 13.4

Consider for following reaction:

$$2HCl + \tfrac{1}{2}O_2 \longrightarrow Cl_2 + H_2O(g)$$

Calculate the ΔG^0_{298} for this reaction.

SOLUTION: First calculate ΔG^0_{298} for the reaction using Equation (13.8).

$$\Delta G^0_{298} = 1\Delta G^0_{f,H_2O} - 2\Delta G^0_{f,HCl}$$

Terms $\Delta G^0_{f,H_2O}$ and $\Delta G^0_{f,HCl}$ can be obtained from Table 13.1.

$$\Delta G^0_{f,H_2O} = -54{,}635 \text{ cal/gmol}$$
$$\Delta G^0_{f,HCl} = -22{,}778 \text{ cal/gmol}$$

Therefore

$$\Delta G^0_{298} = (1)(-54{,}635) - (2)(-22{,}778)$$
$$= -9079 \text{ cal/gmol}$$

The reader is left the exercise of showing that

$$\Delta H^0_{298} = -13{,}672 \text{ cal/gmol}$$ ■

ILLUSTRATIVE EXAMPLE 13.5

Calculate the standard free energy of reaction for

$$CO_2(g) + CH_4(g) = CH_3COOH(1)$$

SOLUTION: For this reaction

$$\Delta G^0_{298} = (\Delta G^0_f)_{CH_3COOH(l)} - (\Delta G^0_f)_{CH_4(g)} - (\Delta G^0_f)_{CO_2(g)}$$

Substituting values from Table 13.1:

$$\Delta G^0_{298} = -93,800 - (-12,140) - (-94,258)$$
$$= -93,800 + 12,140 + 94,258$$
$$= +12,598 \text{ cal/gmol}$$

■

THE CHEMICAL REACTION EQUILIBRIUM CONSTANT, K

Most reactions have an opposing (or reverse) reaction. The rates of the forward and reverse reactions determine the equilibrium distribution of the reactants and products. Consider once again the authors' favorite elementary reversible reaction

$$A + B \rightleftharpoons C + D \tag{13.9}$$

The elementary rate law of the forward reaction for A is

$$r_A = -k_A c_A c_B \tag{13.10}$$

where k_A is the forward reaction velocity constant for the forward reaction. For the reverse reaction

$$r'_A = -k'_A c_C c_D \tag{13.11}$$

where k'_A is the reaction velocity constant for the reverse reaction. At equilibrium, the rate of production of A is just equal to the rate of consumption or disappearance of A so that, at equilibrium, the rates of production and consumption are equal, i.e.,

$$r_A = r'_A \tag{13.12}$$

Combining Equations (13.10), (13.11), and (13.12) gives

$$k_A c_A c_B = k'_A c_C c_D \tag{13.13}$$

or

$$\frac{k}{k'} = \frac{c_C c_D}{c_A c_B} \tag{13.14}$$

The ratio of these rate constants is defined as the *chemical reaction equilibrium* (CRE) constant K based on concentration, i.e.,

$$K_c = \frac{k}{k'} = \frac{c_C c_D}{c_A c_B} \tag{13.15}$$

This equilibrium constant may be measured experimentally, or obtained directly from rate constants. The reader should note that the equilibrium constant above is based on *concentrations*. It is very specific and is limited in application to some liquid phase reactions.

The generally accepted all-purpose equilibrium constant K (also referred to as the *true* equilibrium constant) can be shown to be related to the previously developed standard free energy of reaction.

The following equation is used to calculate the chemical reaction equilibrium constant K at a temperature T:

$$\Delta G_T^0 = -RT \ln K \tag{13.16}$$

The value of this equilibrium constant depends on the temperature at which the equilibrium is established. The effect of temperature on K must now be examined. This is addressed in the next three sections.

ILLUSTRATIVE EXAMPLE 13.6

Calculate the chemical reaction equilibrium constant K for a reaction that has a ΔG^0 value of $-20.0\,\text{kcal/gmol}$ at $70°F$.

SOLUTION: From Equation (13.16)

$$\ln K = -\frac{\Delta G_T^0}{RT}$$

with $T = 70°F = 294K$.
    For $\Delta G^0 = -20.0\,\text{kcal/gmol}$

$$\ln K = -\frac{(-20,000)}{(1.99)(294)}$$

$$= +34.2$$

$$K = 7.02 \times 10^{14}$$

This low value of K suggests an extremely high conversion of reactants. Details on conversion calculations are provided in the next chapter.   ∎

ILLUSTRATIVE EXAMPLE 13.7

Consider the reaction

$$CO_2 = CO + \tfrac{1}{2}O_2$$

Calculate the chemical reaction equilibrium constant, K, at standard conditions.

SOLUTION: As noted earlier,

$$\Delta G^0 = \Delta G^0_{f298} \ (\text{products}) - \Delta G^0_{f298} \ (\text{reactant})$$

For this reaction,

$$\Delta G^0 = \left[\underset{CO}{\Delta G^0_{f298}} + \tfrac{1}{2} \underset{O_2}{\overset{0}{\Delta G^0_{f298}}} \right] - \underset{CO_2}{\Delta G^0_{f298}}$$

From Table 13.1

$$\Delta G^0 = [-32{,}781 + 0 - 94{,}258]$$

$$\Delta G^0 = 61{,}477 \ \text{cal/gmol}$$

Since

$$\Delta G^0 = -RT \ln K$$

$$\ln K = \frac{61{,}477}{-1.987(298)}$$

$$= -103.824$$

$$= 8.125 \times 10^{-46}$$

$$K \cong 0.0000$$

As one would expect, since K is "essentially" zero, the reaction will "essentially" not take place at 298K.   ■

EFFECT OF TEMPERATURE ON ΔG^0 AND K: SIMPLIFIED APPROACH

Starting with the definition of the free energy, it can be shown that[1,2]

$$dG = V \, dP - S \, dT \qquad (13.17)$$

This equation then reduces to (at standard conditions)

$$(dG)_P = -S \, dT$$

$$\left(\frac{\partial G}{\partial T} \right)_P = -S$$

$$\left(\frac{\partial \Delta G}{\partial T} \right)_P = -\Delta S$$

$$\left(\frac{\partial \Delta G^0}{\partial T} \right)_P = -\Delta S^0 \qquad (13.18)$$

In addition, from the definition of G,

$$G = H - TS \tag{13.19}$$

At standard state conditions,

$$G^0 = H^0 - TS^0 \tag{13.20}$$

For two different states,

$$\Delta G^0 = \Delta H^0 - T\,\Delta S^0; \quad \text{constant } T \tag{13.21}$$

Rearranging Equation (13.21)

$$\Delta S^0 = \frac{\Delta H^0 - \Delta G^0}{T} \tag{13.22}$$

Combining Equations (13.18) and (13.22) leads to

$$\frac{d(\Delta G^0)}{dT} - \frac{\Delta G^0}{T} = -\frac{\Delta H^0}{T} \tag{13.23}$$

Equation (13.23) may also be written as

$$\frac{d(\Delta G^0/T)}{dT} = -\frac{\Delta H^0}{T^2} \tag{13.24}$$

This equation provides the dependence of ΔG^0 on temperature. Since

$$-\frac{\Delta G^0}{RT} = \ln K \tag{13.16}$$

substitution into Equation (13.24) ultimately yields

$$\frac{d\ln K}{dT} = \frac{\Delta H^0}{RT^2} \tag{13.25}$$

Equation (13.25) shows the effect of temperature on the equilibrium constant, and hence on the equilibrium yield (to be discussed in the next chapter). If the reaction is exothermic, ΔH^0 is negative and the equilibrium constant decreases with an increase in temperature; for an endothermic reaction, the equilibrium constant increases with an increase in temperature. Note that both Equations (13.24) and (13.25) apply on a no-strings-attached basis.

If the term ΔH^0, which is the standard enthalpy of reaction (heat of reaction defined in Chapter 10), is assumed to be constant with temperature, Equation (13.25) can be integrated between the temperatures T and T_1 to give

$$\ln\frac{K_T}{K_1} = -\frac{\Delta H^0}{R}\left(\frac{1}{T} - \frac{1}{T_1}\right) \tag{13.26}$$

This approximate equation may be used to determine the equilibrium constant K at a temperature T from a known value of K at some other temperature T_1 if the difference between the two temperatures is small. If the standard heat of reaction is known as a function of temperature, however, Equation (13.25) can be integrated rigorously. Information for this situation follows.

ILLUSTRATIVE EXAMPLE 13.8

Prove that the enthalpy of reaction for any chemical reaction is independent of temperature when ΔC_P for the reaction equation is zero.

SOLUTION: As derived in Chapter 10,

$$\Delta H_T^0 = \Delta H_{298}^0 + \int_0^T (\Delta C_P)\, dT$$

Since $\Delta C_P = 0$, the above equation reduces to

$$\Delta H_T^0 = \Delta H_{298}^0$$

In effect, the standard enthalpy of reaction at any temperature T is given by ΔH_{298}^0 and this does not vary with temperature. ∎

EFFECT OF TEMPERATURE ON ΔG^0 AND K: α, β, AND γ DATA

Equation (13.25) can be rearranged and integrated to give

$$\ln K = \frac{1}{R}\int \frac{\Delta H^0}{T^2}\, dT + I \tag{13.27}$$

where I is a constant of integration.

It was shown in Chapter 10 that if the molar heat capacity for each chemical species taking part in the reaction is known and can be expressed as a power series in T (Kelvin) of the form [see Equation (10.13)]

$$C_P = \alpha + \beta T + \gamma T^2$$

then ΔH^0 at a given temperature T becomes [see Equation (10.16)]:

$$\Delta H_T^0 = \Delta H_0 + \Delta\alpha T + \frac{\Delta\beta T^2}{2} + \frac{\Delta\gamma T^3}{3}$$

In this equation, the constant ΔH_0 can be calculated from the standard heat of reaction at 25°C. With ΔH_0 determined, ΔH_T^0 can be substituted into Equation (13.27) and integrated to obtain

$$\ln K = -\frac{\Delta H_0}{RT} + \frac{\Delta \alpha}{R} \ln T + \frac{\Delta \beta}{2RT} + \frac{\Delta \gamma}{6R} T^2 + I \tag{13.28}$$

Here the constant I may be evaluated from a knowledge of the equilibrium constant at one temperature. This is usually obtained from standard free energy of formation data at 25°C. Equation (13.16) is employed to obtain K at this temperature. A similar equation for ΔG_T^0 may be obtained by combining Equations (13.16) and (13.28):

$$\Delta G_T^0 = \Delta H_0 - (\Delta \alpha)T \ln T - \left(\frac{\Delta \beta}{2}\right) T^2 - \left(\frac{\Delta \gamma}{6}\right) T^3 - IRT \tag{13.29}$$

ILLUSTRATIVE EXAMPLE 13.9

Consider the reaction

$$CO_2 \rightleftharpoons CO + \tfrac{1}{2}O_2$$

Using α, β, and γ heat capacity data, develop an expression for $\ln(K)$ as a function of the temperature.

SOLUTION: The chemical reaction equilibrium constant (K) can be evaluated using Equation (13.29) after replacing ΔG_T^0 by $-RT \ln K$.

$$\ln K = -\Delta H_0/RT + (\Delta \alpha)T \ln T + (\Delta \beta/2R) T + (\Delta \gamma/6R) T^2 + I$$

Following the procedure set forth in Chapter 10 for this reaction, the following can be deduced from Table 7.5.

Refer to Table 7.4 to find heat capacity data of the form

$$C_P = \alpha + \beta T + \gamma T^2$$

Substituting

$$\Delta \alpha = \left[\tfrac{1}{2}(6.148) + 6.420\right] - [6.214] = 3.28$$

$$\Delta \beta = \left[\tfrac{1}{2}(3.102 \times 10^{-3}) + (1.665 \times 10^{-3})\right] - \left[10.396 \times 10^{-3}\right] = -0.00718$$

$$\Delta \gamma = \left[\tfrac{1}{2}(-0.923 \times 10^{-6}) + (-0.196 \times 10^{-6})\right] - \left[-3.545 \times 10^{-6}\right] = 2.888 \times 10^{-6}$$

From Table 13.1 or from Illustrative Example 13.7,

$$\Delta G^0_{298} = 61{,}447 \, \text{cal/gmol}$$

Equation (10.18) is employed to calculate the constant, ΔH_0

$$\Delta H_0 = \Delta H_T - \Delta \alpha T - \frac{\Delta \beta}{2} T^2 - \frac{\Delta \gamma}{3} T^3$$

Substituting the above,

$$\Delta H_0 = 67{,}635 \, \text{cal/gmol} - (3.28)(298) - \left(\frac{-0.00718}{2} \right)(298)^2 - \left(\frac{2.888 \times 10^{-6}}{3} \right)(298)^3$$

$$= 66{,}951 \, \text{cal/gmol}$$

K_{298} may now be determined from

$$\ln K_T = -\frac{\Delta G^0_T}{RT}$$

$$= -\frac{61{,}447 \, \text{cal/gmol}}{(1.987 \, \text{cal/gmol} \cdot \text{K})(298 \text{K})} \tag{13.16}$$

$$\ln K_{298\text{K}} = -103.8$$

The constant "I" may now be evaluated in Equation (13.28),

$$I = \ln K + \frac{\Delta H_0}{RT} - \frac{\Delta \alpha}{R} \ln T - \frac{\Delta \beta}{2R} T - \frac{\Delta \gamma}{6R} T^2$$

Substituting

$$I = (-103.8) + \frac{66{,}951}{(1.987)(298)} - \left(\frac{3.28}{1.987} \right) \ln (298)$$

$$- \left(\frac{-0.00718}{2(1.987)} \right)(298) - \left(\frac{2.888 \times 10^{-6}}{6(1.987)} \right)(298)^2$$

$$= 0.4509$$

Thus, the describing equation becomes:

$$\ln(K) = -\frac{33{,}722}{T} + (1.560) \ln T - (0.00181)T + (2.42 \times 10^{-7})T^2 + 0.4509$$

■

ILLUSTRATIVE EXAMPLE 13.10

Refer to Illustrative Example 13.9. Calculate K at 2050°F.

SOLUTION: Convert the temperature to K:

$$2050°F = 1394.3K$$

Substitute $T = 1394.3K$

$$\ln K = -(33{,}722/T) + (1.560)\ln T - (0.00181)T + 2.42 \times 10^{-7}T^2 + 0.4509$$
$$= -14.493$$

$$K = 5.0784 \times 10^{-7}$$

This low value suggests an extremely low conversion of CO_2 to CO.   ∎

EFFECT OF TEMPERATURE ON ΔG^0 AND K: a, b, AND c DATA

The reader is left the exercise of developing companion equations to Equations (13.28) and (13.29) if the heat capacity variation with temperature is given by [see Equation (10.20)]:

$$C_P = a + bT + cT^{-2}$$

For this case, the final results take the form:

$$\ln K = -(\Delta H_0/RT) + (\Delta a/R)\ln T + (\Delta b/2R)T + (\Delta c/2R)T^{-2} + I \qquad (13.30)$$

and

$$\Delta G_T^0 = \Delta H_0 - (\Delta a)T\ln T - (\Delta b/2)T^2 - (\Delta c/2)/T - IRT \qquad (13.31)$$

with (once again)

$$\Delta G_T^0 = -RT \ln K \qquad (13.16)$$

ILLUSTRATIVE EXAMPLE 13.11

The equilibrium constant K for the reaction (at 1 atm)

$$2HCl(g) + 0.5O_2 \longrightarrow Cl_2 + H_2O(g)$$

may be expressed in the form[3-5)

$$K = Ae^{B/T}$$

where $A = 0.229 \times 10^{-3}$

  $B = 7340$

  $T =$ absolute temperature (K)

As part of a thermodynamics project, you are required to develop a more rigorous equation for K as a function of the absolute temperature T.

SOLUTION: The following two results are provided from Illustrative Example 13.4:

$$\Delta G^0_{298} = -9079 \text{ cal/gmol}$$

$$\Delta H^0_{298} = -13{,}672 \text{ cal/gmol}$$

Employ Equation (13.30) so that

$$\ln K = \frac{-\Delta H_0}{RT} + \frac{\Delta \alpha}{R} \ln T + \frac{\Delta b}{2R} T + \frac{\Delta c}{2R} T^{-2} + I$$

Next, ΔH_0 and I must be determined. ΔH_0 is found by a procedure similar to that used in Chapter 10 [see Equation (10.11)]:

$$\Delta H^0_T = \Delta H^0_{298} + \int_{298}^{T} \Delta C_P \, dT$$

For heat capacities of the form [see Equation (10.20)]:

$$C_P = a + bT + cT^{-2}$$

Table 7.4 can be employed to generate the following terms:

$$\Delta a = (7.30 + 8.85) - [(2)(6.27) + (0.5)(7.16)]$$
$$= 0.03$$
$$\Delta b = (2.46 \times 10^{-3} + 0.16 \times 10^{-3}) - [(2)(1.24 \times 10^{-3}) + (0.5)(1.0 \times 10^{-3})]$$
$$= -3.6 \times 10^{-4}$$
$$\Delta c = (0.0 - 0.68 \times 10^5) - [(2)(0.30 \times 10^5) + (0.5)(-0.4 \times 10^5)]$$
$$= -1.08 \times 10^5$$

From this, Equation (10.22) then becomes:

$$\Delta H_T^0 = \Delta H_{298}^0 + \int_{298}^{T} [\Delta a + (\Delta b)T + (\Delta c)T^{-2}]\,dT$$

or

$$\Delta H_T^0 = \Delta H_{298}^0 + \Delta a(T - 298) + \frac{1}{2}\Delta b[T^2 - (298)^2] - \Delta c\left(\frac{1}{T} - \frac{1}{298}\right)$$

Combining the constant terms into ΔH_0 (as in Chapter 10) yields the following:

$$\Delta H_T^0 = \Delta H_0 + (\Delta a)T + \frac{1}{2}(\Delta b)T^2 - (\Delta c)T^{-1}$$

where

$$\Delta H_0 = \Delta H_{298}^0 - 298\Delta a - \frac{1}{2}(298)^2\Delta b + \frac{1}{298}\Delta c$$

$$= 13{,}682 - 298(0.03) - \frac{1}{2}(298)^2(-3.6 \times 10^{-4}) + \frac{1}{298}(-1.08 \times 10^5)$$

$$= 14{,}027 \text{ cal/gmol}$$

From Equation (13.16)

$$\ln K = \frac{-\Delta G_{298}^0}{RT} = \frac{9079}{(1.99)(298)}$$

$$= 15.31$$

Therefore,

$$15.31 = \frac{14{,}027}{(1.99)(298)} + \frac{0.03}{1.99}(\ln 298) + \frac{-3.6 \times 10^{-4}}{(2)(1.99)}(298)$$

$$+ \frac{-1.08 \times 10^5}{(2)(1.99)}(298)^{-2} + I$$

Solving for I,

$$I = -8.09$$

The final form of the equation for K is

$$\ln K = \frac{7048.7}{T} + 0.0151(\ln T) - 9.06 \times 10^{-5}T - 2.714 \times 10^4(T)^{-2} - 8.09 \qquad \blacksquare$$

ILLUSTRATIVE EXAMPLE 13.12

Refer to Illustrative Example 13.11. Calculate K at 500K.

SOLUTION: Substitute into the equation

$$\ln K = \frac{7048.7}{T} + 0.0151(\ln T) - 9.06 \times 10^{-5}T - 2.714 \times 10^{4}(T)^{-2} - 8.09$$

For $T = 500$K

$$\ln K = 14.1 + 0.09 - 0.05 - 0.11 - 8.09$$
$$= 6.05$$
$$K = 424$$

Obviously, the (forward) reaction is favored at low temperatures. ∎

PROCEDURES TO DETERMINE *K*

There are basically nine ways to calcutate the chemical reaction equilibrium constant K_T at a given temperature. Obviously, two pieces of information are required if one employs Equations (13.28) or (13.31); in effect, there is a need to calculate ΔH_0 and I. The most convenient method, and most often used, is by determining ΔH^0 at one temperature, T, and ΔG^0 at another. This is usually accomplished by obtaining values of ΔH^0 and ΔG^0 at 298K. These are, as well as other procedures requiring two pieces of information, summarized in Table 13.2.

Table 13.2 Information Required to Determine *K* at any Temperature

1. ΔH^0_{298} and ΔG^0_{298}
2. ΔH^0_{T} and ΔG^0_{298}
3. ΔH^0_{298} and ΔG^0_{T}
4. ΔH^0_{298} and K_{298}
5. ΔH^0_{298} and K_{T}
6. ΔH^0_{T} and K_{298}
7. ΔH^0_{298} and K_{298}
8. K_{298} and K_{T}
9. K_{T1} and K_{T2}

In addition, K can be evaluated if actual equilibrium concentration (or the equivalent) data is available. This is treated in the next chapter. Finally, an approach is available that is based on the standard entropy of reaction; details for this method are available in the literature.[2]

ILLUSTRATIVE EXAMPLE 13.13

A scientist discovers that there is no heat capacity data for acetaldehyde. In order to derive an empirical relation in the form

$$C_P = \alpha + \beta T + \gamma T^2$$

experiments are carried out using the reaction

$$2C_2H_4 + O_2 \longrightarrow 2C_2H_4O$$

at 298K, where all the usual ΔH^0_{298} and ΔG^0_{298} values are known for all components. Heat capacity data for C_2H_4 and O_2 are also known. By previous experiments, the scientist knows that *K* at 600K and 1 atm equals 1.5×10^{32}. She also knows that $\Delta H_0 = -103,525$ cal/gmol O_2 and $IR = 3.5$ for the above reaction. Calculate α, β, and γ for acetaldehyde and compare with the literature values for this compound. The available heat capacity data for the two reactants are given in Table 13.3.

Table 13.3   Heat Capacity Data

	α	$\beta \times 10^3$	$\gamma \times 10^6$
C_2H_4	2.830	28.601	-8.726
O_2	6.148	3.102	-0.923

SOLUTION: Note that there are three unknowns, and information on three conditions are specified. Calculate the standard enthalpy of reaction, ΔH^0, at 298K in cal/gmol O_2 reacted:

$$\Delta H^0_{298} = 2(-39,760) - 2(12,496)$$
$$= -104,512 \text{ cal/gmol } O_2 \text{ reacted}$$

Calculate the standard free energy of reaction, ΔG^0, at 298K in cal/gmol:

$$\Delta G^0_{298} = -96,484 \text{ cal/gmol } O_2 \text{ reacted}$$

Write the equation for ΔH^0 at 298K in terms of ΔH_0, $\Delta \alpha$, $\Delta \beta$, and $\Delta \gamma$:

$$\Delta H^0_T = \Delta H_0 + (\Delta \alpha)T + (\Delta \beta/2)T^2 + (\Delta \gamma/3)T^3 \qquad (10.18)$$

At $T = 298$K and $\Delta H_0 = -103,525$ cal/gmol

$$-987 = 298(\Delta \alpha) + 44,402(\Delta \beta) + 8.82 \times 10^6(\Delta \gamma)$$

Write the equation for ΔG^0 at 298K in terms of ΔH_0, $\Delta \alpha$, $\Delta \beta$, $\Delta \gamma$, and IR. At $T = 298$K and $IR = 3.5$,

$$\Delta G_T^0 = \Delta H_0 - (\Delta \alpha)(T) \ln T - (\Delta \beta / 2)T^2 - (\Delta \gamma / 6)T^3 - IRT$$
$$-8084 = 1698(\Delta \alpha) + 44{,}402(\Delta \beta) + 4.41 \times 10^6 (\Delta \gamma)$$

(13.29)

Also write the equation for ΔG^0 at 600K in terms of ΔH_0, $\Delta \alpha$, $\Delta \beta$, $\Delta \gamma$, and IR. Since

$$K = 1.5 \times 10^{32} \text{ at } 600K$$

and

$$\Delta G_T^0 = -RT \ln K$$

one obtains

$$\Delta G_{600}^0 = -88{,}350$$

At $T = 600$K,

$$-17{,}275 = 3839(\Delta \alpha) + 1.8 \times 10^5 (\Delta \beta) + 3.6 \times 10^7 (\Delta \gamma)$$

Calculate $\Delta \alpha$, $\Delta \beta$, and $\Delta \gamma$ by solving the above three equations simultaneously.

$$\Delta \alpha = -5.046$$
$$\Delta \beta = 1.017 \times 10^{-2}$$
$$\Delta \gamma = 7.406 \times 10^{-6}$$

Calculate α, β, and γ for acetaldehyde.

For α:

$$\Delta \alpha = -5.046 = 2\alpha - 2(2.830) - 6.148$$
$$\alpha = 3.380$$

For β:

$$\Delta \beta = 10.17 \times 10^{-3} = 2\beta - (2)(28.601 \times 10^{-3}) - 3.102 \times 10^{-3}$$
$$\beta = 35.24 \times 10^{-3}$$

For γ:

$$\Delta \gamma = 7.406 \times 10^{-6} = 2\gamma - (2)(-8.726 \times 10^{-6}) + 0.923 \times 10^{-6}$$
$$\gamma = -5.485 \times 10^{-6}$$

∎

ILLUSTRATIVE EXAMPLE 13.14

Comment on the results of the previous example.

SOLUTION: Experimental values for α, β, and γ are listed below (see Table 7.5)

$$\alpha = 3.364$$
$$\beta = 35.722 \times 10^{-3}$$
$$\gamma = -12.236 \times 10^{-6}$$

The agreement for α and β is excellent; the result is reasonable for γ in view of its sensitivity to T^3. ∎

REFERENCES

1. D. GREEN and R. PERRY, *"Perry's Chemical Engineers' Handbook,"* 8th edition, McGraw-Hill, New York, 2008.
2. J. SMITH, H. VAN NESS and M. ABBOTT, *"Introduction to Chemical Engineering Thermodynamics,"* 6th edition, McGraw-Hill, New York, 2001.
3. L. THEODORE: personal notes, Manhattan College, 1985.
4. J. REYNOLDS, J. JERIS, and L. THEODORE, *"Handbook of Chemical and Environmental Engineering Calculations,"* John Wiley & Sons, Hoboken, NJ, 2004.
5. J. SANTOLERI, J. REYNOLDS, and L. THEODORE, *"Introduction to Hazardous Waste Incineration,"* 2nd edition, John Wiley & Sons, Hoboken, NJ, 2000.

NOTE: Additional problems for each chapter are available for all readers at www. These problems may be used for additional review or homework purposes.

Chapter **14**

Chemical Reaction Equilibrium Applications

William Shakespeare [1564–1616]

Against that time when thou shalt strangely pass,
And scarcely greet me with that sun, thine eye,
When love, converted from the thing it was,
Shall reasons find of settled gravity.

—Sonnet 49

INTRODUCTION

Thermodynamics provides the practicing engineer with two vital pieces of information regarding chemical reactions:

1 the enthalpy (heat) effects associated with the reaction, and
2 the maximum extent to which the reaction may proceed.

Topic (1) received treatment in Part II. This last chapter of Part III addresses subject (2).

Industry is generally interested in the attainment of a product. When the synthesis of this product is demonstrated to be possible, the interest from a thermodynamics perspective centers on what has come to be defined as the extent of reaction, or alternatively, the yield of the product. And yes, thermodynamics provides significant useful information on the extent and/or yield associated with the reaction.

This final chapter of Part III—Equilibrium Thermodynamics—is concerned with Chemical Reaction Equilibrium Applications. As noted in Chapter 13, the presentation provides information on the extent of reactions for equilibrium conditions. The main objective is to apply the principles developed earlier in a manner that provides extent of reaction information in simple and understandable terms so that it can be used to perform additional engineering/thermodynamic calculations.

Thermodynamics for the Practicing Engineer. By L. Theodore, F. Ricci, and T. Van Vliet
Copyright © 2009 by John Wiley & Sons, Inc.

Topics covered in this chapter are:

1 Rate vs Equilibrium Considerations

2 Extent of Reaction

3 The Reaction Coordinate

4 Gas Phase Reactions

5 Equilibrium Conversion Calculations: Simplified Approach

6 Equilibrium Conversion Calculations: Rigorous Approach

7 Other Reactions

It should be noted that the development and applications primarily address gas phase reactions because of the introductory nature of this book. Readers desiring more and/or additional information are referred to the references at the end of this chapter.

RATE vs EQUILIBRIUM CONSIDERATIONS

With regard to chemical reactions, there are two important questions that are of concern to the engineer:

1 how *far* will the reaction go, and

2 how *fast* will the reaction go?

Chemical thermodynamics provides the answer to the first question; however, it provides nothing about the second. Reaction rates fall within the domain of chemical kinetics[1] and will not be treated in this text.

To illustrate the difference and importance of both of the above questions on an engineering analysis of a chemical reaction, consider the following process.[2,3] Substance *A*, which costs 1 cent/ton, can be converted to *B*, which costs $1 million/lb, by the reaction $A \leftrightarrow B$. Chemical thermodynamics will provide information on the maximum amount of *B* that can be formed. If 99.99% of *A* can be converted to *B*, the reaction would then appear to be economically feasible, from a *thermodynamic* point-of-view. However, a *kinetic* analysis might indicate that the reaction is so slow that, for all practical purposes, its rate is vanishingly small. For example, it might take 10^6 years to obtain a $10^{-6}\%$ conversion of *A*. The reaction is then economically unfeasible. Thus, it can be seen that both equilibrium and kinetic effects must be considered in an overall engineering analysis of a chemical reaction. The same principle applies to gaseous mass transfer separation (e.g., absorption), control of gaseous pollutants by combustion,[4] and incineration of hazardous waste.[3]

Equilibrium and rate are therefore both extremely important factors to be considered in the design and the prediction of performance of equipment employed for chemical reactions. The rate at which a reaction proceeds will depend on the departure from equilibrium, with the rate at which equilibrium is established essentially dependent on a host of factors. As expected, this rate process ceases upon the attainment of equilibrium.

As noted in Chapter 13, if a chemical reaction is conducted in which reactants go to products, the products will be formed at a rate governed (in part) by the concentration of the reactants and conditions such as temperature and pressure. Eventually, as the reactants form products and the products react to form reactants, the *net* rate of reaction must equal zero. At this point, equilibrium will have been achieved.

EXTENT OF REACTION

The chemical industry is usually concerned with the attainment of a product or products. Chemical reaction equilibrium principles allow the engineer/scientist to determine the end-products of a chemical reaction for a given set of operating conditions and initial reactant(s) if the final state is at equilibrium. However, from the standpoint of obtaining sufficient product(s) of economic value, a final state of equilibrium is almost always undesirable. It should be noted that the environmental industry centers on the destruction (or removal) of a reactant, often referred to as a waste.

Once a reaction is completed, or comes to rest (equilibrium), interest then centers on the so-called extent of the reaction. The extent can be expressed in terms of either conversion or yield. Numerous definitions appear for both conversion and yield.

In order to define the commonly used conversion term, one of the reactants is chosen as the basis of the calculation and then the other species involved in the reaction are related to that basis. In most instances, it is best to choose the limiting reactant (see Chapter 5) as the basis of calculation.

Consider the authors' favorite reaction equation:

$$aA + bB \longrightarrow cC + dD \tag{14.1}$$

The upper-case letters once again represent chemical species and the lower-case letters represent stoichiometric coefficients. Taking species A as the *basis of calculation*, the reaction expression is divided through by the stoichiometric coefficient of species A in order to arrange the reaction expression in the following form:

$$A + \frac{b}{a}B \longrightarrow \frac{c}{a}C + \frac{d}{a}D \tag{14.2}$$

This essentially places every quantity on a "per mole of A" basis. The conversion variable X_A can then be defined as the number of moles of A that have reacted per mole of A fed to the system, i.e.,

$$X_A = \frac{\text{moles of } A \text{ reacted}}{\text{moles of } A \text{ fed}} \tag{14.3}$$

Consider the two simultaneous and/or multiple reactions

$$aA + bB \longrightarrow cC + dD$$

that are now accompanied by a second (side or consecutive) reaction,

$$gA + fC \longrightarrow eE + fF \tag{14.4}$$

Since species A appears as the reactant in both reactions, it is chosen as the basis of calculation, and both equations are divided by their respective stoichiometric coefficients of A:

$$\text{Reaction 1: } A + \frac{b}{a}B \longrightarrow \frac{c}{a}C + \frac{d}{a}D$$

$$\tag{14.5}$$

$$\text{Reaction 2: } A + \frac{f}{g}C \longrightarrow \frac{e}{g}E$$

There are two conversion variables in this case:

$$X_{A1} = \frac{\text{moles of } A \text{ consumed by Reaction 1}}{\text{initial moles of } A} \tag{14.6}$$

This defines the conversion of A to form C and D in Reaction 1. Similarly, X_{A2} is defined as the conversion of A to form E and F in Reaction 2:

$$X_{A2} = \frac{\text{moles of } A \text{ consumed by Reaction 2}}{\text{initial moles of } A} \tag{14.7}$$

This topic receives additional treatment in the last section of the chapter.

The extent of reaction can also be defined or specified in terms of the yield. The yield is a common term that is employed in industry since it provides information on the "yield" of a product. Although the yield term has been defined in many ways, it is usually expressed as

1 the quantity of a product formed divided by maximum quantity of the product that can be formed, or

2 the quantity of a product formed divided by the quantity one of the reactants.

One of the authors[5] has defined other conversion variables that can be employed to describe the extent of the reaction. These are listed below for reactant/species A without any explanatory details:

1 $\alpha_A = $ moles of A reacted

2 $\alpha_A^* = $ mass of A reacted

3 $X_A^* = \dfrac{\text{mass of } A \text{ reacted}}{\text{initial mass of } A}$

4 $X_A' = \dfrac{\text{mass of } A \text{ reacted}}{\text{initial moles of } A}$

5 $X_A'' = \dfrac{\text{moles of } A \text{ reacted}}{\text{initial mass of } A}$

For equilibrium reactions, as with the case in this chapter, the most convenient approach is to employ a reaction coordinate. This receives treatment in the next section.

Finally, it should be noted that the highest conversion that can be achieved in reversible reactions is the equilibrium conversion (which takes an infinite period of time to achieve). For endothermic reactions, the equilibrium conversion increases with increasing temperature up to a maximum of 1.0; for exothermic reactions the equilibrium conversion decreases with increasing temperature. The reader is cautioned that these equilibrium concentration calculations are, for most intents and purposes, a set of "fake" or "artificial" values. They almost always represent an upper limit on the expected concentration at the temperature in question. Other chemical reactions, kinetic effects, and temperature variations in the system may render these calculations valueless. Nonetheless, these calculations serve a useful purpose since they do provide a reasonable estimate of these concentrations.

THE REACTION COORDINATE

The reaction coordinate provides a development of maximum generality regarding the stoichiometry of a equilibrium reaction. The reaction equation on a general basis may be written as

$$\sum v_i R_i = 0 \tag{14.8}$$

where R_i represents all the participating chemical species and v_i represents the stoichiometric coefficients that are taken as positive for products and negative for reactants. The change in the number of mols, n_i, of each participating species i due to the chemical reaction is given by

$$\frac{dn_1}{v_1} = \frac{dn_2}{v_2} = \cdots = \frac{dn_i}{v_i} = d\varepsilon \tag{14.9}$$

All the ratios are therefore equal and equal to $d\varepsilon$. The term ε is defined as the reaction coordinate or the *extent of reaction*. Thus, ε alone can be used to describe the extent of a reaction, and equilibrium stoichiometric calculations can be performed based on ε.

As an example of this procedure, consider once again the authors' favorite reaction:

$$aA + bB \;\rightleftharpoons\; cC + dD \tag{14.1}$$

This may now be written as [see Equation (14.8)]:

$$v_1 R_1 + v_2 R_2 + v_3 R_3 + v_4 R_4 = 0$$

where $\quad v_1 = -a;\quad R_1 = A$
$\qquad\quad v_2 = -b;\quad R_2 = B$
$\qquad\quad v_3 = c;\quad\;\; R_3 = C$
$\qquad\quad v_4 = d;\quad\;\; R_4 = D$

In accordance with Equation (14.9), any change due to chemical reaction can be represented by

$$\frac{dn_1}{v_1} = \frac{dn_2}{v_2} = \frac{dn_3}{v_3} = \frac{dn_4}{v_4} = d\varepsilon$$

or

$$dn_i = v_i d\varepsilon \tag{14.11}$$

Equation (14.11) may be integrated to give

$$n_i - n_{io} = v_i \varepsilon \tag{14.12}$$

or

$$n_i = n_{io} + v_i \varepsilon \tag{14.13}$$

Note that the reaction coordinate term ε can be applied to both reactants and products.

ILLUSTRATIVE EXAMPLE 14.1

Consider the following:

$$\underset{(1)}{CO_2} + \underset{(2)}{3H_2} \;\longrightarrow\; \underset{(3)}{CH_3OH} + \underset{(4)}{H_2O}$$

If the initial number of moles of the participating species 1, 2, 3, and 4 are n_{10}, n_{20}, n_{30}, and n_{40}, respectively, express the moles of all species at equilibrium in terms of the reaction coordinate, ε.

SOLUTION: For this reaction

$$\begin{aligned} R_1 &= CO_2; & v_1 &= -1 \\ R_2 &= H_2; & v_2 &= -3 \\ R_3 &= CH_3OH; & v_3 &= +1 \\ R_4 &= H_2O; & v_4 &= +1 \end{aligned}$$

Therefore

$$n_1 = n_{10} - \varepsilon$$
$$n_2 = n_{20} - 3\varepsilon$$
$$n_3 = n_{30} + \varepsilon$$
$$n_4 = n_{40} + \varepsilon$$

■

ILLUSTRATIVE EXAMPLE 14.2

Refer to Illustrative Example 14.1. If the initial number of moles of CO_2, H_2, CH_3OH, and H_2O are 3, 1, 0, and 1, respectively, express the equilibrium number of moles of each participating species in terms of ε.

SOLUTION: For this case,

$$n_{10} = R_{10} = 3$$
$$n_{20} = R_{20} = 1$$
$$n_{30} = R_{30} = 0$$
$$n_{40} = R_{40} = 1$$

Thus,

$$n_1 = 3 - \varepsilon$$
$$n_2 = 1 - 3\varepsilon$$
$$n_3 = \varepsilon$$
$$n_4 = 1 + \varepsilon$$

The authors' method to solve this example is to write down the reaction equation and then insert the initial and final (equilibrium) number of moles above and below the reaction equation, respectively. Thus,

$$\overset{3}{CO_2} + \overset{1}{H_2} \longrightarrow \overset{0}{CH_3OH} + \overset{1}{H_2O} \quad \Sigma n_{io} = n_o = 5$$
$$\underset{3-\varepsilon}{} \quad \underset{1-3\varepsilon}{} \quad \underset{\varepsilon}{} \quad \underset{1+\varepsilon}{} \quad \Sigma n_i = n = 5 - 2\varepsilon$$

Note that this approach provides the initial (top) and the equilibrium (bottom) number of moles. ■

ILLUSTRATIVE EXAMPLE 14.3

Refer to Illustrative Example 14.2. Assuming that the reaction occurs at 1 atm pressure and some unknown temperature, express the equilibrium mole fraction of the four species in terms of ε.

SOLUTION: In line with the definition of mole fraction, and noting that the equilibrium number of moles is n, one can write (since $n = 5 - 2\varepsilon$)

$$y_1 = \frac{3 - \varepsilon}{5 - 2\varepsilon}$$

$$y_2 = \frac{1 - 3\varepsilon}{5 - 2\varepsilon}$$

$$y_3 = \frac{\varepsilon}{5 - 2\varepsilon}$$

$$y_4 = \frac{1 + \varepsilon}{5 - 2\varepsilon}$$

Note that temperature plays no role in this calculation. ∎

ILLUSTRATIVE EXAMPLE 14.4

Refer to Illustrative Example 14.3. Calculate the partial pressures of the participating species if the pressure is 0.5 atm. Assume ideal gas behavior and $\varepsilon = 0.3$.

SOLUTION: Based on the definition of partial pressure

$$p_i = y_i P$$

Since $P = 0.5$ atm

$$p_1 = y_1 P$$

$$= \left(\frac{3 - 0.3}{5 - 0.6}\right) 0.5$$

$$= 0.307 \text{ atm}$$

Similarly

$$p_2 = \left(\frac{1 - (3)(0.3)}{4.4}\right) 0.5$$

$$= 0.011 \text{ atm}$$

$$p_3 = \left(\frac{0.3}{4.4}\right) 0.5$$

$$= 0.034 \text{ atm}$$

$$p_4 = \left(\frac{1 + 0.3}{4.4}\right) 0.5$$

$$= 0.148 \text{ atm}$$

In accordance with the ideal gas law,

$$\sum p_i = P$$

The above results satisfy this equation.   ∎

GAS PHASE REACTIONS

This chapter primarily addresses gas phase reactions. Although a rigorous, detailed presentation of this topic is beyond the scope of this chapter, this treatment is presented in the hope that it may provide at least an introduction to chemical reaction equilibrium for gas phase reactions. This topic finds widespread application in engineering calculations. The basic problem that remains is to relate K to understandable physical quantities and, as will soon be shown for gas-phase reactions, the term K may be approximately represented in terms of the equilibrium partial pressures or concentrations of the participating species.

For homogeneous chemical reactions, it has been found to be convenient to introduce a new term referred to as the activity:

$$a_i = \frac{\hat{f}_i}{f_i^0} \tag{14.14}$$

The activity is simply the ratio of the fugacity component in a mixture, $\hat{f}_i$, to its value in the standard state f_i^0. (The fugacity is a measure of the escaping tendency of a liquid and is related to the vapor pressure.) For gases, the activity is equal to the component fugacity, since the fugacity in the standard state is arbitrarily assigned a value of unity. These activities can be shown to be related to the standard free energy of reaction. For the reaction

$$aA + bB \longrightarrow cC + dD \tag{14.1}$$

the relationship takes the form

$$\Delta G^0 = -RT(\ln \hat{a}_C^c + \ln \hat{a}_D^d - \ln \hat{a}_A^a - \ln \hat{a}_B^b) \tag{14.15}$$

or

$$\Delta G^0 = -RT \ln \frac{\hat{a}_C^c \hat{a}_D^d}{\hat{a}_A^a \hat{a}_B^b} \tag{14.16}$$

Equation (14.16) may also be written as

$$\Delta G^0 = -RT \ln K$$

where K is defined as

$$K = \frac{\hat{a}_C^c \hat{a}_D^d}{a_A^a a_B^b} \qquad (14.17)$$

It should be noted that the specification of standard states is arbitrary. However, for the sake of convenience, when using standard of free energy of formation data from Table 13.1, the following standard states apply:

1 *Gases.* The standard state is the pure gas component at the reaction temperature and at a pressure such that the fugacity is 1 atm, or unity.

2 *Liquids and Solids* (to be discussed later). The standard state is the pure substance at the reaction temperature and at a pressure of 1 atm.

Since the standard state is defined in terms of a specified pressure, the properties of a particular species in its standard state do not depend on pressure. Thus, ΔG^0 and K are independent of pressure and depend only on temperature.

It now remains to relate K in Equations (14.16) and (14.17) to measurable quantities. For gas phase reactions, with $f_i^0 = 1.0$,

$$K = \Pi \hat{f}_i^{v_i} \qquad (14.18)$$

or simply

$$K = \frac{\hat{f}_C^c \hat{f}_D^d}{\hat{f}_A^a \hat{f}_B^b} \qquad (14.19)$$

The fugacity in a mixture can also be expressed as[1,2]

$$\hat{f}_i = \hat{\phi}_i y_i P \qquad (14.20)$$

where $\hat{\phi}_i$ is defined as the fugacity coefficient in solution. Substituting Equation (14.20) into Equation (14.19) gives

$$KP^{-v} = \Pi (y_i \hat{\phi}_i)^{v_i} \qquad (14.21)$$

For an ideal solution, $\hat{\phi}_i = \phi_i$, so that

$$KP^{-v} = \Pi (y_i \phi_i)^{v_i} \qquad (14.22)$$

where ϕ_i is the pure component fugacity coefficient. For an ideal gas, $\phi_i = 1.0$, so that

$$KP^{-v} = \Pi (y_i)^{v_i} \qquad (14.23)$$

The above approach can now be applied to the authors' favorite reaction

$$aA + bB \rightleftharpoons cC + dD$$

For this case

$$\hat{f}_A = \hat{\phi}_A \, y_A P$$
$$\hat{f}_B = \hat{\phi}_B \, y_B P$$

(14.24)

$$\vdots$$

etc.

and ultimately, one obtains

$$K = \frac{(\hat{\phi}_C y_C P)^c (\hat{\phi}_D y_D P)^d}{(\hat{\phi}_A y_A P)^a (\hat{\phi}_B y_B P)^b}$$

$$= \left(\frac{\hat{\phi}_C^c \hat{\phi}_D^d}{\hat{\phi}_A^a \hat{\phi}_B^b} \right) \left(\frac{y_C^c y_D^d}{y_A^a y_B^b} \right) P^{c+d-a-b}$$

$$= K_{\hat{\phi}} K_y P^\nu; \quad \nu = c + d - a - b$$

(14.25)

For an ideal solution, Equation (14.25) becomes

$$KP^{-\nu} = K_\phi K_y$$

(14.26)

For an ideal gas, Equation (14.26) reduces to

$$KP^{-\nu} = K_y$$

(14.27)

ILLUSTRATIVE EXAMPLE 14.5

If the gas phase in the reaction in Illustrative Example 14.3 is assumed to be ideal, express K in terms of ε.

SOLUTION: From Equation (14.27)

$$K = K_y P^\nu$$

Since

$$y_1 = \frac{3 - \varepsilon}{5 - 2\varepsilon}$$

$$y_2 = \frac{1 - 3\varepsilon}{5 - 2\varepsilon}$$

$$y_3 = \frac{\varepsilon}{5 - 2\varepsilon}$$

$$y_4 = \frac{1 + \varepsilon}{5 - 2\varepsilon}$$

and

$$\nu = 1 + 1 - 3 - 1 = -2$$

the above equation becomes

$$K = \frac{\left[\left(\dfrac{\varepsilon}{5 - 2\varepsilon}\right)^1 \left(\dfrac{1 + \varepsilon}{5 - 2\varepsilon}\right)^1\right]}{\left(\dfrac{3 - \varepsilon}{5 - 2\varepsilon}\right)^3 \left(\dfrac{1 - 3\varepsilon}{5 - 2\varepsilon}\right)^1} P^{-2}$$

This equation may be further simplified (if one so desires) by canceling out some of the $(5 - 2\varepsilon)$ terms. The reader is again reminded that the approaches presented in this problem set apply only to ideal gases. ∎

For gas-phase reactions, the term K may also be approximately represented in terms of the partial pressures of the components involved since $p_i = y_i P$. This functional relationship can be deduced directly from Equation (14.27) and is given in Equation (14.28).

$$K = K_p \tag{14.28}$$

The term K_p is an equilibrium constant based on partial pressures:

$$K_p = \frac{p_C^c p_D^d}{p_A^a p_B^b} \tag{14.29}$$

where p_A = partial pressure of component A, etc. Thus, this equation may be used to determine the partial pressures of the participating components at equilibrium. It is important to note once again that the componential partial pressures (p_i) are *equilibrium* values. For product gases, p_i usually represents the *maximum* values that can ultimately be achieved; for reactant gases, p_i represents *minimum* values.

Additional explanation details are available in the literature.[6,7]

EQUILIBRIUM CONVERSION CALCULATIONS: SIMPLIFIED APPROACH

The reaction coordinate approach allows for the calculation of equilibrium conversions. The procedure requires the determination of ε as it appears in Equations

(14.3) and (14.4). This calculation can be simplified if ε is small. This indicates a very low conversion so that certain terms can be neglected in the describing equation.[8] For example, consider Illustrative Example (14.3). If ε is small, note that

$$y_1 = \frac{3 - \varepsilon}{5 - 2\varepsilon} = \frac{3}{5} = 0.6$$

$$y_2 = \frac{1 - 3\varepsilon}{5 - 2\varepsilon} = \frac{1}{5} = 0.2$$

$$y_3 = \frac{\varepsilon}{5 - 2\varepsilon} = \frac{\varepsilon}{5}$$

$$y_4 = \frac{1 + \varepsilon}{5 - 2\varepsilon} = \frac{1}{5} = 0.2$$

(14.30)

so that K may be rewritten as

$$K = \frac{(\varepsilon/5)(0.2)}{(0.6)(0.2)^3} P^2$$

$$= 8.33(\varepsilon)(P)^2$$

(14.31)

The above approach may also be applied to solving problems involving the partial pressure.

ILLUSTRATIVE EXAMPLE 14.6

Consider the reaction

$$CO_2 = CO + \tfrac{1}{2}O_2$$

Calculate the mole fraction of CO in equilibrium with initial mole fractions of CO_2 and O_2 of 0.0317 and 0.0584, respectively, at 2050°F and 1.0 atm. (The remaining component mole fractions may be considered as inerts.) The pressure of the system may be assumed constant.

SOLUTION: It was shown in Illustrative Example (13.10) that at 2050°F (or 1394.3K),

$$\ln K = -13.904$$

$$K = 9.156 \times 10^{-7}$$

For this atmospheric ($P = 1$) equilibrium reaction

$$K = K_p = \frac{(p_{CO})(p_{O_2})^{0.5}}{p_{CO_2}}$$

Assuming a low conversion due to the low K, the equilibrium partial pressures may be assumed equal to the initial partial pressure. Thus,

$$p_{CO} = \frac{(K)(p_{CO_2})}{(p_{O_2})^{0.5}}$$

with

$$p_{CO_2} = 0.0317 \,(\text{approximately constant})$$
$$p_{O_2} = 0.0584 \,(\text{approximately constant})$$

The approximate partial pressure (in atm) or mole fraction—since the pressure is 1 atm—of CO is

$$y_{CO} = p_{CO} = \frac{(9.156 \times 10^{-7})(0.0317)}{(0.0584)^{0.5}}$$
$$= 1.207 \times 10^{-7} \text{ at } 2050°F \qquad\blacksquare$$

ILLUSTRATIVE EXAMPLE 14.7

Repeat the calculation of Illustrative Example 14.6 at 250°F and convert the answer to ppm. As shown in the Chapter 13 homework Problem (see homework problem web-address), set $K = 1.015 \times 10^{-33}$ at this temperature.

SOLUTION: Repeat the previous calculation at 250°F. For this condition,

$$K = 1.015 \times 10^{-33}$$

Substituting

$$p_{CO} = \frac{(1.015 \times 10^{-33})(0.0317)}{(0.0584)^{0.5}}$$
$$= 1.4 \times 10^{-34}$$
$$y_{CO} = 1.4 \times 10^{-34}$$
$$= 1.4 \times 10^{-28} \text{ ppm} \qquad\blacksquare$$

ILLUSTRATIVE EXAMPLE 14.8

Refer to Illustrative Example 13.11. The final result took the form.

$$\ln K = \frac{7048.7}{T} + 0.0151 \ln T - 9.06 \times 10^{-5} T - 2.714 \times 10^4 T^{-2} - 8.09$$

for the reaction

$$2HCl + 0.5O_2 \longrightarrow Cl_2 + H_2O$$

If the initial partial pressures of HCl, O_2, Cl_2, and H_2O are 0.146, 0.106, 0.0, and 0.0292, respectively, calculate the equilibrium partial pressure of Cl_2 if the operating conditions are 1.0 atm and 1250K.

SOLUTION: By definition,

$$K = K_p = \frac{p_{Cl_2}\, p_{H_2O}}{p_{HCl}^2\, p_{O_2}^{0.5}}$$

The following procedure[8] is satisfactory for purposes of engineering calculations. At equilibrium,

$$p_{Cl_2} = p_{Cl_2}(\text{initial}) + x$$
$$= x$$

The term x represents the increase in the partial pressure of the chlorine due to this equilibrium reaction. In addition,

$$p_{H_2O} = p_{H_2O}(\text{initial}) + x = 0.0292 + x$$
$$p_{HCl} = p_{HCl}(\text{initial}) - 2x = 0.146 - 2x$$
$$p_{O_2} = p_{O_2}(\text{initial}) - 0.5x = 0.106 - 0.5x$$

K_p can then be expressed as

$$K_p = \frac{(x)(0.0292 + x)}{(0.146 - 2x)^2(0.106 - 0.5x)^{0.5}}$$

Now, calculate K_p at 1250K using the result from the equation above.

$$\ln K = \frac{7048.7}{1250} + 0.015(\ln 1250) - 9.06 \times 10^{-5}(1250) - 2.714 \times 10^4(1250)^{-2} - 8.09$$
$$= -2.475$$
$$K_p = K = 0.0842$$

Therefore,

$$0.0842 = \frac{(x)(0.0292 + x)}{(0.146 - 2x)^2(0.106 - 0.5x)^{0.5}}$$

Solving for x, which is the equilibrium partial pressure of Cl_2, by a trial-and-error calculation yields

$$p_{Cl_2} = x = 0.01050 \text{ atm}$$

Note that approximately 1% of the discharge flue gas is chlorine—a rather sizable amount. ∎

EQUILIBRIUM CONVERSION CALCULATIONS: RIGOROUS APPROACH

When the extent of an equilibrium reaction is above 5%, the calculational procedures set forth in the previous section should not be employed. Another more accurate approach is available[9] and the presentation in this section deals with the equilibrium reactions for which the extent of reaction, or conversion, is not small. For this case, the earlier terms in K containing ε cannot be neglected. However, the assumption of ideal gas behavior still applies.

The reader is referred to the equations developed in the previous section in Illustrative Example 14.5. For the rigorous approach, the equations provided for terms y_1, y_2, y_3, and y_4 must now be retained in their original form, including the denominator where the 2ε term was neglected. Thus, for this development, the equation for K_y would once again take the form

$$K_y = \frac{\left(\dfrac{\varepsilon}{5 - 2\varepsilon}\right)^1 \left(\dfrac{1 + \varepsilon}{5 - 2\varepsilon}\right)^1}{\left(\dfrac{3 - \varepsilon}{5 - 2\varepsilon}\right)^3 \left(\dfrac{1 - \varepsilon}{5 - 2\varepsilon}\right)^1} \tag{14.32}$$

A total of five Illustrative Examples follow. Each addresses a particular topic and/or effect that is of importance in chemical reaction equilibrium. Several (homework) problems and examination questions for this Part are also available.

ILLUSTRATIVE EXAMPLE 14.9

Compute the degree of dissociation (extent of reaction) of bromine at 527°C and 1 atm for the reaction

$$Br_2(g) = 2Br(g)$$

The standard free energy of reaction in cal/gmol Br_2 is given by

$$\Delta G_T^0 = 53{,}424 - 2.6\,T\,\ln(T) + 0.0005T^2 - 5.0T; \quad T = K$$

Assume ideal gas behavior.

SOLUTION: Calculate ΔG_T^0 at a temperature of 527°C:

$$\Delta G_T^0 = 53{,}424 - (2.6)(T)\,\ln T + (0.0005)T^2 - (5.0)T$$
$$T = 527°C = 800K$$

Substituting

$$\Delta G_{527}^0 = +35{,}840$$

Calculate K at 527°C:

$$\Delta G^0 = -RT \ln K$$
$$\ln K = -35,840/(1.987)(800)$$
$$= -22.547$$
$$K = 1.615 \times 10^{-10}$$

Write the equation for K in terms of K_y and P:

$$K = K_y P; \quad P = 1 \text{ atm}$$
$$K = K_y$$

Express K_y in terms of the reaction coordinate variable ε:

$$n_{Br,O} = 0$$
$$n_{Br_2,O} = 1.0 \text{ (assumed)}$$
$$n_{Br} = 2\varepsilon$$
$$n_{Br_2} = 1 - \varepsilon$$
$$n = 1 + \varepsilon$$
$$y_{Br} = 2\varepsilon/(1 + \varepsilon)$$
$$y_{Br_2} = (1 - \varepsilon)/(1 + \varepsilon)$$

Since $P = 1$ atm,

$$K_y = y_{Br}^2/y_{Br}$$
$$= [2\varepsilon/(1 + \varepsilon)]^2/[(1 - \varepsilon)/(1 + \varepsilon)]$$
$$= 4\varepsilon^2/(1 - \varepsilon^2)$$

Solving the quadratic equation analytically yields

$$\varepsilon = 6.354 \times 10^{-6}$$

where ε once again effectively describes the extent of reaction. Note that a negligible amount of bromine is formed, indicating that the simplified approach could have been employed. ∎

ILLUSTRATIVE EXAMPLE 14.10

Refer to Illustrative Example 14.9. What is the effect of increasing the pressure to 10.0 atm? Assume ideal gas behavior to apply.

SOLUTION: At 10 atm,

$$K = K_y P^\nu$$

Since $v = 1 - 2 = -1$,

$$K_y = K/10$$

Resolving for ε gives

$$\varepsilon = 1.96 \times 10^{-6}$$

Once again, a negligible amount of bromine is formed. However, at a pressure of 10 atm, the assumption of ideal gas behavior is questionable. ∎

ILLUSTRATIVE EXAMPLE 14.11

For the gaseous reaction

$$2A + B \rightleftharpoons C$$

conducted at a pressure of 2.0 atm and a temperature of 373K, the equilibrium mixture was composed of 1 mole A, 2.5 moles B, and 2 moles of C. Assuming ideal gas behavior, determine the standard free energy change of this reaction at 273K.

SOLUTION: Based on the stated equilibrium number of moles:

$$\underset{1}{\overset{?}{2A}} + \underset{2.5}{\overset{?}{B}} \longrightarrow \underset{2}{\overset{?}{C}} \quad \Sigma n_i = n = 5.5$$

Therefore, at equilibrium

$$y_A = \frac{1.0}{5.5} = 0.182$$

$$y_B = \frac{2.5}{5.5} = 0.454$$

$$y_C = \frac{2}{5.5} = 0.364$$

For this reaction,

$$K = \left[\frac{y_C}{(y_A)^2 (y_B)} \right] P^v$$

where

$$v = 1 - 2 - 1 = -2$$

Therefore,

$$K = \frac{0.364}{(0.182)^2(0.454)}(2)^{-2}$$

$$K = 6.0512$$

$$\ln(K) = 1.8003$$

Substituting into Equation (13.16)

$$\ln(K) = \frac{-\Delta G^0}{RT}$$

and solving for ΔG^0_{273} gives

$$1.8003 = \frac{-\Delta G^0}{(273\text{K})(1.987)}$$

so that

$$\Delta G^0_{273} = 976.57 \, \text{cal/gmol}$$

■

ILLUSTRATIVE EXAMPLE 14.12

Nitrogen peroxide dissociates into nitrogen dioxide according to the following reaction:

$$N_2O_4(g) = 2NO_2(g)$$

Calculate the equilibrium composition of a mixture formed by the dissociation at 500K and 1 atm if

$$\Delta G^0_T = -13,600 + 4.16T; \quad \text{cal/gmol of } N_2O_4, \; T = K$$

Comment on the following with regard to this reaction. Discuss the effect on the true equilibrium constant and the equilibrium conversion of

1 increasing the temperature
2 increasing the pressure
3 introducing an inert gas
4 nitrogen dioxide initially present
5 liquid nitrogen present

SOLUTION: For this reaction

$$\Delta G^0_T = -13,600 + 4.16T$$

For $T = 500K$

$$\Delta G_{500}^0 = -11,520 \text{ cal/gmol}$$

Calculate K at this temperature:

$$\Delta G_{500}^0 = -RT \ln K$$

$$\ln K = -(-11,520)/(1.987)(500)$$
$$= 11.6$$
$$K = 1.09 \times 10^5$$

Set up the reaction coordinates ε:

$$\underset{1-\varepsilon}{\overset{1}{N_2O_4}} = \underset{2\varepsilon}{\overset{0}{2NO_2}} \quad \overset{1}{1+\varepsilon}$$

$$K = K_y = 3.31 \times 10^6$$
$$K_y = (2\varepsilon/1+\varepsilon)^2/[(1-\varepsilon)/(1+\varepsilon)] = 1.09 \times 10^5$$

Solving for ε,

$$\varepsilon = 0.9999 \text{ (as expected)}$$

or near 100% conversion.

Note that the notation $\uparrow$, $\longrightarrow$, and $\downarrow$ represent an increase, no change, and a decrease, respectively. The following conclusions may thus be drawn:

1 As $T \uparrow$, $K \downarrow$, conversion (ε) $\downarrow$

2 As $P \uparrow$, $K \longrightarrow$, $\varepsilon \downarrow$

3 As inerts $\uparrow$, $K \longrightarrow$, $\varepsilon \uparrow$

4 As $NO_2 \uparrow$, $K \longrightarrow$, $\varepsilon \downarrow$

5 Liquid N_2 does not exist at these conditions. However, it will vaporize and serve as an inert ... so that $K \longrightarrow$, $\varepsilon \uparrow$ ■

ILLUSTRATIVE EXAMPLE 14.13

Ethanol is manufactured according to the vapor phase reaction below:

$$C_2H_4 + H_2O \rightleftharpoons C_2H_5OH$$

If the reaction is conducted at 150°C and 1.5 atm, and the feed stream consists of 90 mole % H_2O and 10% mole % C_2H_4, determine the equilibrium product mole fraction composition(s):

Data: $\Delta G^0(150°C) = 2375 \text{ cal/gmol}$

$$\Delta H^0(298K) = -10,100 \text{ cal/gmol}$$

Also discuss the effect of increasing the pressure on the equilibrium composition. Justify your answer.

SOLUTION: Assumes 10 moles initially present so that

$$\overset{1}{C_2H_4}(g) + \overset{9}{H_2O}(g) = \overset{0}{C_2H_5OH} \qquad \overset{10}{}$$
$$\underset{1-\varepsilon}{} \qquad \underset{9-\varepsilon}{} \qquad \underset{\varepsilon}{} \qquad \underset{10-\varepsilon}{}$$

$T = 150°C = 423K$; $P = 1.5$ atm; assume ideal gas.

$$\int_{1}^{n_{C_2H_4}} \frac{dn_{C_2H_4}}{-1} = \int_{9}^{n_{H_2O}} \frac{dn_{H_2O}}{-1} = \int_{0}^{n_{C_2H_5OH}} \frac{dn_{C_2H_5OH}}{1} = \int_{0}^{\varepsilon} d\varepsilon; \quad \varepsilon = \varepsilon_e$$

Therefore

$$n_{C_2H_4} = 1 - \varepsilon \qquad y_{C_2H_4} = \frac{1-\varepsilon}{10-\varepsilon}$$

$$n_{H_2O} = 9 - \varepsilon \qquad y_{H_2O} = \frac{9-\varepsilon}{10-\varepsilon}$$

$$n_{C_2H_5OH} = \varepsilon \qquad y_{C_2H_5OH} = \frac{\varepsilon}{10-\varepsilon}$$

$$n = 10 - \varepsilon$$

For an ideal gas

$$K_y = KP^{-\nu}; \quad \nu = 1 - 1 - 1 = -1$$

Therefore,

$$\frac{y_{C_2H_5OH}}{y_{C_2H_4}y_{H_2O}} = KP$$

Substituting,

$$\frac{\dfrac{\varepsilon}{10-\varepsilon}}{\left(\dfrac{1-\varepsilon}{10-\varepsilon}\right)\left(\dfrac{9-\varepsilon}{10-\varepsilon}\right)} = K(1.5)$$

$$K = \frac{(\varepsilon)(10-\varepsilon)}{(2)(1-\varepsilon)(9-\varepsilon)}$$

Use Equation (13.16) to calculate K:

$$\ln K = -\Delta G^0/RT$$
$$= (-2375)/(1.987)(423)$$
$$= -2.8257$$
$$K = 0.0595 \equiv 0.06$$

Therefore

$$\frac{\varepsilon(10 - \varepsilon)}{2(1 - \varepsilon)(9 - \varepsilon)} = 0.06$$

$$1.12\varepsilon^2 - 11.2\varepsilon + 1.08 = 0$$

Solving for ε

$$\varepsilon = \frac{11.2 \pm \sqrt{(11.2)^2 - (4)(1.12)(1.08)}}{(2)(1.12)}$$

$$= 0.0966$$

$$y_{C_2H_4} = \frac{1 - 0.0966}{10 - 0.0966} = 0.0913 = 9.13\%$$

$$y_{H_2O} = \frac{9 - 0.0966}{10 - 0.0966} = 0.899 = 89.9\%$$

$$y_{C_2H_5OH} = \frac{0.0966}{10 - 0.0966} = 0.018 = 1.8\%$$

Note that

$$\sum y_i = 0.0913 + 0.899 + 0.018$$

$$= 1.008, \text{ OK}$$

Since $v = -1$ (negative), an increase in pressure at constant temperature will cause ε to increase. Therefore, the reaction would shift (favorably) to the right.   ■

OTHER REACTIONS

No attempt will be made to treat the other classes of reaction(s). Rather, this section only addresses consecutive/simultaneous reactions, and superficially examines liquid reactions and reactions involving species in the solid phase.

ILLUSTRATIVE EXAMPLE 14.14

Consider the following simultaneous gas reactions:

$$1 \quad aA + bB \longrightarrow cC + dD$$

$$2 \quad gC + eE \longrightarrow fF$$

If ε_1 and ε_2 are the reaction coordinates for reactions 1 and 2, respectively, express the equilibrium mole fractions of all the participating species if all components, other than F, are initially present.

SOLUTION: Assign ε_1 as the reaction coordinates for (1) and ε_2 for reaction (2). Apply ε_1 first to reaction (1)

$$\underset{\substack{n_{Ao} \\ n_{Ao}-a\varepsilon_1}}{aA} + \underset{\substack{n_{Bo} \\ n_{Bo}-b\varepsilon_1}}{bB} = \underset{\substack{n_{Co} \\ n_{Co}+c\varepsilon_1}}{cC} + \underset{\substack{n_{Do} \\ n_{Do}+d\varepsilon_1}}{dD}$$

Note that species C has already participated in reaction 1. Therefore, the initial number of moles that reaction (2) "senses" includes this equation

$$\underset{n_{Co}}{gC} + \underset{n_{Eo}}{eE} = \underset{0}{fF}$$

Now, apply ε_2 to reaction (2)

$$\underset{\substack{n_{Co} \\ n_{Co}+c\varepsilon_1-g\varepsilon_2}}{gC} + \underset{\substack{n_{Eo} \\ n_{Eo}-e\varepsilon_2}}{eE} = \underset{\substack{fF \\ f\varepsilon_2}}{fF}$$

Rewrite the above two equations:

$$\underset{\substack{n_{Ao} \\ n_{Ao}-a\varepsilon_1}}{aA} + \underset{\substack{n_{Bo} \\ n_{Bo}-b\varepsilon_1}}{bB} = \underset{\substack{n_{Co} \\ n_{Co}+c\varepsilon_1-g\varepsilon_2}}{cC} + \underset{\substack{n_{Do} \\ n_{Do}+d\varepsilon_1}}{dD}$$

$$\underset{\substack{n_{Co} \\ n_{Co}+c\varepsilon_1-g\varepsilon_2}}{gC} + \underset{\substack{n_{Eo} \\ n_{Eo}-e\varepsilon_2}}{eE} = \underset{\substack{fF \\ f\varepsilon_2}}{fF}$$

To determine the total number of moles present at equilibrium, simply obtain the sum of all the species, keeping in mind that component C can only contribute to the total once. Therefore,

$$n_A = n_{Ao} - a\varepsilon_1$$
$$n_B = n_{Bo} - b\varepsilon_1$$
$$n_C = n_{Co} + c\varepsilon_1 - g\varepsilon_2$$
$$n_D = n_{Do} + d\varepsilon_1$$
$$n_E = n_{Eo} \qquad\quad - e\varepsilon_2$$
$$n_F = \qquad\qquad\quad + f\varepsilon_2$$
$$\text{Total } n = n_o + (c+d-a-b)\varepsilon_1 + (f-g-e)\varepsilon_2$$

The corresponding mole fraction are given by

$$y_i = \frac{n_i}{n}$$

■

ILLUSTRATIVE EXAMPLE 14.15

Refer to Illustrative Example 14.14. Calculate ε_1 and ε_2 for the following conditions:

$$K_1 = 152$$
$$K_2 = 665$$
$$P = 1 \text{ atm}$$

$n_{Ao} = 1;$	$a = 1$
$n_{Bo} = 3;$	$b = 4$
$n_{Co} = 0;$	$c = 2, g = 1$
$n_{Do} = 0;$	$d = 2$
$n_{Eo} = 1;$	$e = 2$
$n_{Fo} = 0;$	$f = 1$

SOLUTION: For 1 atm and ideal conditions

$$K_1 = \frac{y_C^c \, y_D^d}{y_A^a \, y_B^b} = 152$$

$$K_2 = \frac{y_F^f}{y_C^c \, y_E^e} = 665$$

For this case,

$$n_o = 1 + 3 + 1 = 5$$
$$n = 5 + (2 + 2 - 1 - 4)\varepsilon_1 + (1 - 1 - 2)\varepsilon_2$$
$$= 5 - \varepsilon_1 - 2\varepsilon_2$$

with

$$y_A = \frac{1 - \varepsilon_1}{n}$$

$$y_B = \frac{3 - 4\varepsilon_1}{n}$$

$$y_C = \frac{2\varepsilon_1 - \varepsilon_2}{n}$$

$$y_D = \frac{2\varepsilon_1}{n}$$

$$y_E = \frac{1 - 2\varepsilon_2}{n}$$

$$y_F = \frac{\varepsilon_2}{n}$$

The reader is left the exercise of solving for the two unknown ε_1 and ε_2. Note also that there are two equations: one for K_1 and one for K_2.

Hint: The following three constraints apply:

$$\varepsilon_1 < 0.5\varepsilon_2$$
$$\varepsilon_1 < 0.75$$
$$\varepsilon_2 < 0.5$$

Answers: $\varepsilon_1 = 0.622$, $\varepsilon_2 = 0.402$, $n = 3.574$

Therefore

$$y_A = 0.106,$$
$$y_B = 0.143,$$
$$y_C = 0.236,$$
$$y_D = 0.348,$$
$$y_E = 0.055,$$
$$y_F = 0.112$$

and

$$\sum y_i = 1.0$$

■

For liquid phase reactions, K retains its original form, i.e.

$$K = \Pi \hat{a}_i^{\nu_i} \tag{14.33}$$

with

$$\hat{a}_i = \frac{\hat{f}_i}{f_i^0}; \quad f_i^0 = \text{pure } i \text{ at } T \text{ and } 1\,\text{atm} \tag{14.34}$$

Except for $P \gg 1\,\text{atm}$, Equation (14.33) can be approximated by

$$K = \Pi(x_i)^{\nu_i}; \quad x_i = \text{mole fraction} \tag{14.35}$$

For ideal solutions, $y_i = 1.0$ so that

$$K = \Pi x_i^{\nu_i} \tag{14.36}$$

For liquid-phase reactions, K is approximately given by

$$K = K_C \tag{14.37}$$

where $K_C = C_C^c C_D^d / C_A^a C_B^b$

$C_i = $ concentration of component i (consistent units)

The reader is referred to the literature for a more detailed treatment.[6]

The reader should again note that Equation (14.37) is valid for systems approaching constant volume conditions. Liquid reactions fit this description; gas phase reactions usually do not.

For solid phase reaction or when a solid is one of the participating species, one may assume its activity is approximately unity since it is "pure." The solid phase component may therefore be "melded" into the equilibrium expression for K. Thus, K for the reaction

$$C(s) + \tfrac{1}{2} O_2 = CO \tag{14.38}$$

is approximately given by

$$K = \frac{y_{CO}}{(y_{O_2})^{0.8}} \tag{14.39}$$

REFERENCES

1. S. Fogler, "*Elements of Chemical Reaction Engineering*," 4th edition, Prentice-Hall, Upper Saddle River, NJ, 2006.
2. J. Reynolds, J. Jeris, and L. Theodore, "*Handbook of Chemical and Environmental Engineering Calculations*," John Wiley & Sons, Hoboken, NJ, 2004.
3. J. Santoleri, J. Reynolds, and L. Theodore, "*Introduction to Hazardous Waste Incineration*," 2nd edition, John Wiley & Sons, Hoboken, NJ, 2000.
4. L. Theodore, "*Air Pollution Control Equipment*," John Wiley & Sons, Hoboken, NJ, 2009.
5. L. Theodore, personal notes, Manhattan College, 1967.
6. J. Smith, H. Van Ness, and M. Abbott, "*Introduction to Chemical Engineering Thermodynamics*," 6th edition, McGraw-Hill, New York, 2001.
7. D. Green and R. Perry, "*Perry's Chemical Engineers' Handbook*," 8th edition, McGraw-Hill, New York, 2008.
8. L. Theodore, personal notes, Manhattan College, 1968.
9. L. Theodore, personal notes, Manhattan College, 1978.

NOTE: Additional problems for each chapter are available for all readers at www. These problems may be used for additional review or homework purposes.

Part IV

Other Topics

The Bible: Old Testament

He passed by on the other side.

—Luke X, 31

Walter Savage Landor [1775–1864]

Elegance in prose composition is mainly this: a just admission of topics and of words; neither too many nor too few of either.

—Imaginary Conversations: Chesterfield and Chatham

This last part, Part IV, has been provided the generic title "OTHER TOPICS". Its main purpose is to address all subjects that have been deemed important by ABET (Accreditations Board for Engineering Technology). These subjects include:

1 Economic Considerations—Chapter 15

2 Open-Ended Problems—Chapter 16

3 Other ABET Topics—Chapter 17

Two additional topics, Fuel Options and Exergy are treated in Chapters 18 and 19, respectively.

The following four topics are addressed in Chapter 17:

1 Environmental Management

2 Health, Safety, and Accident Management

3 Numerical Methods

4 Ethics

Obviously, a detailed presentation of each of these subjects is beyond the scope of this text. Because of this, the reader should note that only a limited number of Illustrative Examples are included. However, the reader is encouraged to refer to the cited references for additional details.

Thermodynamics for the Practicing Engineer. By L. Theodore, F. Ricci, and T. Van Vliet
Copyright © 2009 John Wiley & Sons, Inc.

Chapter **15**

Economic Considerations

Franklin Delano Roosevelt [1882–1945]

We have always known that heedless self-interest was bad morals; we know now that it is bad economics.

—Second Inaugural Address [January 20, 1937]

Publilius Syrus

Money alone sets all the world in motion.

—Maxim 656

INTRODUCTION

Every engineer should be able to execute an economic evaluation of a proposed project. If the project is not profitable, it should obviously not be pursued, and the earlier such a project can be identified, the fewer are the resources that will be wasted.

Before the cost of a process or facility can be evaluated, the factors contributing to the cost must be recognized. There are two major contributing factors: capital costs and operating costs; these are discussed in the next two sections. Once the total cost has been estimated, the engineer must determine whether or not the project will be profitable. This usually involves converting all cost contributions to an annualized basis, a method that is discussed in a later section. If more than one project proposal is under study, this method provides a basis for comparing alternative proposals and for choosing the best proposal. Project optimization, including a brief description of a perturbation analysis is also presented. The chapter concludes with applications that illustrate the material presented.

Detailed cost estimates are beyond the scope of this text. Such procedures are capable of producing accuracies in the neighborhood of $\pm 5\%$; however, such estimates generally require many months of engineering work. This chapter is designed to give the reader a basis for a *preliminary cost analysis* only, with an expected accuracy of approximately $\pm 20\%$.

CAPITAL COSTS

Equipment cost is a function of many variables, one of the most significant of which is *capacity*. Another important variable includes the degree of equipment sophistication. Preliminary estimates are often made from simple cost–capacity relationships that are valid when the other variables are confined to narrow ranges of values; these relationships can be represented by approximate linear (on log–log coordinates) cost equations of the form[1]

$$C = \alpha q^{\beta} \tag{15.1}$$

where C = cost

q = some measure of equipment capacity

α, β = empirical "constants" that depend mainly on equipment type

It should be emphasized that this procedure is suitable for rough estimation only; actual estimates from vendors are more preferable. Only major pieces of equipment are included in this analysis; smaller peripheral equipment such as pumps and compressors are often not included. However, methods for estimating the costs of storage tanks, conveyors, pumps, compressors, site development, and so on are available in the literature.[1] If greater accuracy is needed, however, actual quotes from vendors should be used.

Again, equipment cost estimation models are useful for a very preliminary estimation. If more accurate values are needed and if old price data are available, the use of an indexing method is better, although a bit more time consuming. The method consists of adjusting the earlier cost data to present values using factors that correct for inflation. A number of such indices are available; some of the most commonly used are:

1 *Chemical Engineering Fabricated Equipment Cost Index* (FECI),[2] earlier values of which are listed in Table 15.1.

2 *Chemical Engineering Plant Cost Index,*[3] listed in Table 15.2.

3 *Marshall and Swift (M&S) Equipment Cost Index,*[4] listed in Table 15.3.

Table 15.1   Fabricated Equipment Cost Index[2]

Year	Index
1999	434.1
1998	435.6
1997	430.4
1996	425.5
1995	425.4
1994	401.6
1993	391.2
1957–1959	100

Table 15.2   Plant Cost Index[3]

Year	Index
2004	457.4
2003	402.0
2002	395.6
2001	394.3
2000	394.1
1999	390.6
1998	389.5
1997	385.5
1996	381.7
1957–1959	100

Table 15.3   Marshall & Swift Equipment Cost Index[4]

Year	Index
2004	1124.7
2003	1123.6
2002	1104.2
2001	1093.9
2000	1089.0
1999	1068.3
1998	1061.9
1997	1056.8
1996	1039.1
1926	100

    Cost data given as of a specific date can be converted to more recent costs through the use of these cost indices. In general, the indices are based upon constant dollars in a base year and actual dollars in a specific year. In this way, with the proper application of the index, the effect of inflation (or deflation) and price increases are included in the analysis by multiplying the historical cost by the ratio of the present cost index divided by the index applicable in the historical year.

    Other indices for construction, labor, buildings, engineering, and so on are also available in the literature.[2–4] Generally, it is not wise to use past cost data older than 5–10 years, even with the use of the cost indices. Within that time span, the technologies used in the processes may have changed drastically. The use of the indices could cause the estimates to be significantly different to the actual costs. Such an error might lead to the choice of an alternative proposal other than the least costly. Note that the three indices listed in Tables 15.1–15.3 range over many years. The base years for Tables 15.1 and 15.2 are 1957–1959 with an index of 100. One may compare the fabricated equipment index to costs available from the time frame between 1993 and 1999. Another comparison would be the plant cost index from 1989 to 1999. The Marshall and Swift index uses 1926 as a base year with a cost index of 100. Comparing the ratios for any 2 years, calculated from Tables 15.2 and 15.3, one notes that the change in index varies from 1 to 3%. Since these methods are for preliminary cost estimates with an expected accuracy of approximately $\pm 20\%$, use of either index will not significantly impact the final estimate.[5]

The usual technique for determining the *capital costs* (i.e., *total* capital costs, which include equipment design, purchase, and installation) for a process or facility is based on the *factored method* of establishing direct and indirect installation costs as a function of the known equipment costs. This is basically a *modified Lang method*, whereby cost factors are applied to known equipment costs.[6,7]

The first step is to obtain from vendors (or, if less accuracy is acceptable, from one of the estimation techniques previously discussed) the purchase prices of the primary and auxiliary equipment. The total base price, designated by X, which should include instrumentation, control, taxes, freight costs, etc., serves as the basis for estimating the direct and indirect installation costs. The installation costs are obtained by multiplying X by the cost factors, which are available in the literature.[6–11] For more refined estimates, the cost factors can be adjusted to more closely model the proposed system by using adjustment factors that take into account the complexity and sensitivity of the system.[6,7]

The second step is to estimate the direct installation costs by summing all the cost factors involved in the direct installation costs, which include piping, insulation, foundation supports, etc. The sum of these factors is designated as the DCF (*direct installation cost factor*). The direct installation costs are then the product of the DCF and X.

The third step consists of estimating the indirect installation costs. The procedure here is the same as that for the direct installation cost; i.e., all the cost factors for the indirect installation costs (engineering and supervision, startup, construction fees, etc.) are added; the sum is designated by ICF (indirect installation cost factor). The indirect installation costs are then the product of ICF and X.

Once the direct and indirect installation costs have been calculated, the *total capital cost* (TCC) may be evaluated as

$$\text{TCC} = X + (\text{DCF})(X) + (\text{ICF})(X) \tag{15.2}$$

This cost is then converted to *annualized* capital costs with the use of the *capital recovery factor* (CRF), which is described later. The *annualized capital cost* (ACC) is the product of the CRF and TCC and represents the total installed equipment cost distributed over the lifetime of the facility.

Some guidelines in purchasing equipment are listed below:

1 Do not buy or sign any documents unless provided with certified independent test data.

2 Previous clients of the vendor company should be contacted and their facilities visited.

3 Payment schedules should be established based on targeted goals, e.g., 10% on receipt of approved engineering drawings, 15% upon receipt of major fabrication materials, etc.

4 Finally, 10–15% of the cost should be withheld until the installation is completed.

5 Permit approval from state and local regulatory officials should be obtained. This may sometimes require federal approvals as well.

6 A process and mechanical guarantee from the vendors involved should be required. Startup assistance should be included and assurance of prompt technical assistance should be obtained in writing. A complete and coordinated operating manual should be provided.

7 Vendors should provide key spare parts necessary at startup and replacement parts for a minimum of the first 6 months to 1 year of operation. This should also consider long-term delivery items.

OPERATING COSTS

Operating costs can vary from site to site since these costs, in part, reflect local conditions (e.g. staffing practices, labor, and utility costs). Operating costs, such as capital costs, may be separated into two categories: direct and indirect costs. *Direct* costs are those that cover material and labor and are directly involved in operating the facility. These include labor, materials, maintenance labor and maintenance supplies, replacement parts, waste treatment and disposal fees, utilities, and laboratory costs. *Indirect* costs are those operating costs associated with but not directly involved in operating the facility; costs such as overhead (e.g. building–land leasing and office supplies), administrative fees, local property taxes, and insurance fees fall into this category.

The major direct operating costs are usually those associated with labor and materials. *Materials* costs for most systems involve the cost of chemicals needed for the operation of the process.[12] *Labor* costs differ greatly.[13] Salary costs vary from state to state and depend significantly on the location of the facility. The cost of *utilities* generally consists of that for electricity, water, fuel, compressed air, and steam. The annual costs are estimated with the assistance of material and energy balances. Costs for environmental management should also be estimated. Annual *maintenance* costs can be estimated as a percentage of the capital equipment cost.

The *indirect* operating costs consist of overhead, local property tax, insurance, administration, less any credits. The overhead comprises payroll, fringe benefits, social security, unemployment insurance, and other compensation that is indirectly paid to the plant personnel. This cost can be estimated as 70–90% of the operating labor, supervision, and maintenance costs.[11,12] Local *property taxes* and *insurance* can be estimated as 2–3% of the TCC while administration costs can be estimated as 2% of the TCC.

The total operating cost is the sum of the direct operating costs and the indirect operating costs less any credits from the process. Unlike capital costs, operating costs are always calculated on an annual basis.

PROJECT EVALUATION

In comparing alternative processes or different options of a particular process from an economic point-of-view, it is recommended that the total capital cost be converted to an annual basis by distributing it over the projected lifetime of the facility. The sum of both the annualized capital costs (ACC) and the *annual operating costs* (AOC) is known as the *total annualized cost* (TAC) for the facility. The economic merit of the proposed facility, process, or scheme can be examined once the total annual cost is available. Alternative facilities or options may also be compared. Note, a small flaw in this procedure is the assumption that the operating costs remain constant throughout the lifetime of the facility. However, since the analysis is geared to comparing different alternatives, the changes with time should be somewhat uniform among the various alternatives, resulting in little loss of accuracy.

The conversion of the total capital cost to an annualized basis involves an economic parameter known as the capital recovery factor (CRF). These factors can be found in any standard economics text[13,14] or can be calculated directly from

$$CRF = \frac{(i)(1 + i)^n}{(1 + i)^n - 1} \tag{15.3}$$

where n = projected lifetime of the system (years)

i = annual interest rate (expressed as a fraction)

The CRF is a positive, fractional number. The ACC is computed by multiplying the TCC by the CRF. The annualized capital cost reflects the cost associated in recovering the initial capital outlay over the depreciable life of the system. Investment and operating costs can be accounted for in other ways, such as a *present worth* analysis. However, the capital recovery method is often preferred because of its simplicity and versatility. This is especially true when comparing systems having different depreciable lives. There are usually other considerations in such decisions besides the economics, but if all the other factors are equal, the proposal with the lowest total annualized cost should be the most viable.

If a plant is under consideration for construction, the total annualized cost should be sufficient to determine whether or not the proposal is economically attractive as compared to other proposals. If, however, a commercial process is being considered, the profitability of the proposed operation becomes an additional factor. For the sake of simplicity, some methods assume a facility lifetime of 10 years and that the land is already available.

One difficulty in this analysis is estimating the revenue generated from the facility because both technology and costs can change from year to year. If a reasonable estimate as to the revenue that will be generated from the facility can be made, a rate of return can be calculated. This method of analysis is known as the *discounted cash flow method using an end-of-year convention*, i.e., the cash flows are assumed to be generated at the end of the year, rather than throughout the year (the latter obviously being the real case). An expanded explanation of this method can be found in any engineering economics text.[13]

PERTURBATION STUDIES IN OPTIMIZATION

Once a particular process scheme has been selected, it is common practice to optimize the process from a capital cost and O&M (operation and maintenance) standpoint. There are many optimization procedures available, most of them too detailed for meaningful application to a major facility. These sophisticated optimization techniques, some of which are routinely used in the design of conventional chemical and petrochemical plants, invariably involve computer calculations. However, use of these techniques is often not warranted.

One simple optimization procedure that is recommended is the *perturbation study*. This involves a systematic change (or *perturbation*) of variables, one by one, in an attempt to locate the optimum design from a cost and operation viewpoint. To be practical, this often means that the engineer must limit the number of variables by assigning constant values to those process variables that are known beforehand to play an insignificant role. Reasonable guesses and simple or short-cut mathematical methods can further simplify the procedure. Much information can be gathered from this type of study since it usually identifies those variables that significantly impact the overall performance of the process and also helps identify the major contributors to the total annualized cost.

ILLUSTRATIVE EXAMPLE 15.1

Plans for the construction of a small electric generating station at a plant were initiated in 1993. The cost for this plant at that time was determined to be $245,000. Estimate the cost of the station in terms of both 1995 and 1999 dollars employing the FECI factors.

SOLUTION: This problem involves the use of the CE (*Chemical Engineering*) Fabricated Equipment Cost Index to obtain the FECI factors for the years 1993, 1995, and 1999. The cost for a particular year is the product of the cost in 1993 and the ratio of the FECI factor for that year to that for 1993. From Table 15.1:

$$1993 \text{ FECI} = 391.2$$
$$1995 \text{ FECI} = 425.4$$
$$1999 \text{ FECI} = 434.1$$

The cost in 1995 dollars is therefore

$$\begin{aligned}\text{Cost}(1995) &= \text{cost}(1993)(1995 \text{ FECI}/1993 \text{ FECI}) \\ &= (\$245{,}000)(425.4/391.2) \\ &= \$266{,}419\end{aligned}$$

Similarly, the cost in 1999 is

$$\begin{aligned}\text{Cost}(1999) &= \text{cost}(1993)(1999 \text{ FECI}/1993 \text{ FECI}) \\ &= (\$245{,}000)(434.1/391.2) \\ &= \$271{,}867 \qquad \blacksquare\end{aligned}$$

ILLUSTRATIVE EXAMPLE 15.2

A stream of 100,000 acfm of flue gas from a utility facility is to be cooled in an air preheater. You have been requested to find the best unit to install to cool the flue gas and preheat the combustion air feed to the boiler. A reputable vendor has provided information on the cost of three units, as well as installation, operating, and maintenance costs. Table 15.4 gives all the data you have collected. Determine what preheater you would select in order to minimize costs on an annualized basis.

SOLUTION: The first step is to convert the equipment, installation, and operating costs to total costs by multiplying each by the total gas flow, 100,000 acfm. Hence, for the finned exchanger,

Table 15.4 Preheater Cost Data

	Finned	4-pass	2-pass
Equipment cost	$3.1/acfm	$1.9/acfm	$2.5/acfm
Installation cost	$0.80/acfm	$1.4/acfm	$1.0/acfm
Operating cost	$0.06/acfm-yr	$0.06/acfm-yr	$0.095/acfm-yr
Maintenance cost	$14,000/yr	$28,000/yr	$9500/yr
Lifetime of equipment	20 yr	15 yr	20 yr

Costs are based on comparable performance. Interest rate is 10% and there is zero salvage value.

the total costs are

$$\text{Equipment cost} = 100,000 \text{ acfm } (\$3.1/\text{acfm}) = \$310,000$$
$$\text{Installation cost} = 100,000 \text{ acfm } (\$0.80/\text{acfm}) = \$80,000$$
$$\text{Operating cost} = 100,000 \text{ acfm } (\$0.06/\text{acfm} \cdot \text{yr}) = \$6000/\text{yr}$$

Note that the operating costs are on an annualized basis. The equipment cost and the installation cost must then be converted to an annual basis using the CRF. From Equation (15.3)

$$\text{CRF} = (0.1)(1 + 0.1)^{20}/[(1 + 0.1)^{20} - 1]$$
$$= 0.11746$$

The annual costs for the equipment and the installation is given by the product of the CRF and the total costs of each:

$$\text{Equipment annual cost} = \$310,000(0.11746)$$
$$= \$36,412/\text{yr}$$

$$\text{Installation annual cost} = \$80,000(0.11746)$$
$$= \$9397/\text{yr}$$

The calculations for the 4-pass and the 2-pass exchangers are performed in the same manner. The three preheaters can be compared after all the annual costs are added. The tabulated results are provided in Table 15.5. According to the analysis, the 2-pass exchanger is the most economically attractive device since the annual cost is the lowest.   ■

Table 15.5 Preheater Cost Calculations

	Finned	4-pass	2-pass
Equipment cost	$310,000	$190,000	$250,000
Installation cost	$80,000	$140,000	$100,000
CRF	0.11746	0.13147	0.11746
Annual equipment	$36,412	$24,980	$29,365
Annual installation	$9397	$18,405	$11,746
Annual operating	$6000	$6000	$9500
Annual maintenance	$14,000	$28,000	$9500
Total annual	$65,809	$77,385	$60,111

ILLUSTRATIVE EXAMPLE 15.3

Plans are underway to construct and operate a commercial hazardous waste facility in Trashtown in the state of Ecoarkana. The company is still undecided as to whether to install a liquid injection or rotary kiln incinerator at the waste site. The liquid injection unit is less expensive to purchase and operate than a comparable rotary kiln system. However, projected waste treatment income from the rotary kiln unit is higher since it will handle a larger variety and quantity of wastes. In addition, the rotary kiln will be treating solids as well as liquid wastes.

Based on the economic and financial data provided in Table 15.6, select the incinerator that will yield the higher annual profit.[15]

Table 15.6   Incinerator Economic Data

Costs/Credits	Liquid injection	Rotary kiln
Capital ($)	2,625,000	2,975,000
Installation ($)	1,575,000	1,700,000
Operation ($/yr)	400,000	550,000
Maintenance ($/yr)	650,000	775,000
Income ($/yr)	2,000,000	2,500,000

Calculations should be based on an interest rate of 12% and a process lifetime of 12 years for both incinerators.

SOLUTION: For both units:

$$CRF = (0.12)(1 + 0.12)^{12}/[(1 + 0.12)^{12} - 1]$$
$$= 0.1614$$

Annual capital and installation costs for the liquid injection (LI) unit are

$$LI\,costs = (2,625,000 + 1,575,000)(0.1614)$$
$$= \$677,880/yr$$

Annual capital and installation costs for the rotary kiln (RK) unit are

$$RK\,costs = (2,975,000 + 1,700,000)(0.1614)$$
$$= \$754,545/yr$$

A comparison of costs and credits for both incinerators is given in Table 15.7. A rotary kiln incinerator is recommended. ■

Table 15.7   Incinerator Results

	Liquid injection	Rotary kiln
Total installed ($/yr)	678,000	755,000
Operation ($/yr)	400,000	550,000
Maintenance ($/yr)	650,000	775,000
Total annual cost ($/yr)	1,728,000	2,080,000
Income credit ($/yr)	2,000,000	2,500,000
Profit ($/yr)	272,000	420,000

ILLUSTRATIVE EXAMPLE 15.4

Stacey Shaefer, a recent graduate from Manhattan College's prestigious chemical engineering program was given an assignment to design the most cost-effective heat exchanger to recover energy from a hot flue gas at 500°F. The design is to be based on pre-heating 100°F incoming air (to be employed in the boiler) to a temperature that will result in the maximum annual profit to the utility. A line diagram of the proposed countercurrent exchanger is provided in Fig. 15.1.

Having just completed a heat transfer course with Dr. Flynn and a thermodynamics course with the infamous Dr. Theodore, Stacey realizes that their are two costs that need to be considered:

1 the heat exchanger employed for energy recovery, and

2 the "quality" (from an entropy perspective) of the recovered energy.

Refer to Chapters 6 and 19 for additional details.

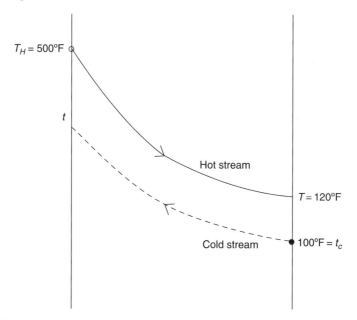

Figure 15.1   Proposed countercurrent exchanger.

She also notes that the higher the discharge temperature of the heated air, t, the smaller will be the temperature difference driving force, and the higher the area requirement of the exchanger and the higher the equipment cost. Alternatively, with a higher t, the "quality" of the recovered energy is higher, thus leading to an increase in recovered profits (by reducing fuel costs).

Based on similar system designs, O'Brien Consultants (OBC) has provided the following annual economic models:

$$\text{Recovered energy profit:} \quad A(t - t_c); \quad A = \$/\text{yr} \cdot {}^\circ F$$
$$\text{Exchange cost:} \quad B/(T_H - t); \quad B = \$ \cdot {}^\circ F/\text{yr}$$

For the above system, OBC suggests a value for the coefficients in the cost model be set at:

$A = 10$

$B = 100,000$

Employing the above information, Stacey has been asked to calculate a t that will

1 provide breakeven operation

2 maximize profits

She is also required to perform the calculation if $A = 10$, $B = 4000$, and $A = 10$, $B = 400,000$. Finally, an analysis of the results is requested.

SOLUTION: Since there are two contributing factors to the cost model, one may write the following equation for the profit, P

$$P = A(t - t_c) - B/(T_H - t); \quad T_H = 500 \quad \text{and} \quad t_c = 100$$

For breakeven operation, set $P = 0$ so that

$$(t - t_c)(T_H - t) = B/A$$

This may be rewritten as

$$t^2 - (T_H + t_c)t + [(B/A) + T_H t_c] = 0$$

The solution to this quadratic equation for $A = 10$ and $B = 100,000$ is

$$t = \frac{600 \pm \sqrt{(600)^2 - (4)(1)(10,000 + 50,000)}}{2}$$
$$= \frac{600 \pm 346}{2}$$
$$= 473^\circ F, \ 127^\circ F$$

To maximize the profit, take the first deriative of P with respect to t and set it equal to zero, i.e.,

$$\frac{dP}{dt} = A - \frac{B}{(T_H - t)^2} = 0$$

Solving,

$$(T_H - t)^2 = B/A$$
$$T_H - t = \sqrt{B/A}$$
$$= \sqrt{10,000}$$
$$= 100$$
$$t = 400°F$$

Upon analyzing the first derivative with t values greater than and less than 400 °F, one observes that the derivative changes sign from $+ \rightarrow -$ at about $t = 400$, indicating a relative maximum.
Similarly, for $A = 10$, $B = 4000$,

$$t_{BE} = 499°F, \ 101°F$$
$$t_{max} = 480°F$$

For $A = 10$, $B = 400,000$,

$$t_{BE} = 300°F$$
$$t_{max} = 300°F$$

Graphical results for the three scenarios is shown in Fig. 15.2.   ∎

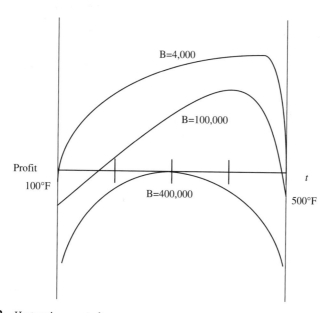

Figure 15.2   Heat exchanger results.

REFERENCES

1. R. J. McCormick and R. J. DeRosier, *"Capital and O&M Cost Relationships for Hazardous Waste Incineration,"* Acurex Corp., Cincinnati, OH, EPA Report 600/2-84-175, Oct. 1984.
2. Economic Indicators, "Chemical Engineering Plant Cost Equipment Index," *Chem. Eng.*, 99, 178, July (1992); **101**, 206, May (1994); **102**, 180, April (1995); **102**, 162, May (1995); **103**, 168, May (1996); **104**, 229, May (1997); **105**, 189, June (1998); **107**, 164, May (2000).
3. Economic Indiators, "Chemical Engineering Plant Cost Index," *Chem. Eng.*, 103, 172, Nov. (1996); **107**, **164**, May (2000).
4. Economic Indicators, "Marshall and Swift Equipment Cost Index," *Chem. Eng.*, 103, 172, Nov. (1996); *Chemical Engineering*, **164**, May 2000.
5. W. M. Vatavuk, *"Escalation Indexes for Air Pollution Control Costs,"* EPA-452/R-95-006; NTIS # PB 96-100862, Sept. 1995.
6. R. B. Neveril, *"Capital and Operating Costs of Selected Air Pollution Control Systems,"* Gard, Inc., Niles, IL, EPA Report 450/5-80-002, Dec. 1978.
7. W. M. Vatavuk and R. B. Neveril, "Factors for Estimating Capital and Operating Costs," *Chem. Eng.*, 157–162, Nov. 3, 1980.
8. G. A. Vogel and E. J. Martin, "Hazardous Waste Incineration, Part 1—Equipment Sizes and Integrated-Facility Costs," *Chem. Eng.*, 143–146, Sept. 5, 1983.
9. G. A. Vogel and E. J. Martin, "Hazardous Waste Incineration, Part 2—Estimating Costs of Equipment and Accessories," *Chem. Eng.*, 75–78, Oct. 17, 1983.
10. G. A. Vogel and E. J. Martin, "Hazardous Waste Incineration, Part 3—Estimating Capital Costs of Facility Components," *Chem. Eng.*, 87–90, Nov. 28, 1983.
11. G. D. Ulrich, *A Guide to Chemical Engineering Process Design and Economics*, John Wiley & Sons, Hoboken, NY, 1984.
12. G. A. Vogel and E. J. Martin, "Hazardous Waste Incineration, Part 4—Estimating Operating Costs," *Chem. Eng.*, 97–100, Jan. 9, 1984.
13. E. P. DeGarmo, J. R. Canada, and W. G. Sullivan, *"Engineering Economy,"* 6th ed., Macmillan, New York, 1979.
14. C. Hodgman, S. Selby, and R. Weast, eds., *"CRC Standard Mathematical Tables,"* 12th ed., Chemical Rubber Company, Cleveland, OH, 1961 (presently CRC Press, Boca Raton, FL).
15. L. Theodore and E. Moy, *"Hazardous Waste Incinerators,"* A Theodore Tutorial, Theodore Tutorials, East Williston, NY, 1999.

NOTE: Additional problems for each chapter are available for all readers at www. These problems may be used for additional review or homework purposes.

Chapter **16**

Open-Ended Problems

The Koran

Let not thy hand be tied up to thy neck; neither open it with an unbounded expansion, lest thou become worthy of reprehension, and be reduced to poverty.

—Chap. 17

INTRODUCTION

The education literature provides frequent references to individuals, particularly engineers and scientists, that have different learning styles. And, in order to successfully draw on these different styles, a variety of approaches can be employed. One such approach involves the use of open-ended problems.

The term "open-ended problem" has come to mean different things to different people. It basically describes an approach to the solution of a problem and/or situation for which there is usually not a unique solution. One of the authors of this text has applied this somewhat unique approach by including numerous open-ended problems in several course offerings at Manhattan College. Student comments for the graduate course "Accident and Emergency Management" were recently tabulated. Student responses (unedited) to the question, "What aspects of this course were most beneficial to you?" are listed below:

1 "The open-ended questions gave engineers a creative license. We don't come across many of these opportunities."

2 "Open-ended questions allowed for candid discussions and viewpoints the class may not have been otherwise exposed to."

3 "The open-ended questions gave us an opportunity to apply what we were learning in class with subjects we have already learned which gave us a better understanding of the course."

4 "Much of the knowledge that was learned in this course can be applied to everyday situations and in our professional lives."

5 "Open-ended problems made me sit down and research the problem to come up with ways to solve them."

6 "I thought the open-ended problems were inventive and got me to think about problems in a better way."

7 "I felt that the open-ended problems were challenging. I, like most engineers, am more comfortable with quantitative problems vs qualitative."

There is a tale[1] that appeared in print many years ago that dissected the value of an open-ended approach to a particular problem. The story is presented below.

Some time ago, I received a call from a colleague who asked if I would be the referee on the grading of an examination question. He was about to give a student a zero for his answer to a physics question, while the student claimed he should receive a perfect score and would if the system were not set up against the student. The instructor and the student agreed to submit this to an impartial arbiter, and I was selected.

I went to my colleague's office and read the examination question: "Show how it is possible to determine the height of a tall building with the aid of a barometer."

The student had answered: "Take the barometer to the top of the building, attach a long rope to it, lower the barometer to the street, and then bring it up, measuring the length of the rope. The length of the rope is the height of the building."

I pointed out that the student really had a strong case for full credit, since he had answered the question completely and correctly. On the other hand, if full credit were given, it could well contribute to a high grade for the student in his physics course. A high grade is supposed to certify competence in physics, but the answer did not confirm this. I suggested that the student have another try at answering the question. I was not surprised that my colleague agreed, but I was surprised that the student did.

I gave the student six minutes to answer the question, with the warning that his answer should show some knowledge of physics. At the end of five minutes, he had not written anything. I asked if he wished to give up, but he said no. He had many answers to this problem; he was just thinking of the best one. I excused myself for interrupting him, and asked him to please go on. In the next minute, he dashed off his answer which read:

"Take the barometer to the top of the building and lean over the edge of the roof. Drop the barometer, timing its fall with a stopwatch. Then, using the formula $S = \frac{1}{2}at^2$, calculate the height of the building."

At this point, I asked my colleague if *he* would give up. He conceded, and I gave the student almost full credit.

In leaving my colleague's office, I recalled that the student had said he had other answers to the problem, so I asked him what they were. "Oh, yes," said the student. "There are many ways of getting the height of a tall building with the aid of a barometer. For example, you could take the barometer out on a sunny day and measure the height of the barometer, the length of its shadow, and the length of the shadow of the building, and by the use of a simple proportion, determine the height of the building."

"Fine," I said. "And the others?"

"Yes," said the student. "There is a very basic measurement method that you will like. In this method, you take the barometer and begin to walk up the stairs. As you climb the stairs, you mark off the length of the barometer along the wall. You then count the number of marks, and this will give you the height of the building in barometer units. A very direct method.

"Of course, if you want a more sophisticated method, you can tie the barometer to the end of a string, swing it as a pendulum, and determine the value of 'g' at the street level and at the top of the building. From the difference between the two values of 'g,' the height of the building can, in principle, be calculated."

"Finally," he concluded, "there are many other ways of solving the problem." "Probably the best," he said, "is to take the barometer to the basement and knock on the superintendent's door. When the superintendent answers, you speak to him as follows: 'Mr. Superintendent, here I have a fine barometer. If you will tell me the height of this building. I will give you this barometer.'"

At this point. I asked the student if he really did not know the conventional answer to this question. He said that he did, but added that he was fed up with high school and college instructors trying to teach him how to think, to use the "scientific method," and to explore the deep inner logic of the subject in a pedantic way, as is often done in the new mathematics, rather than teaching him the structure of the subject. With this in mind, he decided to revive scholasticism as an academic lark to challenge the Sputnik-panicked classrooms of America.

The remainder of this chapter addresses a host of topics involved with open-ended problems. The remaining sections are entitled:

Developing Students' Power of Critical Thinking

Creativity

Brainstorming

Inquiring Minds

The chapter concludes with open-ended applications primarily in the thermodynamics field. Applications in other areas can be found in the literature.[2,3]

DEVELOPING STUDENTS' POWER OF CRITICAL THINKING[4]

It has often been noted that we are living in the middle of an information revolution. For nearly two decades, that revolution has had an effect on teaching and learning. Teachers are hard-pressed to keep up with the advances in their fields. Often their attempts to keep the students informed are limited by the difficulty of making new material available.

The basic need of both teacher and student is to have useful information readily accessible. Then comes the problem of how to use this information properly. The objectives of both teaching and studying such information are: to assure comprehension of the material and to integrate it with the basic tenets of the field it represents; and, to use comprehension of the material as a vehicle for critical thinking, reasoning, and effective argument.

Information is valueless unless it is put to use; otherwise it becomes mere data. To use information most effectively, it should be taken as an instrument for understanding. The process of this utilization works on a number of incremental levels. Information can be: absorbed, comprehended, discussed, argued in reasoned fashion, written about, and integrated with similar and contrasting information.

The development of critical and analytical thinking is the key to the understanding and use of information. It is what allows the student to discuss and argue points of opinion and points of fact. It is the basis for the student's formation of independent ideas. Once formed, these ideas can be written about and integrated with both similar and contrasting information.

CREATIVITY

Engineers bring mathematics and science to bear on practical problems, molding materials and harnessing technology for human benefit. Creativity is often a key component in this synthesis; it is the spark that motivates efforts to devise solutions to novel problems, design new products, and improve existing practices. In the competitive marketplace, it is a crucial asset in the bid to win the race to build better machines, decrease product delivery times, and anticipate the needs of future generations.[5]

Although one of the keys to success for an engineer or a scientist is the ability to generate fresh approaches, processes, and products, they need to be creative. Gibney[5] has detailed how some schools and institutions are attempting to use certain methods that essentially share the same objective: to open students' minds to their own creative potential.

Gibney[5] provides information on "The Art of Problem Definition" developed by Rensselaer Polytechnic Institute. To stress critical thinking, they teach a seven step methodology for creative problem development. These steps are:

1 Define the problem

2 State objective(s)

3 Establish functions

4 Develop specifications

5 Generate multiple alternatives

6 Evaluate alternatives

7 Build

In addition, Gibney[5] identified the phases of the creative process set forth by psychologists. They essentially break the process down into five basic stages:

1 Immersion

2 Incubation

3 Insight

4 Evaluation

5 Elaboration

Psychologists have ultimately described the creative process as recursive. At any one of these stages, a person can double back, revise ideas, or gain new knowledge that reshapes his or her understanding. For this reason, being creative requires patience, discipline, and hard work.

Finally, Delle Femina[6] recently outlined five secrets regarding the creative process:

1 Creativity is ageless

2 You don't have to be Einstein

3 Creativity is not an eight hour job

4 Failure is the mother of all creativity

5 Dead men don't create

The reader is left with a thought from Theodore[7]: Creativity usually experiences a quick and quiet death in rooms that house large conference tables.

BRAINSTORMING

Panitz[8] has demonstrated how brainstorming strategies can help engineering students generate an outpouring of ideas. Brainstorming guidelines include:

1 Carefully define the problem upfront

2 Allow individuals to consider the problem before the group tackles it

3 Create a comfortable environment

4 Record all suggestions

5 Appoint a group member to serve as a facilitator

6 Keep brainstorming groups small

A checklist for change was also provided:

1 Adapt

2 Modify

3 Magnify

4 Minify

5 Put to other uses

6 Substitute

7 Rearrange

8 Reverse

9 Combine

INQUIRING MINDS

In an exceptional and well-written article by Lih[9] titled *Inquiring Minds*, Lih commented on Inquiring Minds saying "You can't transfer knowledge without them." His thoughts (which have been edited) on the inquiring or questioning process follow.

1 Inquiry is an attitude—a very important one when it comes to learning. It has a great deal to do with curiosity, dissatisfaction with the status quo, a desire to dig deeper, and having doubts about what one has been told.

2 Questioning often leads to believing. There is a saying that has been attributed to Confucius: "Tell me, I forget. Show me, I remember. Involve me, I understand." It might also be fair to add: "Answer me, I believe."

3 Effective inquiry requires determination to get to the bottom of things.

4 Effective inquiry requires wisdom and judgment. This is especially true for a long-range intellectual pursuit that is at the forefront of knowledge.

5 Inquiry is the key to successful life-long learning. If one masters the art of questioning, independent learning is a breeze.

6 Questioning is good for the questionee as well. It can help clarify issues, uncover holes in an argument, correct factual and/or conceptual errors, and eventually lead to a more thoughtful outcome.

7 Teachers and leaders should model the importance of inquiry. The teacher/ leader must allow and encourage questions and demonstrate a personal thirst for knowledge.

ILLUSTRATIVE EXAMPLE 16.1

Using terms a liberal arts major could understand, briefly explain the concept of:

1 Unit conversion(s) or conversion factors

2 Dimensional analysis

3 Dimensionless numbers

4 Kinetic energy

5 Potential energy

6 Energy conservation

SOLUTION: Qualitative questions like these are often difficult to answer. Perhaps (1) is the easiest of the six. For this case, one might draw an analogy between 5 pennies and 1 nickel, or 2 nickels and 1 dime, or 12 inches and 1 foot, etc. ■

ILLUSTRATIVE EXAMPLE 16.2

The Alaska Department of Environmental Protection (DEP) has requested your consulting firm to provide "the best estimate" of the oil requirements to sustain a proposed remote outpost in the middle of a pristine region. If both the heating and electrical requirements for the site are known, *outline* how to calculate the average daily oil needs. Assume the energy value of the oil is given. Indicate if any additional information is required. Note that no calculations are required.[10]

SOLUTION: When calculating the average daily oil needs for the outpost, several factors come into play. The most important is knowing the heating and electrical requirements for the site and the heating value of the oil. It is also important to take into account how the oil will be transported to the outpost. It is one thing to purchase the oil and another entirely different thing to transport and store the oil at the outpost. As this nation knows, oil prices are governed by international and national events as well as the demand for the oil. Thus, the fluctuations in the oil prices also need to be accounted for. The thermal conversion efficiencies are also required to be known. The following four-step approach is suggested.

Step 1: Break down the given heating and electrical requirements into a per month basis and from this the average daily oil needs can be obtained. (Take the total monthly requirements and divided it by the number of days in a month to get the average daily requirements.) One reason for this is that winter months tend to use more electricity, heating and oil than summer months. Also, one must know when the "peak hours are." This is what is referred to as "demand."

Step 2: Use conversion factors to convert the heating and electrical units into oil units. Oil is usually sold by the barrel. Once all the units are converted into Btu units, the heating value of oil can be used (such as one barrel $= 6.287$ million Btu) as a way to see how many barrels of oil are required.

Step 3: Find out the cost of oil after calculating how much the daily and monthly consumption will be.

Step 4: Decide how the oil will be transported to the outpost and how/where it will be stored and include all these factors in the analysis.   ■

ILLUSTRATIVE EXAMPLE 16.3

The P, V, T data for a recently synthesized gas is given in Table 16.1 ($M = 609$ gmol). Based on either statistical or theoretical principles (or both), develop an equation describing the P, V, T behavior of this gas. State the limitations of the approach and/or final equations.

Table 16.1 P, V, T Data

P, atm	V, m^3	T, K	z
1.0	10.0	200	1.0
5.0	4.0	400	1.0
20.0	2.3	800	1.15
40.0	1.3	800	1.3
100.0	1.6	1000	3.2

SOLUTION: The sky is the limit regarding what approach to employ. See Illustrative Example 16.5 for suggestion. ∎

ILLUSTRATIVE EXAMPLE 16.4

It has been proposed to use the sensible heat from the flue gas of a boiler to reduce the energy need of the combustion process. A young engineer, recently graduated from Manhattan College's prestigious #2 ranked Chemical Engineering department has proposed to recover the heat in a heat exchanger and use the recovered heat to preheat the air required for combustion. Comment on the advantages and disadvantages of the young engineer's proposal.

SOLUTION: The total energy content of the gas should be taken into consideration since the mass involved is (relatively speaking) small (i.e., the density of gas is low). ∎

ILLUSTRATIVE EXAMPLE 16.5

The heat capacity variation with temperature of a fluid are given in Table 16.2. As part of a simulation project, your immediate supervisor has requested that you develop an algebraic equation describing the C_p variation with T. Outline how to solve the problem and justify your approach.

SOLUTION: Regression analysis is a useful statistical technique for developing a quantitative relationship between a dependent variable and one or more independent variables. It usually utilizes experimental data on the pertinent variable to develop a numerical equation showing the influence of the independent variable(s) on the dependent variable of the system. A simple correlation problem can arise when one asks whether there is any relationship between smoking and heart ailments, between beauty and brains, between kinetic energy and velocity, between internal drag force and pressure, etc. For example, in a thermal combustion unit involving a complex chemical reaction, regression methods have been used to develop an equation relating the conversion to entering concentrations, temperature, pressure, and residence time.[11,12]

The investigation of the relationship between two variables, based on a set of n pairs of measurements (x_1, y_1), (x_2, y_2), ..., (x_n, y_n), usually begins with an attempt to discover the approximate form of the relationship by graphing the data as n points in the x, y-plane. Such a graph is called a scatter diagram. Use of a scatter diagram allows one to quickly discern whether there is any pronounced relationship and, if so, whether the relationship may be treated as approximately linear.

Table 16.2   Heat Capacity Variation with Temperature

C_p, cal/gmol · K	T, K
0.203	200
0.238	300
0.258	400
0.274	500

One of the problems of fitting a curve to a set of points in some efficient manner is essentially that of estimating the parameters of a proposed equation. Although there are numerous methods for performing the estimation of such parameters, the best known and most popular method is referred to as the method of least squares.[12]

The least-squares regression development can be applied to several types of polynomials. Four of these functions are:

$$\text{0th degree polynomial: } y = a_0$$
$$\text{1st degree polynomial: } y = a_0 + a_1 x_1$$
$$\text{2nd degree polynomial: } y = a_0 + a_1 x_1 + a_2 x_2^2$$
$$\text{nth degree polynomial: } y = a_0 + a_1 x_1 + \cdots + a_n x_n^n$$

Applying these equations to the four data points is left as an exercise for the reader. However, additional details are provided in the Numerical Methods section of the next chapter.

The correlation coefficient is a measure of the relationship between variables. A correlation coefficient of 1.0 indicates a perfect association between the variables; a correlation coefficient of 0.0 indicates a completely random relation. It provides no information about the dependence or independence of the variables, and nothing about the nature of the relation between the variables.

Although the correlation coefficient provides an answer as to how well the model fits the data, it does not provide an answer as to whether it is the best and/or correct model. In fact, it can often provide misleading information. For example, if one were interested in fitting a model to the four data points provided, one could apply a zero-, first-, second-, or third-order model to the data. As one would suppose, the zero-order polynomial would provide the worst correlation coefficient while the third-order polynomial, which exactly passes through each data point, would correlate perfectly with a corresponding correlation coefficient of 1.0. Obviously, the latter is almost certainly *not* the "correct" or "best" model.[12]

To obtain information on the "best" model, one must resort to a statistical technique referred to as ANOVA, an acronym referring to ANalysis Of VAriance. This topic is beyond the scope of this text; however, information is available in the literature.[12] ∎

ILLUSTRATIVE EXAMPLE 16.6

Some individuals have claimed (and several books have been written) that Earth and the humans that inhabit Earth are all doomed. The basis of their argument is that the entropy of the universe is constantly on the rise so that a time will come when there will be no useful energy available for doing useful work. Comment on this thesis.

SOLUTION: The reader should first address the question: "How long will it take for Earth to reach a state when there will be no useful energy available?" ∎

ILLUSTRATIVE EXAMPLE 16.7

Based on what you have learned in this text, provide a layman's definition of entropy.

SOLUTION: The reader in referred to Chapter 6 for one possible answer to this question. ∎

ILLUSTRATIVE EXAMPLE 16.8

Based on what you have learned in reading this text, provide a layman's definition to the following terms:

1 Vapor pressure

2 Fugacity coefficient

3 Activity

4 Activity coefficient

5 Phase equilibrium constant

SOLUTION: Explaining (1) as relatively easy. Explaining (2)–(5) is more difficult. Regarding (1), vapor pressure can be related to the escaping tendency of vapor from a liquid. A good example would be perfume in an open bottle. ∎

ILLUSTRATIVE EXAMPLE 16.9

Discuss some key design specifications for combustion devices.

SOLUTION: A key design specification for any combustion device, including boilers and incinerators, is the operating temperature. Materials of construction must withstand the operating temperature without experiencing any damage for safe and efficient plant operation. In addition, the volumetric flow rate is a strong function of temperature (see Chapter 5) and plays an important role in properly sizing combustion/reactor equipment. ∎

REFERENCES

1. www.lhup.edu/~dsimanek/angelpin.htm.
2. A. M. FLYNN, J. REYNOLDS, and L. THEODORE, "*A Course on Health Safety and Accident Management,*" Proceedings of the AWMA Meeting, Baltimore, 2002.
3. P. ABULENCIA and L. THEODORE, "*Fluid Flow for the Practicing Engineer,*" John Wiley & Sons, Hoboken, NJ, 2009.
4. Manhattan College Center for Teaching, "*Developing Students' Power of Critical Thinking,*" Bronx, NY, January, 1989.
5. K. GIBNEY, "*Awakening Creativity,*" ASEE Promo, March, 1988.
6. J. DELLA FEMINA, "*Jerry's Rules,*" Modern Maturity, March–April, 2000.
7. L. THEODORE, personal notes, Manhattan College, 2000.
8. B. PANITZ, "*Brain Storms,*" ASEE Promo, March, 1998.
9. M. LIH, "*Inquiring Minds,*" ASEE Promo, December, 1998.
10. A. BAEZ, Adapted from: homework assignment submitted to L. Theodore, 2007.
11. B. ANDERSON, "Regression Analysis Correlates Relationships between Variables," *Chem. Eng.*, May 13, 1963.
12. S. SHAEFER and L. THEODORE, "*Probability and Statistics Applications for Environmental Science,*" CRC Press/Taylor & Francis, Boca Raton, FL, 2007.

NOTE: Additional problems for each chapter are available for all readers at www. These problems may be used for additional review or homework purposes.

Chapter **17**

Other ABET Topics

Publilius Syrus [Circa 42 B.C.]
We are interested in others when they are interested in us.

—Maxim 16

INTRODUCTION

As noted in the Introduction to Part IV, ABET now requires the inconclusion of certain topics in most engineering courses. Six key topics were introduced at that time. Economic Considerations received treatment in Chapter 15 while Open-Ended Problems were addressed in the Chapter 16. This chapter reviews the remaining four topics, which are

1 Environmental Management
2 Health, Safety, and Accident Management
3 Numerical Methods
4 Ethics

A brief introduction to each of the four subjects is followed by two applications. References for additional reading and an expanded treatment of these topics is also provided.

ENVIRONMENTAL MANAGEMENT

In the past four decades, there has been an increased awareness of a wide range of environmental issues covering all sources: air, land, and water. More and more people are becoming aware of these environmental concerns, and it is important that professional people, many of whom do not possess an understanding of environmental problems, have the proper information available when involved with environmental issues. All professionals should have a basic understanding of the technical and scientific terms related to these issues. In addition to serving the needs of the professional, this section

---

qualitatively examines how one can increase his or her awareness of and help solve the environmental problems facing society. An introduction to seven key environmental management areas is given below. This section concludes with two applications.

1 Air issues looks into several different areas related to air pollutants and their control. Atmospheric dispersion of pollutants can be mathematically modeled to predict where pollutants emitted from a particular source, such as a stack at a combustion facility, will fall to the ground and at what concentration. Pollution control equipment can be added to various sources to reduce the amount of pollutants before they are emitted into the air. One topic that few people are aware of is the issue of indoor air quality. Inadequate ventilation systems in homes and businesses directly affect the quality of health of the people within the buildings. For example, the episode of Legionaires' disease that occurred in Philadelphia in the 1970s was related to microorganisms which grew in the cooling water of the air-conditioning system. Finally, there is the issue of global warming.

2 Pollutant dispersion in water systems and wastewater treatment are important the water issues. Pollutants entering rivers, lakes, and oceans come from a wide variety of sources, including stormwater runoff, industrial discharges, and accidental spills. It is important to understand how these substances disperse in order to determine how to control them. Municipal and industrial wastewater treatment systems are designed to reduce or eliminate problem substances before they are introduced into natural water systems, industrial use systems, drinking water supplies, and other water systems. Often, wastewater from industrial plants must be pretreated before it can be discharged into a municipal treatment system.

3 Important solid waste issues include treatment and disposal methods for municipal, medical, and radioactive wastes. Programs to reduce and dispose of municipal waste include reuse, reduction, recycling, and composting, in addition to incineration and landfilling. Potentially infectious waste generated in medical facilities must be specially packaged, handled, stored, transported, treated, and disposed of to ensure the safety of both the waste handlers and the general public. Radioactive waste may have far more serious impacts on human health and the environment, and treatment and disposal requirements for radioactive substances must be strictly adhered to. Superfund is also included in this area.

4 Pollution prevention generally covers domestic and industrial means of reducing pollution. This can be accomplished through

   1 proper residential and commercial building design

   2 proper heating, cooling, and ventilation systems

   3 energy conservation

   4 reduction of water consumption, and

   5 attempts to reuse or reduce materials before they become wastes.

Domestic and industrial solutions to environmental problems result by considering ways to make homes and workplaces more energy-efficient as well as ways to reduce the amount of wastes generated within them.

5 Noise pollution, although not traditionally an important topic, is included in this list. The effects of noise pollution are not generally noticed until hearing is impaired. Although impairment of hearing is a commonly known result of noise pollution, few people realize that stress is also a significant result of excessive noise exposure. The human body enacts its innate physiologic defensive mechanisms under conditions of loud noise and the fight to control these physical instincts causes tremendous stress on the individual.

6 Managers also need to be informed on how to make decisions about associated risks and how to communicate these risks and their effects on the environment to the public. This can include short-term and long-term threats to human health and the environment.

7 Other environmental areas of interest include electromagnetic fields that emanate from power distribution systems, acid rain, greenhouse effects, underground storage tanks, the general subject of metals, questions concerned with asbestos, architectural environmental considerations, ISO 4000, sustainable development, etc.

This section is not intended to be all-encompassing. Rather, it is to be used as a starting point. Little is presented on environmental regulations because of the enormity of the subject matter; in a very real sense, it is a moving target that is beyond the scope of this text.[1] Although much of the material above is qualitative in nature, quantitative material and calculations are available in the literature.[1,2]

ILLUSTRATIVE EXAMPLE 17.1

Describe the health risk process.

SOLUTION: Health risk assessments provide an orderly, explicit way to deal with scientific issues in evaluating whether a problem exists and what the magnitude of the problem may be. This evaluation typically involves large uncertainties because the available scientific data are limited, and the mechanisms for adverse health impacts or environmental damage are only imperfectly understood. When examining risk, how does one decide how safe is "safe" or how clean is "clean"? To begin with, one has to look at both sides of the risk equation, that is, both the toxicity of a pollutant and the extent of public exposure. Information is required at both the current and the potential exposure, considering all possible exposure pathways. In addition to human health risks, one needs to look at potential ecological or other environmental effects. It should be remembered that there are always uncertainties in conducting a comprehensive risk assessment and these assumptions must be included in the analysis.

In recent years, several guidelines and handbooks have been produced to help explain approaches for doing health risk assessments. As discussed by a special National Academy of Sciences committee convened in 1983, most human or environmental health problems can be evaluated by dissecting the analysis into four parts; hazard identification, dose-response assessment or hazard assessment, exposure assessment, and risk characterization (see Fig. 17.1). For some perceived hazards, the risk assessment might stop with the first step, hazard identification, if no adverse effect is identified, or if an agency elects to take regulatory

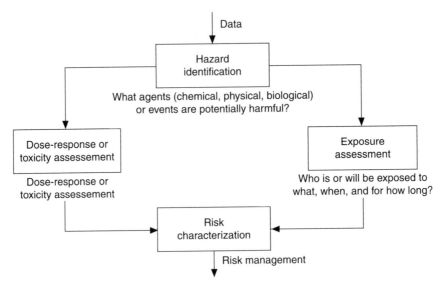

Figure 17.1   The health risk evaluation process.[2]

action without further analysis.[3] Regarding hazard identification, a hazard is defined as a toxic agent or a set of conditions that has the potential to cause adverse effects to human health or the environment. Hazard identification involves an evaluation of various forms of information in order to identify the different hazards. Dose-response or toxicity assessment is required in an overall assessment; responses/effects can vary widely since all chemicals and contaminants vary in their capacity to cause adverse effects. This step frequently requires that assumptions be made to relate experimental data from animals and humans. Exposure assessment is the determination of the magnitude, frequency, duration, and routes of exposure of human populations and ecosystems. Finally, in risk characterization, toxicology, and exposure data/information are combined to obtain qualitative or quantitative expression of risk.

Risk assessment also involves the integration of the information and analysis associated with the above four steps to provide a complete characterization of the nature and magnitude of risk and the degree of confidence associated with this characterization. A critical component of the assessment is a full elucidation of the uncertainties associated with each of the major steps. Under this broad concept of risk assessment are encompassed all of the essential problems of toxicology. Risk assessment takes into account all of the available dose-response data. It should treat uncertainty not by the application of arbitrary safety factors, but by stating them in quantitatively and qualitatively explicit terms, so that they are not hidden from decision-makers. Risk assessment, defined in this broad way, forces an assessor to confront all the scientific uncertainties and to set forth in explicit terms the means used in specific cases to deal with these uncertainties.[4] An expanded presentation on each of the four health risk assessment steps is provided below. In effect, risk characterization is the process of estimating the incidence of a health effect under the various conditions of human or animal exposure described in the exposure assessment. It is performed by combining the exposure and dose-response assessments. The summary effects of the uncertainties in the preceding steps should also be described in this step.

There are two major types of risk: maximum individual risk and population risk. Maximum individual risk is defined exactly as it implies, i.e., the maximum risk to an individual person. This person is considered to have a 70-yr lifetime of exposure to a process or a chemical. Population risk is basically the risk to a population. It is expressed as a certain number of deaths per thousand or per million people. These risks are often based on very conservative assumptions that may yield too high a risk. ∎

ILLUSTRATIVE EXAMPLE 17.2

Provide a more detailed presentation on dose-response and/or toxicity.

SOLUTION: Dose-response assessment is the process of characterizing the relation between the dose of an agent administered or received and the incidence of an adverse health effect in exposed populations, and estimating the incidence of the effect as a function of exposure to the agent. This process considers such important factors as intensity of exposure, age pattern of exposure, and other possible variables that might affect response, such as sex, lifestyle, and other modifying factors. A dose-response assessment usually requires extrapolation from high to low doses and extrapolation from animals to humans, or one laboratory animal species to a wildlife species. A dose-response assessment should describe and justify the methods of extrapolation used to predict incidence, and it should characterize the statistical and biological uncertainties in these methods. When possible, the uncertainties should be described numerically rather than qualitatively.

Toxicologists tend to focus their attention primarily on extrapolations from cancer bioassays. However, there is also a need to evaluate the risks of lower doses to see how they affect the various organs and systems in the body. Many scientific papers focus on the use of a safety factor or uncertainty factor approach, since all adverse effects other than cancer and mutation-based developmental effects are believed to have a threshold—a dose below which no adverse effect should occur. Several researchers have discussed various approaches to setting acceptable daily intakes or exposure limits for developmental and reproductive toxicants. It is thought that an acceptable limit of exposure could be determined using cancer models, but today they are considered inappropriate because of thresholds.[2]

For a variety of reasons, it is difficult to precisely evaluate toxic responses caused by acute exposures to hazardous materials. First, humans experience a wide range of acute adverse health effects including irritation, narcosis, asphyxiation, sensitization, blindness, organ system damage, and death. In addition, the severity of many of these effects varies with intensity and duration of exposure. Second, there is a high degree of variation in response among individuals in a typical population. Third, for the overwhelming majority of substances encountered in industry, there is insufficient data on toxic responses of humans to permit an accurate or precise assessment of the substance's hazard potential. Fourth, many releases involve multiple components. There are presently no rules on how these types of releases should be evaluated. Fifth, there are no toxicology testing protocols that exist for studying episodic releases on animals. In general, this has been a neglected area of toxicology research. There are many useful measures available to employ as benchmarks for predicting the likelihood that a release event will result in serious injury or death.

Dangers are not necessarily defined by the presence of a particular chemical, but rather by the amount of that substance one is exposed to, also known as the dose. A dose is usually expressed in milligrams of chemical received per kilogram of body weight per day. For toxic substances other than carcinogens, a threshold dose must be exceeded before a health effect will occur, and for many substances, there is a dosage below which there is no harm, i.e., a health effect will occur or at least will be detected at the threshold. For carcinogens, it is assumed that there is no threshold, and, therefore, any substance that produces cancer is assumed to produce cancer at any concentration. It is vital to establish the link to cancer and to determine if that risk is acceptable. Analyses of cancer risks are much more complex than those for noncancer risks.

Not all contaminants or chemicals are equal in their capacity to cause adverse effects. Thus, clean-up standards or action levels are based in part on the compounds' toxicological properties. Toxicity data employed are derived largely from animal experiments in which the animals (primarily mice and rats) are exposed to increasingly higher concentrations or doses. As described above, responses or effects can vary widely from no observable effect to temporary and reversible effects, to permanent injury to organs, to chronic functional impairment, to, ultimately, death.   ■

HEALTH, SAFETY, AND ACCIDENT MANAGEMENT[5,6]

Accidents are a fact of life, whether they are a careless mishap at home, an unavoidable collision on the freeway, or a miscalculation at a chemical plant. Even in prehistoric times, long before the advent of technology, a club-wielding caveman might have swung at his prey and inadvertently toppled his friend in what can only be classified as an "accident." As Man progressed, so did the severity of his misfortunes. The "Modern Era" has brought about assembly lines, chemical manufacturers, nuclear power plants, etc., all carrying the capability of disaster. To keep pace with the changing times, safety precautions must constantly be upgraded. It is no longer sufficient, as with the caveman, to shout the warning, "Watch out with that thing!" Today's problems require more elaborate systems of warnings and controls to minimize the chances of serious accidents.

Industrial accidents occur in many ways—a chemical spill, an explosion, a nuclear power plant melt-down, etc. There are often problems in transport with trucks overturning, trains derailing, or ships capsizing. There are "acts of God" such as earthquakes and storms. The one common thread through all of these situations is that they are rarely expected and frequently mismanaged.

Most industrial process plants are safe to be around. Plant management, aided by reliable operators, who are in turn backed up by still-more-reliable automatic controls, does its best to keep operations moving along within the limits usually considered reasonably safe to man and machine. Occasionally, however, there is a whoosh or a bang that is invariably to the detriment of the operation, endangering investment and human life, and rudely upsetting the plant's loss expectancy.

Accidents have occurred since the birth of civilization. Anyone who crosses a street, rides in a car, or swims in a pool runs the risk of injury through carelessness, poor judgment, ignorance, or other circumstances. This has not changed throughout history.

Risk evaluation of accidents serves a dual purpose. It estimates the probability that an accident will occur and also assesses the severity of the consequences of an accident. Consequences may include damage to the surrounding environment, financial loss, or injury to life.

This section is primarily concerned with the methods used to identify hazards and the causes and consequences of accidents. Issues dealing with health risks have been explored in the previous section. Risk assessment of accidents provides an effective way to help ensure either that a mishap does not occur or reduces the likelihood of an accident. The result of the risk assessment also allows concerned parties to take precautions to prevent an accident before it happens.

Regarding definitions, the first thing an individual needs to know is what exactly is an accident. An accident is defined by one of the authors as an unexpected event that has undesirable consequences. The causes of accidents have to be identified in order to help prevent accidents from occurring. Any situation or characteristic of a system, plant, or process that has the potential to cause damage to life, property, or the environment is considered a hazard. A hazard can also be defined as any characteristic that has the potential to cause an accident. The severity of a hazard plays a large part in the potential amount of damage a hazard can cause if it occurs.

Risk may be broadly defined as the probability that human injury, damage to property, damage to the environment, or financial loss will occur. An acceptable risk is a risk whose probability is unlikely to occur during the lifetime of the plant or process. An acceptable risk can also be defined as an accident that has a high probability of occurring, but with negligible consequences. Risks can be ranked qualitatively in categories of high, medium, and low. Risk can also be ranked quantitatively as the annual number of fatalities per million affected individuals. This is normally denoted as a number times one millionth. For example 3×10^{-6} indicates that, on average, 3 (workers) will die every year for every million individuals.

Another quantitative approach that has become popular in industry is the Fatal Accident Rate (FAR) concept. This determines or estimates the number of fatalities over the lifetime of 1000 workers. The lifetime of a worker is defined as 10^5 hours, which is based on a 40-hr work week for 50 years. A reasonable FAR for a chemical plant is 3.0, with 4.0 usually taken as a maximum. A FAR of 3.0 means that there are 3 deaths for every 1000 workers over a 50-year period.[2] The FAR for an individual at home is approximately 3.0.

There are several steps in evaluating the risk of an accident (see Fig. 17.2). These are detailed below if the system in question is a chemical plant.

1 A brief description of the equipment and chemicals used in the plant is needed.

2 Any hazard in the system has to be identified. Hazards that may occur in a chemical plant include fire, toxic vapor release, slippage, corrosion, explosions, rupture of a pressurized vessel, and runaway reactions.

3 The event or series of events that will initiate an accident has to be identified. An event could be a failure to follow correct safety procedures, improperly repaired equipment, or failure of a safety mechanism.

4 The probability that the accident will occur has to be determined. For example, if a chemical plant has a given life, what is the probability that the temperature

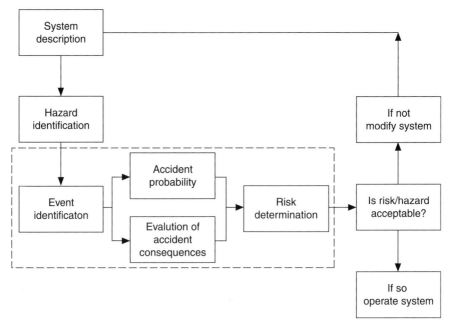

Figure 17.2   Hazard risk assessment flowchart.

in a reactor will exceed a specified temperature range? The probability can be ranked from low to high. A low probability means that it is unlikely for the event to occur in the life of the plant. A medium probability suggests that there is a possibility that the event will occur. A high probability means that the event will probably occur during the life of the plant.

5 The severity of the consequences of the accident must be determined.

6 If the probability of the accident and the severity of its consequences are low, then the risk is usually deemed acceptable and the plant should be allowed to operate. If the probability of occurrence is too high or the damage to the surroundings is too great, then the risk is usually unacceptable and the system needs to be modified to minimize these effects.

The heart of the hazard risk assessment algorithm provided is enclosed in the dashed box (see Fig. 17.2). The algorithm allows for re-evaluation of the process if the risk is deemed unacceptable (the process is repeated starting with either step 1 or 2).

ILLUSTRATIVE EXAMPLE 17.3

Determine if the 50 psig design pressure suggested by the American Petroleum Institute (API) will contain an explosive mixture of air and *n*-pentane with an initial temperature of 25°C and 5 psig initial pressure. Assume a stoichiometric concentration of pentane in air and that the combustion reaction proceeds to completion. The internal energy of reaction of n-pentane at 25°C is 777.46 kcal/gmol. The heat capacity constants (at constant pressure) are given for the flue gas components in Table 17.1.[7] The specific heat at constant pressure, in turn, can be expressed by the following equation:

$$C_p = a + bT + cT^2 + dT^3$$

where C_p = heat capacity at constant pressure, cal/gmol · K
        R = ideal gas law constant, cal/gmol · K

Table 17.1   Heat Capacity Constants

	a	b	c	d
N_2	6.529	0.149×10^{-2}	-0.0227×10^{-5}	0
CO_2	5.316	1.429×10^{-2}	-0.8362×10^{-5}	1.784×10^{-9}
H_2O	6.970	0.345×10^{-2}	-0.0483×10^{-5}	0

SOLUTION: First, write the combustion reaction equation for 1 mole of pentane with stoichio-metric air:

$$C_5H_{12} + 8O_2 + [30.1N_2] \longrightarrow 5CO_2 + 6H_2O + [30.1N_2]$$

Note that

$$n_{N_2} = (0.79/0.21)n_{O_2}$$
$$= (0.79/0.21)(8)$$
$$= 30.1$$

Then, determine the number of moles initially and finally present, and the change in the number of moles:

$$n_{in} = 1 + 8 + 30.1$$
$$= 39.1$$
$$n_{out} = 5 + 6 + 30.1$$
$$= 41.1$$

Express the constant volume heat capacity as a function of the constant pressure heat capacity (see Table 17.2):

$$C_v = C_p - R; \quad R = 1.987 \, \text{cal/gmol} \cdot \text{K}$$

Express the change in internal energy by integrating the internal energy change equation, i.e.,

$$dU = C_v \, dT$$

Table 17.2 Heat Capacity Constants (Modified for C_v)

	a	$a - R$	b	c	d	n_1
N_2	6.529	4.542	0.149×10^{-2}	-0.0227×10^{-5}	0	30.1
CO_2	5.316	3.329	1.429×10^{-2}	-0.8362×10^{-5}	1.784×10^{-9}	5
H_2O	6.970	4.983	0.345×10^{-2}	-0.0483×10^{-5}	0	6

The integrated form of the left-hand side below is provided on the right-hand side

$$
\begin{aligned}
\int_{T_{in}}^{T_{out}} \Sigma(n_i C_{v,i})_{out}\, dT = \int_{T_{in}}^{T_{out}} & [(4.542)(30.1) + (3.329)(5) + (4.893)(6)] \\
& + [(0.149 \times 10^{-2})(30.1) + (1.429 \times 10^{-2})(5) \\
& + (0.345 \times 10^{-2})(6)]T \\
& + [(-0.0227 \times 10^{-5})(30.1) + (-0.8362 \times 10^{-5})(5) \\
& + (-0.0483 \times 10^{-5})(6)]T^2 \\
& + [(1.784 \times 10^{-9})(5)]T^3\, dT \\
= & [(4.542)(30.1) + (3.329)(5) + (4.893)(6)](T_{out} - 298) \\
& + [(0.149 \times 10^{-2})(30.1) + (1.429 \times 10^{-2})(5) \\
& + (0.345 \times 10^{-2})(6)](T_{out}^2 - 298^2)/2 \\
& + [(-0.0227 \times 10^{-5})(30.1) + (-0.8362 \times 10^{-5})(5) \\
& + (-0.0483 \times 10^{-5})(6)](T_{out}^3 - 298^3)/3 \\
& + [(1.784 \times 10^{-9})(5)](T_{out}^4 - 298^4)/4
\end{aligned}
$$

Solve the equation obtained on the RHS of the above equation for T_{out} by trial-and-error until the equation has the value of the internal energy of reaction at 25°C given previously:

$$
\begin{aligned}
777.46 \times 10^3 = & [(4.542)(30.1) + (3.329)(5) + (4.893)(6)](T_{out} - 298) \\
& + [(0.149 \times 10^{-2})(30.1) + (1.429 \times 10^{-2})(5) \\
& + (0.345 \times 10^{-2})(6)](T_{out}^2 - 298^2)/2 \\
& + [(-0.0227 \times 10^{-5})(30.1) + (-0.8362 \times 10^{-5})(5) \\
& + (-0.0483 \times 10^{-5})(6)](T_{out}^3 - 298^3)/3 \\
& + [(1.784 \times 10^{-9})(5)](T_{out}^4 - 298^4)/4
\end{aligned}
$$

By trial-and-error calculation, $T_{out} = 2870K$.

Calculate the final pressure in the vessel:

$$P_f = (14.7 + 5)(41.1/39.1)(2870/298)$$
$$= 200.9 \, \text{psia}$$
$$= 186.2 \, \text{psig}$$

So, can the knockout drum withstand the explosion? Since the final pressure of 186.2 psig is less than the burst pressure of 200 psig, the vessel can withstand the explosion.

In a real application, the vessel pressure should never be near the burst pressure. Typically, two-thirds of the burst pressure is recommended as the maximum pressure that should be allowed. If the assumption that the reaction will proceed to completion does not hold, the above calculations should be performed first to determine the equilibrium concentration for an adiabatic reaction. ∎

ILLUSTRATIVE EXAMPLE 17.4

The firefighters of the local community are evaluating the potential fire hazard from a propane storage tank used to preheat a combustion system that is located on the site of a refinery. If the tank were to rupture, it is expected that 2200 kg of propane would escape and finding an ignition source, would form a fireball. Estimate the size and duration of the fireball. Also determine the area within which structural damage would occur due to radiation effects (intensity of radiation greater than 10 kW/m²) and the area within which a person would sustain a first degree burn (intensity of radiation greater than 4 kW/m²).

The size and the duration of the fireball may be estimated from the following equations:

$$D = 9.56W^{0.325}$$
$$T = 0.196W^{0.349}$$

where D = diameter of the fireball in ft
$\quad\quad T$ = duration of the fireball in s
$\quad\quad W$ = mass of the propane, including oxygen (not air), in lb

The intensity of the radiation can be estimated from the following equation:

$$I = e\sigma T^4 (a/r)^2$$

where I = intensity of radiation, kW/m²
$\quad\quad e$ = emissivity = 0.5
$\quad\quad \sigma$ = Stefan–Boltzmann constant = 5.67×10^{-8}, J/m² · K⁴ · s
$\quad\quad T$ = temperature of the fireball, K
$\quad\quad a$ = radius of the fireball, m
$\quad\quad r$ = distance from the fireball, m

Although it is difficult to predict the temperature of the fireball, assume that the propane reacts with 100% excess air and that the temperature of the fireball is equivalent to the corresponding

Table 17.3   Thermodynamic Data

	ΔH_f^0 (kJ/gmol)	$\overline{C}_p$ (J/gmol · K)
Propane	− 103.85	
Oxygen	0	33.635
Nitrogen	0	31.840
Carbon dioxide	− 393.51	50.919
Water	− 241.826	39.672

adiabatic flame temperature (AFT). Thermodynamic data is provided in Table 17.3. Assume that standard conditions (1 atm, 25°C) exist in the atmosphere.[7]

SOLUTION: Calculate the amount of oxygen consumed. With pure stoichiometric oxygen

$$C_3H_8 + 5O_2 \longrightarrow 3CO_2 + 4H_2O$$

With 100% excess air,

$$C_3H_8 + 10O_2 + [(10)(79/21)N_2] \longrightarrow 3CO_2 + 4H_2O + 5O_2 + [(10)(79/21)N_2]$$

and

$$
\begin{aligned}
\text{mass } O_2 &= (2200\,\text{kg})(2.2\,\text{lb/kg})(1\,\text{lbmol}\,C_3H_8/44\,\text{lb}\,C_3H_8) \\
&\quad \times (10\,\text{lbmol}\,O_2/\text{lbmol}\,C_3H_8)(32\,\text{lb}\,O_2/\text{lbmol}\,O_2)(2) \\
&= 35{,}200\,\text{lb}\,O_2
\end{aligned}
$$

Determine the mass of the reactants in the fireball:

$$
\begin{aligned}
\text{mass } O_2 + \text{mass}\,C_3H_8 &= 35{,}200 + (2200\,\text{kg})(2.2\,\text{lb/kg}) \\
&= 40{,}040\,\text{lb}
\end{aligned}
$$

Estimate the size and duration of the fireball using the equation provided.

$$
\begin{aligned}
D &= 9.56W^{0.325} = 9.56(40{,}040)^{0.325} \\
&= 299.4 \\
&\approx 300\,\text{ft} \\
T &= 0.196W^{0.349} \\
&= 7.9\,\text{s}
\end{aligned}
$$

Estimate the standard heat (enthalpy) of reaction:

$$\Delta H^0 = [(3)\Delta H^0_{f,CO_2} + (4)\Delta H^0_{f,H_2O} - (5)\Delta H^0_{f,O_2} - \Delta H^0_{f,C_3H_8}]$$
$$= [3(-393.51) + 4(-241.826) - 5(0) - 1(-103.5)]$$
$$= -2044 \text{ kJ/gmol } C_3H_8$$

Converting to J,

$$\Delta H^0 = (-2044 \text{ kJ/gmol})(1000 \text{ J/kJ})(1 \text{ gmol}/44 \text{ g})(1000 \text{ g/kg})$$
$$= (-4.645 \times 10^7) \text{ J/kg}$$
$$= (-4.645 \times 10^7)(2200)$$
$$= -1.022 \times 10^{11} \text{ J}$$

Calculate the number of kgmols of both the propane and the products.

$$C_3H_8 = (2200 \text{ kg})/(44 \text{ kg/kgmol})$$
$$= 50 \text{ kgmol}$$
$$O_2 = 5(50 \text{ kgmol})$$
$$= 250 \text{ kgmol}$$
$$CO_2 = 3(50 \text{ kgmol})$$
$$= 150 \text{ kgmol}$$
$$H_2O = 4(50 \text{ kgmol})$$
$$= 200 \text{ kgmol}$$
$$N_2 = (\text{mol } O_2 \text{ in feed})(0.79/0.21)$$
$$= (500)(3.76)$$
$$= 1880 \text{ kgmol}$$

Define and calculate the sensible heat in terms of the temperature change (increase) of the products of combustion employing average heat capacity values.

$$\Delta H = \Sigma n_f \overline{C}_{p_i} \Delta T$$
$$= (\Delta T) \Sigma n_f C_{p_i}$$
$$= \Delta T[(150)(50.919) + (200)(39.672) + (250)(33.635)$$
$$+ (1880)(31.84)]10^3$$
$$= 8.384 \times 10^7 \Delta T$$

Estimate the adiabatic flame temperature using the above results

$$8.384 \times 10^7 \Delta T = 1.022 \times 10^{11}$$
$$\Delta T = 1219°C$$
$$T = 1219 + 298$$
$$= 1517K$$

Finally, calculate the distance where the radiation intensity is 10 kW/m^2

$$I = e\sigma T^4 (a/r)^2$$

$$10 \text{ kW/m}^2 = (0.5)(5.67 \times 10^{-8} \text{ J/m}^2 \cdot \text{K}^4 \text{ s})(1517\text{K})^4$$

$$\times \, [\{(150 \text{ ft})/(3.048 \text{ ft/m})\}/r]^2$$

$$r^2 = 3.636 \times 10^7 \text{ J} \cdot \text{m}^2/\text{kW} \cdot \text{s}$$

Converting to consistent units

$$r^2 = (3.636 \times 10^7 \text{ J} \cdot \text{m}^2/\text{kW} \cdot \text{s})(2.7778 \times 10^{-7} \text{ kW/h})(3600 \text{ s/h})$$

$$= 36,360 \text{ m}^2$$

$$r = 191 \text{ m}$$

$$= 582 \text{ ft}$$

Also, calculate the distance where the radiation intensity is 4 kW/m^2

$$4 \text{ kW/m}^2 = 43.68 \times 10^7 \text{ J}/r^2 \cdot \text{s}$$

$$r^2 = 9.090 \times 10^7 \text{ J} \cdot \text{m}^2/\text{kW} \cdot \text{s}$$

$$= 90,900 \text{ m}^2$$

$$r = 301.5 \text{ m}$$

$$= 919 \text{ ft}$$
∎

The potential for fire hazards is rather high in the chemical industry. The scale of a fire hazard can be determined by assessing the following factors.

1 *Inventory.* The larger the inventory of material, the greater the loss potential.

2 *Energy.* The more energy available for release, such as the stored energy in a material state or a chemical reaction, the greater the potential.

3 *Time.* A higher rate of release of a hazard and a long warning time (i.e., period before emergency countermeasures can be taken), increases the loss potential.

4 *Exposure.* The intensity of the hazard and the distance over which it may cause injury or damage also directly affects the potential.

NUMERICAL METHODS

Early in one's career, the engineer/scientist learns how to use equations and mathematical methods to obtain exact answers to a large range of relatively simple problems. Unfortunately, these techniques are often not adequate for solving real-world problems. The reader should note that one rarely needs exact answers in technical practice. Most real-world applications are usually inexact because they have been generated from data or parameters that are measured, and hence represent only approximations. What one is likely to require in a realistic situation is not an exact answer but rather one having reasonable accuracy from an engineering point-of-view.

The solution to an engineering or scientific problem usually requires an answer to an equation or equations, and the answer(s) may be approximate or exact. Obviously, an exact answer is preferred but because of the complexity of some equations, exact solutions may not be attainable. Furthermore, to engineers, an answer that is precise is often not necessary and can waste time. For this condition, one may resort to another method that has come to be defined as a numerical method. Unlike the exact solution, which is continuous and in closed form, numerical methods provide an inexact (but reasonably accurate) solution. The numerical method leads to discrete answers that are almost always acceptable.

The numerical methods referred to above provide a step-by-step procedure that ultimately leads to an answer and a solution to a particular problem. The method usually requires a large number of calculations and is therefore ideally suited for digital computation.

High-speed computing equipment has had a tremendous impact on engineering design, scientific computation, and data processing. The ability of computers to handle large quantities of data and to perform mathematical operations at tremendous speeds permits the examination of many more cases and more engineering variables than could possibly be handled on the slide rule—the trademark of engineers of yester-year (including one of the authors). Scientific calculations previously estimated in life-times of computation time are currently generated in seconds, and in some instances, nanoseconds.

Today, many powerful commercial mathematical applications are available and widely used in academia and industry. These programs include MathCad, Matlab, Mathematica, etc. In addition, new programs such as Visual Basic, .NET, JAVA, C++, etc., are constantly evolving. These user-friendly programs allow engineers and scientists to perform mathematical calculations without knowing any program-ming or the numerical method involved in the solution.

With commercial mathematical application's powerful computing ability, learn-ing the numerical procedures involved does little for the practicing engineer in terms of interpreting and analyzing answers. For example, instead of performing a long-hand numerical integration using the trapezoid rule, one can use one of the pro-grams mentioned above to do the calculation. However, the job of engineer then becomes that of assessing if the answer given by the computer program seems reason-able. If it does, one can be fairly certain the numbers are correct (providing the rest of the input was correct), but if the answer seems unreasonable, the engineer's job is to figure out what technique the program is using and what might cause it to return an incorrect answer (or, more precisely, a correct answer to the equation, but not to the problem). These software packages are the modern day slide rule for the 21st century engineer and the numerical methods listed below provide the instructions on how to properly use this new slide rule.

1 Simultaneous Linear Algebraic Equations (SLAE)

2 Nonlinear Algebraic Equations

3 Numerical Integration

4 Numerical Differentiation

5 Ordinary Differential Equations (ODE)

6 Partial Differential Equations (PDE)

7 Regression Analysis

8 Optimization

Details on these methods are available in the literature.[8]

Two applications follow. The first is concerned with (7)—regression analysis; the second is concerned with (2)—nonlinear algebraic equations.

ILLUSTRATIVE EXAMPLE 17.5

Outline the least squares method of obtaining the coefficients in a second-order polynomial.

SOLUTION: The method of least squares is based on the assumption that the sum of the squares of the deviations between the assumed function and the given data (i.e., the sum of these errors) must be a minimum. If the ordinate of the data point is designated as y_i and the value of the approximating function at the same value of x_i is $f(x_i)$, the error at x_i can be written as

$$E_i = y_i - f(x_i)$$

Mathematically, the method of least squares requires that

$$\sum_{i=1}^{n} [y_i - f(x_i)]^2 = \text{minimum}$$

where n is the number of data points.

For a second-order polynomial, such as

$$f(x) = a_0 + a_1 x + a_2 x^2$$

the error, E_i, at each data point, i, is therefore

$$E_i = a_0 + a_1 x + a_2 x^2 - y_i$$

The sum of the errors squared, S, is

$$S = \sum E_i^2 = \sum (a_0 + a_1 x + a_2 x^2 - y_i)^2$$

The requirement for this function, S, to be a minimum is given by

$$\left. \frac{\partial S}{\partial a_2} \right|_{a_1, a_0} = 0; \quad \left. \frac{\partial S}{\partial a_1} \right|_{a_2, a_0} = 0; \quad \left. \frac{\partial S}{\partial a_0} \right|_{a_2, a_1} = 0$$

For a second-order polynomial, these three equations can be shown to reduce to

$$\sum_{i=1}^{n} 2x_i^2(a_2x_i^2 + a_1x_i + a_0 - y_i) = 0$$

$$\sum_{i=1}^{n} 2x_i(a_2x_i^2 + a_1x_i + a_0 - y_i) = 0$$

$$\sum_{i=1}^{n} 2(a_2x_i^2 + a_1x_i + a_0 - y_i) = 0$$

These three linear algebraic equations can then be solved simultaneously for a_0, a_1, and a_2.

As noted in the previous chapter, the least-squares regression development can be applied to several types of polynomials. Four of these functions are:

0th degree polynomial: $y = a_0$

1st degree polynomial: $y = a_0 + a_1x_1$

2nd degree polynomial: $y = a_0 + a_1x_1 + a_2x_2^2$

nth degree polynomial: $y = a_0 + a_1x_1 + \cdots + a_nx_n^n$ ∎

ILLUSTRATIVE EXAMPLE 17.6

The chromatographic analysis of the liquid in the top tray of a distillation column is given in Table 17.4. The temperature recorder at the top tray reads 231°F and the pressure is one atm. Comment on the consistency of the chromatographic analysis and temperature reading. Also calculate the composition of the vapor leaving the top tray. The equilibrium constant XK for the components at atmospheric pressure are given by

$$XK(I) = A(I) + B(I) \cdot T + C(I) \cdot T^2; \quad T = °F$$

and the values of A, B, and C are given in Table 17.5.

Table 17.4 Liquid Concentration in Top Tray of Column

Component	Mole Fraction
Benzene	0.37
Toluene	0.23
o-Xylene	0.15
m-Xylene	0.13
p-Xylene	0.12

Table 17.5   Values of A, B, and C

No.	Component	A	B	C
1	Benzene	0.538	−0.0179	0.000107
2	Toluene	0.635	−0.0186	0.000083
3	o-Xylene	0.578	−0.0155	0.000058
4	m-Xylene	0.595	−0.0161	0.000062
5	p-Xylene	0.597	−0.0162	0.000063

SOLUTION: The problem calls for a bubble point temperature calculation. The bubble point of the mixture given by the chromatographic analysis can be calculated and (hopefully) should closely agree with 231°F. This involves a trial-and-error process using the following two equations (see also Chapter 12):

$$Y(I) = XK(I) \cdot X(I) \tag{17.1}$$

and

$$\sum_{I=1}^{5} Y(I) = 1.0 \tag{17.2}$$

The calculations proceed as follows. A temperature is assumed and the equilibrium constant for each component is calculated using the equation:

$$XK(I) = A(I) + B(I) \cdot T + C(I) \cdot T^2 \tag{17.3}$$

The mole fraction in the gas phase, $Y(I)$, is then calculated for each component using Equation (17.1). In the next step, Equation (17.2) is used as a check. If

$$\sum_{I=1}^{5} Y(I) - 1.0 \neq 0$$

the assumed temperature is incorrect and must be updated. This is accomplished by using the Newton iterative technique. If $T(N)$ and $T(N + 1)$ are the original and the updated temperatures, respectively, then

$$T(N + 1) = T(N) - \frac{F[T(N)]}{F_{\text{deriv}}[T(N)]} \tag{17.4}$$

where $F[T(N)]$ is obtained by Equation (17.2) and is given by

$$F[T(N)] = \sum_{I=1}^{5} [XK(I) \cdot X(I)] - 1.0 = 0$$

and $F_{\text{deriv}}[T(N)]$ is obtained by differentiating the above equation with respect to temperature

$$F_{\text{deriv}}[T(N)] = \sum_{I=1}^{5} \{[B(I) + 2 \cdot C(I) \cdot T(N)] \cdot X(I)\}$$

The development of Equation (17.4) is available in the literature.[8] This procedure is continued until a temperature is obtained such that Equation (17.2) is satisfied, i.e., when the LHS of Equation (17.2) is within a given tolerance, ε, of 1.0, i.e.,

$$\left| \sum_{I=1}^{5} Y(I) - 1.0 \right| \leq \varepsilon$$

The reader is left the exercise of demonstrating that the temperature generated using the Newton technique is in reasonable agreement with that given by the recorder. The vapor phase compositions can be obtained directly from Equation (17.1). ∎

ETHICS

The primary responsibility of an engineering professional is to protect public health and safety. However, engineering professionals also have a responsibility to their employers or clients, to their families and themselves, and to the environment. Meeting these responsibilities will challenge the practicing engineer to draw upon a system of ethical values.

Well, what about ethics? Ethics means "doing the right thing" as opposed to "what you have the right to do." But doing the right thing is not always obvious or easy. In fact, ethical decisions are often difficult and may involve a certain amount of self-sacrifice. Doing the right thing for a practicing engineer can be especially challenging. Furthermore, the corporate and government world has confused this concept by developing ethics programs that emphasize only what you have the right to do. An organization, for example, may have a list—often called a Code of Ethics or Code of Conduct—of what an employee can and cannot get away with. Employees are required to sign an acknowledgement that they have read and understood the list. The company unfortunately calls this "ethics training".[9]

One difficulty in some situations is recognizing when a question of ethics is involved. Frequently, in the area of environmental management, a breach of ethics involves a practice that endangers public health and safety or covers up a violation of a rule or regulation. Occasionally, however, a breach may involve a case of the exact opposite. This might seem an unlikely scenario. How can someone be too honest, too caring or too professional?

Regarding the above, one example is lying to save a life. Suppose you are standing on a street and a woman runs past you chased by two men. She screams, "They are trying to attack me!" as she dashes into the entry of a building around a corner. The men ask you, "Which way did she go?" What do you tell them? Clearly, the right thing is to lie. In this case, the value of caring overrides the value of honesty. This situation is exaggerated to illustrate that sometimes it is appropriate to violate certain values to protect public health and safety. In doing the right thing, ideally one should not have to make snap decisions and should take the time to investigate all of the facts (e.g., whether or not the woman was a thief and the men were police).

Sometimes, one must decide how much to sacrifice to ensure public health and safety. In establishing environmental regulations, the regulating agency must decide how safe and how stringent to make the regulations. For example, in the case of air toxic regulations, one standard may result in 10 cancer cases per one million people. But why isn't it one or none? Who should decide?[10]

Two applications in the "thermodynamics" arena follow. These case studies have been primarily drawn from the work of Wilcox and Theodore.[11]

ILLUSTRATIVE EXAMPLE 17.7

Your firm has been contracted by the government to work on a public utility project. Specifically, the firm must complete the first part of the project, which includes determining components of construction to meet design parameters. The results of the firm's work will be turned over to various governmental agencies.

The firm happens to be going through some financial troubles at this time, and they have been overcharging the government in order to help keep the firm going. There are several smaller projects being financed by the overcharges; they are being billed as work related to the government project. So, in addition to being overcharged, the government is being billed for services not rendered.

Your firm has provided good results based on conscientious work; they are therefore meeting their contractual obligations. The firm simply is padding the budget.

There is minimal risk to your firm of getting caught because they have worked under this kind of arrangement with the government several times before and are highly trusted. Compared to the total cost of the project, the extra charges are rather small. However, if your firm does get caught, the professional reputations of several of your friends and colleagues could be ruined; they could be fined and maybe even imprisoned. Finally, if the firm does get caught, chances are it will be forced to fold, and you will be unemployed.

Questions for Discussion

1 What are the facts in this case?

2 What is your responsibility in the matter?

3 Is it allowable because "everybody rips off the government"?

4 What course of action would you take?

5 If you choose to say nothing, then are you equally as guilty as those directly responsible?

6 Would your decision change if your wife/husband worked for the same firm?

7 What could be the repercussions of your decision, whatever it may be?

ILLUSTRATIVE EXAMPLE 17.8

Bangladesh Power Development Board (BPDB) is a semi-government organization under the ministry of electricity, government of Bangladesh. It is responsible for development of power generation as well as its transmission and distribution. Pacific Consultant Ltd (PCL), with long experience in power system engineering, has been working as a prequalified consultant to BPDB. Development International (DI) is a Canadian engineering firm with fifty years of experience throughout the world. PCL and DI have jointly entered into a contract with BPDB for all the engineering required to build a new steam power plant in Bangladesh.

As a part of planning and design, PCL and DI have surveyed the existing electrical loads and those predicted for twenty years; they have collected all necessary data for load flow studies. After those studies, and after considering the load center, communication and transportation facilities, supply of water, fuel source (gas), and other environmental factors, they select a site for the establishment of a new thousand-megawatt power station. The proposal for the new site is submitted to the ministry through BPDB.

One day, Mr. Kader, project coordinator of PCL, who is responsible to the client, receives a telephone call from the chairman of BPDB informing him, "Honorable Minister desires that the site should be shifted by three miles, to be near his native village, Taojan."

Mr. Kader, accompanied by Mr. Hepburn, the Canadian team leader, visit the new site and find some disadvantages. It's hilly, far from both the river (for water as a source and transportation) and the highway. Transportation of heavy equipment would be difficult. Only the fuel source (gas) is available. Mr. Hepburn assesses that the new site will elevate the total estimated cost of the power plant by 10%. The operation and maintenance costs will also be affected.

Mr. Hepburn, team leader of an internationally reputed firm, does not want to recommend the new site. But Mr. Kader, project coordinator, is worried that, if they don't, both companies will lose their contracts or face difficulty in receiving future contracts with BPDB.

Questions for Discussion

1 What are the facts in this case?

2 What are the ethical issues in this case?

3 What are the possible solutions?

4 What would you recommend in such a case?

REFERENCES

1. L. STANDER and L. THEODORE, "*Environmental Regulatory Calculations Handbook*," John Wiley & Sons, Hoboken, NJ, 2008.
2. G. BURKE, B. SINGH, and L. THEODORE, "*Handbook of Environmental Management and Technology*", 2nd edition, John Wiley & Sons, Hoboken, NJ, 2000.
3. D. PAUSTENBACH, "*The Risk Assessment of Environmental and Human Health Hazards: A Textbook of Cast Studies*," John Wiley & Sons, Hoboken, NJ, 1989.
4. J. RODRICKS and R. TARDIFF, "*Assessment and Management of Chemical Risks*," American Chemical Society, Washington D.C., 1984.
5. L. THEODORE, J. REYNOLDS, and F. TAYLOR, "*Accident and Emergency Management*," John Wiley & Sons, Hoboken, NJ, 1989.
6. A. M. FLYNN and L. THEODORE, "*Health, Safety and Accident Management in the Chemical Process Industries*," CRC Press/Taylor & Francis, Boca Raton, 2002 (originally published by Marcel-Dekker).

7. Adapted from: source unknown, location unknown, date unknown.
8. R. KETTLER and S. PRAWEL, "*Modern Methods of Engineering Computation*," McGraw-Hill, New York, 1969.
9. H. TABACK, "Ethics Corner: Doing the Right Thing," *Environmental Management*, Pittsburgh, January, 2002.
10. H. TABACK, "Ethics Corner: When Ego Gets in the Way," *Environmental Management*, Pittsburgh, January, 2002.
11. J. WILCOX and L. THEODORE, "*Engineering and Environmental Ethics: A Case Study Approach*," John Wiley & Sons, Hoboken, NJ, 1998.

NOTE: Additional problems for each chapter are available for all readers at www. These problems may be used for additional review or homework purposes.

Chapter **18**

Fuel Options

Sir Walter Scott [1771–1832]

The sun never sets on the immense empire of Charles V.

—Life of Napoleon (1827)

INTRODUCTION

An evaluation methodology should be established for the comparison of energy resources. Its purpose would be to provide an answer to the question: "Can a procedure be developed that can realistically and practically quantify the overall advantages and disadvantages of the various energy resource options?" A baker's dozen of these energy resource options are provided below[1]:

1 Natural Gas

2 Liquid Fuels (oil)

3 Coal

4 Shale

5 Tar Sands

6 Solar

7 Nuclear (fission)

8 Hydroelectric

9 Wind

10 Geothermic

11 Hydrogen

12 Bioenergy

13 Waste

Details of the first three energy resources are presented in this chapter.

A list of 12 parameters/categories that affect the answer to the question posed in the first paragraph of this chapter for each option has been prepared by one of the

Thermodynamics for the Practicing Engineer. By L. Theodore, F. Ricci, and T. Van Vliet
Copyright © 2009 John Wiley & Sons, Inc.

authors.[1] These are listed below. An attempt to perform this analysis on the various energy resource analyses has also been initiated for different sectors of the world, including the U.S.[2]

1 Resource Quantity

2 Resource Availability

3 Energy Quality (Exergy Analysis)

4 Economic Considerations

5 Conversion Requirements

6 Transportation Requirements

7 Delivery Requirements

8 Operation and Maintenance

9 Regulatory Issues

10 Environmental Concerns

11 Consumer Experience

12 Public Acceptance

Energy is central to all current and future human activities and, as such, society is at a crisis stage. Historically, the primary reason for the emergence of the U.S. as a global leader has been the ability to satisfy its energy needs independently. Without a strategy for determining the most beneficial path to achieving future independence, the future is resigned to be at the mercy of those that control energy resources.[1]

The current energy policy being pursued by the U.S. is disjointed, random, and fraught with vested interests. Achieving energy independence by pursuing current approaches will be costly, inefficient and disruptive, especially as resources diminish. Environmental impacts will become major concerns and the consequences of continuing on the current course may be surprisingly difficult to correct at a later date.[1]

A viable U.S. depends on energy independence and the time to evaluate all options is now. Foreign interests will diminish this country's strengths with potentially little concern for the future or the critical issues that provide global sustainability.[1]

FUEL PROPERTIES

The properties of the fuels most often used in a thermal combustion applications are reviewed here. Fuels burned may be gaseous, liquid, or solid, and are listed below.

- *Gaseous fuels* are principally natural gas (80–95% methane, with the balance ethane, propane, and small quantities of other gases). Light hydrocarbons obtained from petroleum or coal treatment may also be used.

- *Liquid fuels* are mainly hydrocarbons obtained by distilling crude oil (petroleum). The various grades of fuel oil, gasoline, shale oil, and various petroleum cuts and residues are considered in this category.

- *Solid fuels* consist principally of coal (a mixture of carbon, water, noncombustible ash, hydrocarbons, and sulfur). Coke, wood, and solid waste (garbage) may also be employed.

These classifications are not mutually exclusive and necessarily overlap in some areas. The stoichiometric and thermochemical analyses of problems involving the different classes of fuels are similar. Consequently, the actual physical form of the fuel is not important in any calculation study.

In the following sections, fuels are discussed primarily in terms of their usefulness and economic utilization in assisting the thermal combustion process. Properties of fuels considered to be of prime importance are composition and heating value. The heating value refers to the quantity of heat released during combustion of a unit amount of fuel gas. Thus, the heating value of a fuel is one of its most important properties. As described in Chapters 5 and 10, the major products of the complete combustion of a fuel are CO_2 and H_2O. Because of the presence of varying amounts of water in different fuels, two methods for expressing heating value are in common use. Once again, the *gross* or *higher heating value* (HHV) is defined as the amount of heat evolved in the complete combustion of a fuel with any water present in the liquid state; the *net* or *lower heating value* (LHV) is defined as the amount of heat evolved in the combustion of a fuel when all the products, including the water, are in the gaseous state. Thus, the gross heating value is greater than the net heating value by the latent heat of vaporization of the total amount of water originally present and that formed by oxidation of the hydrogen in the fuel.

In dealing with the combustion of some fuels, it is occasionally convenient to express the composition in terms of a single hydrocarbon, even though the fuel is a mixture of many hydrocarbons. For example, gasoline is usually considered to be octane (C_8H_{18}) and diesel fuel is considered to be dodecane ($C_{12}H_{26}$). However, the composition of most fuels employed in combustion calculation is given in terms of its componential (or ultimate) analysis.

Approximate componential analyses of the three types of fuels most often employed at thermal facilities are given in Table 18.1. Estimated values for the net heating value are also included.

Table 18.1   Componential Analysis for Common Fuels

	Mass fraction (lb element/lb fuel)		
Element	Residual fuel oil (No. 6)	Distillate fuel oil (No. 2)	Natural gas
C	0.866	0.872	0.693
H	0.102	0.123	0.227
N	—	—	0.08
S	0.03	0.005	—
Net heating value	19,500 Btu/lb	18,600 Btu/lb	1000 Btu/scf (60°F)

NATURAL GAS

Natural gas is perhaps the closest approach to an ideal fuel because it is practically free from noncombustible gas or particulate residue. Of the many gaseous fuels, natural gas is the most important. No fuel preparation is necessary because gases are easily mixed with air and the combustion reaction proceeds rapidly once the ignition temperature is reached.

Natural gas is found compressed in porous rock or cavities which are sealed between strata under the Earth's surface. When these gas-bearing pools are tapped by drilling wells, the gas is found to be under pressure, which may be as high as 2000 psig. As gas is withdrawn, this pressure gradually decreases until it eventually becomes so low that the field must be abandoned. Thus, the natural gas from a well will usually not remain constant in its elemental composition and/or heating value.

Natural gas consists mainly of methane, with smaller quantities of other hydrocarbons, particularly ethane, and trace amounts of propane. Carbon dioxide and nitrogen are usually present in small amounts and sometimes there can be appreciable amounts of hydrogen sulfide present; this is generally removed at the field before transmission.

The physical and chemical characteristics of natural gas are influenced by the underground conditions existing in the localities where it is found. The gas may contain heavy saturated hydrocarbons, which are liquid at ordinary pressures and temperatures. *Dry* natural gas contains <0.1 gal of gasoline vapor$/1000$ ft^3; >0.1 gal$/1000$ ft^3 is termed *wet*. Note that the terms *wet* and *dry*, when dealing with natural gas, refer to its gasoline content, not to its moisture content. Natural gases are also classified as either *sweet* or *sour*. A *sour* gas is one that contains some mercaptans and a high percentage of hydrogen sulfide, while the *sweet* gas is one in which these compounds are present in trace quantities. Most natural gases employed are *dry* and *sweet*.

Environmental concerns with natural gas are minimal. Most gaseous fuels, with the possible exception of some waste gases, are considered to be clean fuels. Pipeline-grade natural gas is virtually free of sulfur and particulates. Its flue products do not pollute water. In addition, natural gas transportation and distribution facilities have a minimal adverse ecological impact. However, leakage of natural gas can pose a very serious explosion hazard.

The principal air contaminants from gaseous fuels, which are affected by the thermal combustion system design and operation, are the oxidizable materials—carbon monoxide, carbon, and unburned hydrocarbons. Carbon dioxide has recently been added to this list since it is classified as a greenhouse gas. Burner design operating temperature and excess air also affect the production of the oxides of nitrogen. In some instances, the impurity in a gaseous fuel may be hydrogen sulfide. Another important sulfur impurity is carbon disulfide. In most gases, however, the total organic sulfur content is relatively small. Other sulfur compounds that may be present in trace amounts are the thiophenes, carbon oxysulfide, mercaptans, thioesters, and organic sulfide.

Proper operation of a thermal combustion facility employing natural gas requires that the fuel rate be controlled in relation to the demand. The air supply must be appropriate to the fuel supply. This is usually accomplished by automatic controls.

The incoming gas supply is regulated at a constant pressure upstream of the control valve. This valve can be used to control the gas flow. Combustion air regulation is achieved through manipulating dampers or by a special draft controller. Large installations are likely to use more elaborate systems where the fuel and air flows are metered with automatic adjustment to compensate for any changes or disturbances.

Natural gas passes through one or more fixed orifices before entering the combustion chamber. Since flow through an orifice is proportional to the square root of the pressure drop across it, small fluctuations of the upstream pressure will not have a very significant effect on the gas flow rate. However, should it be necessary to reduce the firing rate to 50% of its normal value (2-to-1 *turndown*), a fourfold decrease in gas pressure would be required, with the airflow rate adjusted accordingly. This factor can present a control problem. Before initiating firing with natural gas, the gas injection orifices should be inspected to verify that all passages are unobstructed. Filters and moisture traps should be clean, in place, and operating effectively to prevent any plugging of gas orifices. Proper location and orientation of diffusers should also be confirmed.

LIQUID FUELS

Fuel oil may be defined as petroleum or any of its liquid residues remaining after the more volatile constituents have been removed. Thus, the term *fuel oil* may conveniently cover a wide range of petroleum products. It may be applied to crude petroleum, to a light petroleum fraction similar to kerosene or gas oil, or to a heavy residue left after distillation. The principal industrial liquid fuels are therefore the by-products of natural petroleum. These fuel oils are marketed in two principal classes: distillates and residuals. The principal industrial boiler fuel is residual oil, known as No. 6 or *Bunker C*. The residual oil, as the name implies, is left over after the more valuable products are distilled off. As an industrial and utility fuel, it competes with coal, although its price has fluctuated wildly over the years. It is specified mainly by viscosity. Grades No. 1 and 2 are sometimes designated as *light* and *medium* domestic fuel oils and are specified mainly by the temperature of the distillation range and specific gravity.

Despite the multiplicity of chemical compounds found in fuel oils, there are some important physical and chemical properties that deserve comment:

Specific Gravity   For fuel oils, specific gravity is usually taken as the ratio of the density of the oil at 60°F to the density of water at 60°F. Gravity determinations are readily made by immersing a specific gravity meter into the oil and reading the scale at the point to which the instrument sinks. The specific gravity is then read directly. The American Petroleum Institute (API), the U.S. Bureau of Mines, and the U.S. Bureau of Standards agreed to recommend that only one scale be used in the petroleum industry and that it be known as the API scale. This relationship is given by Equation (2.11).

Viscosity   The Saybolt Universal viscosity is expressed as the amount of time (in seconds) that it takes to run 60 cm^3 of the oil through a standard size orifice at any desired temperature. Viscosity is commonly measured at 100, 150, and 210°F. Fuel oil (particularly No. 6) is very viscous, and it takes a long time to make a determination with the Saybolt Universal viscometer. For this reason, the viscosity of fuel oil may be measured with a Saybolt Furol viscometer, which is the same as the Saybolt Universal viscometer except that the orifice is larger. Interestingly, the viscosity of fuel oil decreases as the temperature rises and becomes nearly constant above about 250°F. Therefore, when fuel oil is heated to reduce the viscosity for good atomization, there is little gain in heating the oil beyond 250°F.

Heating Value   This may be expressed in either British thermal units per gallon or British thermal units per pound (at 60°F). The heating value per gallon increases with specific gravity because there is more weight for a given volume; values range from ∼140,000 to 150,000 Btu/gal. The heating value per pound of fuel oil ranges in value from 18,600 to 19,500 Btu/lb (see Table 18.1).

Flash and Fire Point   The flash point of fuel oil is the lowest temperature at which sufficient vapor is given off to form a momentary flash when flame is brought near the surface of the oil in a small container called a Cleveland Cup. The fire point is the lowest temperature at which the oil gives off enough vapor to burn continuously. Stored fuel oils (as well as liquid wastes) should be maintained significantly below their flash points.

The rate of thermal combustion of fuel oil is limited by vaporization. Light distillate oils readily vaporize; other (heavier) fuel oils, because of their heavier composition, require additional equipment to assure vaporization and complete combustion. In order to achieve complete combustion, oils are atomized into small droplets for rapid vaporization. The rate of evaporation is dependent on surface area, which increases as the atomized droplet size becomes smaller. The desired shape and droplet size are influenced adversely if the fuel viscosity is too high. At ambient temperature, No. 2 fuel oil may be atomized with little difficulty, but typically No. 6 fuel oil must be heated to around 210°F to assure proper atomization. Dirt and foreign matter suspended in the oil may cause wear in the oil pump and blockage of the atomizing nozzles. Strainers or replaceable filters are required in both the oil suction line and the discharge line. During combustion of a distillate fuel oil, the droplet becomes uniformly smaller as it vaporizes. By contrast, a residual oil droplet undergoes thermal and catalytic cracking, and its composition and size undergo various changes with time. Vapor bubbles may form, grow, and burst within a droplet in such a way as to shatter the droplet as it is heated in the combustion zone.

The physical and chemical properties of the oil and the characteristics of the thermal combustion equipment influence the air pollution emissions from a facility. Some of these properties are:

- *Sulfur Content*   Sulfur is a very undesirable element in fuel oil because its products of combustion are acidic and cause corrosion in the ducting, valves, downstream equipment, etc.

- *Solid Impurities*   Fuel oil usually contains all the solid impurities originally present in the crude oil. If these solids contain a large proportion of salt, they are very fusible and can stick to the ducting, valves, waste heat boiler, etc.

- *Vanadium and Sodium Content* The vanadium content in fuel oil may be deposited in the ash on surfaces. These deposits act catalytically in converting SO_2 to SO_3, thereby creating acid dew point problems. Both sodium and vanadium from fuel oil may form sticky ash compounds having low melting temperatures. These compounds increase the deposition of ash and are corrosive.

Some advantages of using No. 2 fuel oil as the fuel additive at a thermal combustion facility are (1) the fuel oil requires less storage space than coal, (2) is not subject to spontaneous combustion, (3) is the least difficult to ignite of the liquid fuels, (4) is not subject to deterioration, and (5) is free from expensive manual handling.

COAL

There is no satisfactory definition of coal. It is a mixture of organic, chemical, and mineral materials produced by a natural process of growth and decay, accumulation of both vegetal and mineral debris, and accomplished by chemical, biological, bacteriological, and metamorphic action. The characteristics of coal vary considerably with location, and even within a given mine some variation in composition is usually encountered. Coals are classified according to rank, which refers to the degree of conversion from one form of coal to another (e.g., lignite to anthracite). The following ranking of coals is employed today: anthracite, bituminous, subbituminous, and lignite.

Solid fuels, including coal, consist of free carbon, moisture, hydrocarbons, oxygen (mostly in the form of oxygenated hydrocarbons), small amounts of sulfur and nitrogen, and nonvolatile noncombustible materials designated as ash. Two types of analyses are in common use for expressing the composition of these fuels: the ultimate analysis and the proximate analysis.

In an *ultimate* analysis, determinations is made of the carbon, moisture (water), ash, nitrogen, sulfur, net hydrogen, and perhaps combined water in the fuel. Ash content is usually determined as a whole (one quantity), although a separate analysis can be made on the ash. Since the hydrogen content is always in excess of the amount needed to form water with all the oxygen present, it is assumed that the oxygen is completely united with the available hydrogen, and this combination may be reported as *combined water*. Note that these two elements do react in the combustion process to form water. The hydrogen in excess of that necessary to combine with the oxygen in the fuel is often termed *net hydrogen*, which usually reacts with the oxygen in the combustion air, halogens (to from acids), etc. Another common approximate method for expressing the composition of a solid fuel is to report it as moisture, volatile combustible matter, fixed carbon, and ash. This is known as a *proximate analysis*. The *moisture* in the proximate analysis is the same as that in the ultimate analysis. The *volatile combustible matter* consists of a large amount of the carbon in the fuel, which is lost as volatile hydrocarbons, plus the hydrogen and combined water, which are given off when combusted. The *fixed carbon* is the carbon left in the fuel after the volatile combustible matter has been removed. *Ash* is the noncombustible residue after complete combustion of the coal. The weight of ash is usually slightly less than that of the mineral matter originally present before burning. However, for high calcium content coals, the ash can be higher than the mineral matter due to the retention of some of the sulfur present in the coal. When coal is heated, it becomes soft and sticky, and as the temperature in a combustor continues to rise, it becomes fluid. Coals with ash that softens and fuses at comparatively low temperatures are likely to deposit the ash within the system and create problems. The sulfur is separately measured, and its amount is useful in judging the air pollution potential of the combustion of the coal. Combustion of sulfur forms oxides, which usually combine with water to form acids that may condense when the flue gas is cooled below its dew point temperature.

Use of coal as a fuel in a thermal combustion system results in particulate (ash) carryover and sulfur oxides (from sulfur) emissions. Fly ash is not considered a major problem; equipment employed for its control is available in the literature.[3] Sulfur emissions can constitute a major and additional problem. The sulfur in coal is found in both organic and inorganic iron pyrites and/or marcasite. There appears to be no evidence that sulfur occurs in coal in its elemental state. The amount of sulfur found in coals varies significantly. During combustion, about 80% of the sulfur in the coal appears in the flue gas in the form of the oxides of sulfur; most of this is SO_2; <5% is SO_3. Coal cleaning at the mine reduces the ash content and simultaneously reduces the sulfur content by removing some of the iron pyrites. Cleaning can be accomplished by gravimetric separation, which is a successful method because pyrite is nearly five times denser than coal. Unfortunately, methods to reduce organic (bound) sulfur generally are not economically viable.

In selecting coal as the fuel at a thermal combustion site, its storage must be considered. Coal slowly deteriorates when exposed to weathering. Attention must be given to the manner in which the coal is stockpiled; large piles loosely formed can ignite spontaneously. This problem is most severe with smaller sizes of coal and high sulfur content. Where large amounts must be stored, such as at power stations, stockpiles are created by using heavy equipment to form piles several hundred feet wide, several thousand feet long, and approximately 20 ft high. When smaller quantities are employed and turnover is fairly rapid, conical piles with a 12-ft depth or less are used. Where open piles are not permitted, silos can be used for coal storage; these should be equipped with fugitive dust control for use during loading.

FUEL SELECTION

No set rule applies to the selection of a given fuel for a particular type of application. However, some general comments and guidelines can be offered. Natural gas and No. 2 distillate oil are the usual options available for many domestic and utility applications.

Number 6 Bunker oil is often not used with smaller units because of its storage and heating requirements. Other units can use natural gas, No. 2 or No. 6 oil. Natural gas appears to be the preferred choice today because of economic and environmental considerations. Due to today's economics, natural gas is most often used in large units. Coal appears to be ideally suited for the US because of its near unlimited supply, but natural gas and fuel oils have been widely used.

If a significant amount of fuel is to be employed at a thermal combustion application, the design engineer may require a thorough survey of the various fuels available at the location of the plant. The results of this study will assist in the fuel selection process. The factors in choosing a fuel for a proposed facility are

1 Cost of fuel

2 Delivered cost of fuel as fired; transportation should be included in this cost.

3 Storage, heating, and handling costs.

4 Added cost of ash handling and removal, particularly with coal.

5 Market price history and trends; these are needed to serve as a guide for future price relationships.

6 Resources or reserves; these should be extensive to assure future supplies.

7 Relative availability; this should be considered under both normal and unusual circumstances or demands.

8 Long-range government conservation; this may be applied to limited reserves, especially when those reserves are in demand for specific essential uses.

STOICHIOMETRIC CALCULATIONS

In thermal combustion applications, one of the key engineering calculations is the prediction of the flue gas flow rate and composition following combustion of an air–fuel mixture. Unfortunately, no simple generalized procedure is available to accomplish this task. The resulting flue gas flow rate and composition are obviously strong functions of both the amount of air and the composition of the fuel. However, this flow rate and composition also depend, to a lesser extent, on the operating temperature in the unit because of chemical reaction equilibrium effects (see Chapters 13 and 14).

One of the difficulties in developing a comprehensive, overall calculational procedure to determine flue gas flows and compositions is coping with the different methods that are in use to characterize fuels. As described above and in Chapter 5, the fuel characterization often takes the form of an ultimate analysis, but there are other ways in which the fuel may be described. The four categories of characterization include:

1 Ultimate or elemental analysis—mass basis, fraction or percentage

2 Ultimate analysis—mole basis

3 Componential or compound analysis—mass basis

4 Componential analysis—mole basis

If an ultimate analysis on a mass basis is provided for the air–fuel mixture, stoichiometric (or theoretical) air mass requirements per unit mass of mixture (m_{st}) may be calculated from

$$m_{st} = 11.5n_C + 34.5n_H - 4.29n_O - 0.97n_{Cl} + 4.29n_S \qquad (18.1)$$

where m_{st} = stoichiometric air requirement per unit mass of fuel (lb air/lb fuel)

n_C = mass fraction of carbon (lb C/lb mixture)

n_H, n_O, n_{Cl}, n_S = mass fractions of hydrogen, oxygen, chlorine, and sulfur, respectively.

The composition of the flue gas produced on a mass basis (for *stoichiometric* combustion only) is then

$$n_{CO_2} = 3.67n_C \qquad (18.2)$$

$$n_{H_2O} = 9.0n_H - 0.25n_{Cl} + \{n_{H_2O,w}\} \qquad (18.3)$$

$$n_{SO_2} = 2.00n_S \qquad (18.4)$$

$$n_{N_2} = 8.78n_C + 26.3n_H - 3.29n_O - 0.74n_{Cl} + 3.29n_S + n_{N,w} \qquad (18.5)$$

$$n_{HCl} = 1.03n_{Cl} \qquad (18.6)$$

where n_{CO_2} = mass CO_2 in the flue gas (lb CO_2/lb mixture)

n_{H_2O}, n_{SO_2}, n_{N_2}, n_{HCl} = masses of each product compound per unit mass of the fuel

$n_{N,w}$ = mass fraction of nitrogen in the fuel (lb N/lb fuel)

$n_{H_2O,w}$ = mass fraction of water in the fuel (lb H_2O/lb fuel); used only when part of the weight fraction of the fuel is expressed as water or moisture

This set of equations assumes that there is enough hydrogen available in the fuel to completely convert the chlorine to hydrogen chloride. The bracketed water term accounts for any water present in the fuel if a mass fraction of water or moisture is included in the analysis. The $m_{N,w}$ term accounts for any nitrogen that may be present in the fuel. If the ultimate analysis is on a *mole* basis, a balanced stoichiometric equation of the following form may be used:

$$C_zH_yO_xCl_wS_vN_u + [z + \phi + v - \tfrac{1}{2}x]O_2 + \tfrac{79}{21}(z + \phi + v - \tfrac{1}{2}x)N_2 \longrightarrow zCO_2$$
$$+ 2\phi H_2O + wHCl + vSO_2 + \left[\tfrac{1}{2}u + \tfrac{79}{21}(z + \phi + v - \tfrac{1}{2}x)\right]N_2 \qquad (18.7)$$

where z, y, x, w, v, u = number of moles (or mole fractions) of C, H, O, Cl, S, N present in the fuel, respectively, and

$$\phi = \tfrac{1}{4}(y - w) \quad \text{when } y > w$$
$$= 0 \quad \text{when } y \leq w \qquad (18.8)$$

Note that the parameter ϕ is included if its value is positive and ignored if it is zero or negative. For this approach, the mole fraction for carbon in the fuel (x_C) is given by

$$x_C = \frac{z}{z + y + x + w + v + u} \qquad (18.9)$$

with similar equations applicable to hydrogen, oxygen, etc. This assumes that the fuel consists only of C, H, O, Cl, S, and N. Finally, the above equations are valid provided:

1 Cl_2, SO_3, and NO_x formation is neglected.

2 Oxygen in the fuel is available for combustion and is therefore treated as a *credit* in calculating stoichiometric air requirements.

3 The air essentially consists of 79% nitrogen and 21% oxygen (by mole or volume).

4 The air essentially consists of 76.8% nitrogen and 23.3% oxygen (by weight or mass).

5 Combustion is complete.

Whether the calculations are performed on a mass or a mole basis, the *excess* air usage requirements for a particular application may easily be included in this analysis (see Chapters 5 and 10 for more details). For example, for 100% excess air, twice the stoichiometric air requirement would be employed; half of this air would be used in stoichiometrically combusting the fuel and the remaining half would be carried through the process in the flue gas as the *extra* or *excess* air.

The preferred choice of the fuel characterization method in performing these calculations is often dictated by the contents of the fuels. If the fuel contains an agglomeration of orgainics, ultimate analysis would make the calculations easier; if the fuel consists of one hydrocarbon (e.g., benzene) in water, the componential analysis is preferred. The choice as to whether the analysis is on a mass or mole basis may also be dictated by the calculational procedure employed.

REFERENCES

1. L. THEODORE: personal notes, Manhattan College, 2006.
2. L. THEODORE: personal notes, Manhattan College, 2007.
3. L. THEODORE, "*Air Pollution Control Equipment*," John Wiley & Sons, Hoboken, NJ, 2009.

NOTE: Additional problems for each chapter are available for all readers at www. These problems may be used for additional review or homework purposes.

Chapter **19**

Exergy: The Concept of "Quality Energy"

Henry Wadsworth Longfellow [1807–1882]
Great is the art of beginning, but greater the art is of ending.

—Elegiac Verse

"While conceptualizing the rather complex notion of exergy, I have been constantly reminded of a rather useful analogy. That of a chemical engineering student after one of Dr. Theodore's exams: he/she may still contain some energy, but certainly none which will be converted into useful work."

—Frank Ricci, 2008

INTRODUCTION

The *First Law of Thermodynamics* is often referred to as the "law of conservation of energy." It states that energy may neither be created nor destroyed, but may be converted from one form into another. It is important to note that, while the first law allows one to construct an energy balance around a given process, it does not provide any useful information with regards to how energy is transferred (e.g., energy always naturally flows from a hot reservoir to a cold reservoir). Abiding by the first law alone, a process may be mathematically described that violates the laws of nature. The *Second Law of Thermodynamics* accounts for directionality, and explains the irreversibility of any naturally occurring process via the aforementioned thermodynamic property of entropy introduced in Chapter 6. Hence, the second law is often referred to as the "limiting law."

THE QUALITY OF HEAT VS WORK

It can be stated that, while both heat and work are dimensionally consistent in terms of units, work can more readily be put to use in order to perform a task. This usage may

Thermodynamics for the Practicing Engineer. By L. Theodore, F. Ricci, and T. Van Vliet
Copyright © 2009 John Wiley & Sons, Inc.

be in lifting a weight, turning a generator, moving a piston, etc., resulting in a decrease in the mechanical, electrical, or internal energy of the "working" system. By contrast, whenever energy in the form of heat is transferred to a system, only a portion of the energy may be converted into useful work.[1]

An example of the limitation of heat to perform work is that of the steam in a steam engine. The hot vapor in a steam engine's cylinder contains a significant amount of thermal energy. The random thermal motion of the water molecules contribute to the unified motion of the rigid metallic piston, thereby transforming some of the steam's internal energy into useful pressure-volume work. However, the Kelvin–Planck statement, mentioned loosely in Chapter 6, decrees that a process may never be created such that all of the steam's energy content can be transformed completely into useful work. The reader is referred to Chapter 6 for a more detailed discussion of heat vs work.

The steam in the above example contained a finite amount of energy which could be converted into useful work. While the exhausted steam exiting the engine still contains internal energy, its *useful* energy has been reduced, and it may no longer contain enough useful energy for the process to continue. It should also be noted that no energy was either created or destroyed during the process. However, the original energy content of the steam has been *dispersed* during the process, and therefore lost much of its *quality energy*, such that it can no longer be used to perform the same work in the engine. This concept of energy degradation is essential to exergy analysis.

ILLUSTRATIVE EXAMPLE 19.1

Explain in theoretical terms why the steam engine mentioned in the preceding paragraphs could never convert all of the heat it receives into work?

SOLUTION: A Carnot engine operates in a completely reversible manner, and represents the maximum amount of work which can be produced for a given heat engine. The Carnot efficiency is defined (fractional basis) as,

$$E_{\text{Carnot}} \equiv 1 - \frac{T_C}{T_H} \tag{19.1}$$

If the steam engine converted all of its heat into work, the efficiency of the engine would be unity. This may only occur either when $T_C = 0K$ or $T_H \to \infty$. Neither of these options is physically possible on Earth. Moreover, actual heat engines are all irreversible, and therefore could never reach the theoretical Carnot efficiency. Thus, the steam engine may never have an efficiency equal to unity.   ∎

EXERGY

Exergy is defined as the maximum amount of work that may be produced by a system as it comes into equilibrium with a reference environment.[2] As such, exergy may be viewed as the equivalent of a systems "quality energy." Exergy is not subject to conservation laws, and is destroyed during any real process as a result of the second law. Another statement regarding exergy is that in all processes which convert energy to useful work, some energy is converted from a higher to lower quality form (e.g., exergy is consumed).[2] It should be noted that the terms *exergy*, *quality energy*, *availability*, and *availability energy* are all synonymous, with exergy being the most systemic.

Exergy destruction is directly proportional to entropy generation. Since all real processes are irreversible, all real processes must realize a net increase in entropy, thereby consuming exergy. As a corollary, it may also be said that exergy degradation accounts for the irreversibility of any real process.

By definition, a reversible process is in equilibrium with its surroundings at every infinitesimal stage of the process path. Since a system's exergy is null when it is at equilibrium with a chosen reference environment, exergy consumption is zero for a reversible process. In the aforementioned example regarding steam, the system's (engine's) surroundings are simply chosen as the reference environment, as is commonly practiced. However, a more thorough discussion of the selection of a reference environment is presented below.

A concise definition of the reference environment is of utmost importance in exergy analysis, particularly with respect to the environment's temperature, pressure, and chemical composition. The reference environment is generally the local surroundings and is chosen by the exergy "analyst" based on convenience. The main criteria of the reference environment are that it must be:

1 Assumed to be in equilibrium.

2 Large enough that it is virtually unaffected by the system of interest as the system moves towards equilibrium with the environment.

Since the great majority of man-made processes take place on the surface of the Earth, the local natural atmosphere is usually taken as the reference environment in most exergy analysis applications, particularly in environmental impact analysis. However, the aforementioned criteria regarding reference selection are not completely satisfied with respect to the Earth's atmosphere. For example, the natural environment is not in a stable equilibrium, as certain properties such as temperature and barometric pressure can exhibit extreme variance over time and location. Therefore, in the strictest sense, the exergy of the Earth's atmosphere is not zero. In order to account for the deviations from theoretical reference environment behavior, several reference environment models have been developed, each with specific protocols, assumptions, and applications. The most important of which have been compiled by Cengel et al.[1]

QUANTITATIVE EXERGY ANALYSIS

Exergy serves to measure the potential of a system to produce work because it is not in equilibrium with its reference environment. The exergy of a system is not a thermodynamic property such as enthalpy or entropy; rather, it is a co-property of both the system and selected reference environment.[1] Thermodynamicists have quantitatively defined exergy for a closed system as

$$X = U + P_O V - T_O S - \sum_i \mu_{i,O} n_i \qquad (19.2)$$

where X = exergy of the system
$\quad\quad\ U$ = internal energy of system
$\quad\quad P_O$ = absolute pressure of reference environment
$\quad\quad\ V$ = volume of system
$\quad\quad T_O$ = absolute temperature of reference environment
$\quad\quad\ S$ = entropy of the system
$\quad\quad \mu_{i,O}$ = chemical potential of the ith component at reference conditions
$\quad\quad\ n_i$ = moles of the ith component in the system

Chemical potential, which may be defined as *partial molar free energy*, is not thoroughly treated in this text. For a more complete development, the reader is referred to Smith et al.[3]

For an open system, where there is flow across the boundary, the relation $H = U + PV$ is substituted into Equation (19.2)

$$X = H - T_O S - \sum_i \mu_{i,O} n_i \qquad (19.3)$$

where H = enthalpy of the system

Equations (19.2) and (19.3) illustrate the co-dependence of exergy in that exergy depends on both intensive properties of the environment (P_O, T_O, $\mu_{i,O}$), as well as extensive properties of the system (U, H, V, S, n_i).[4] These equations may also be modified to include the effects of kinetic and potential energy, electricity, magnetism, gravity, and radiation.[5]

When the differential forms of the preceding equations are integrated between the appropriate limits, they take a more applicable form as shown below

Closed System:

$$X = (U - U_{eq}) + P_O(V - V_{eq}) - T_O(S - S_{eq}) - \sum_i \mu_{i,O}(n_i - n_{i,eq}) \qquad (19.4)$$

Open System:

$$X = (H - H_{eq}) - T_O(S - S_{eq}) - \sum_i \mu_{i,O}(n_i - n_{i,eq}) \qquad (19.5)$$

Note that exergy is not represented as a difference term when integrated from equilibrium conditions to process conditions because, by definition, $X_{eq} = 0$. In these working forms, the exergy of a given system may be calculated at a particular state, which is chosen based on the nature of the process. Internal energy or enthalpy, volume, entropy, and moles of each species must be calculated for the system at process conditions. The subscript "eq" denotes "in thermodynamic equilibrium with the reference environment."[4] Therefore, these thermodynamic properties must be calculated as if the system was in equilibrium with the reference environment (e.g., calculated at the reference temperature, pressure, and as if any chemical reactions and/or changes have reached equilibrium).

Equations (19.4) and (19.5) illustrate how exergy may simply be thought of as the measure of a system's displacement from equilibrium with its surroundings. The equations are also particularly useful in that they mathematically represent exergy as the sum of several thermodynamic quantities. It may be shown through dimensional analysis that each term in Equation (19.4) has units of energy, while the terms in Equation (19.5) are expressed in energy per time. However, in many applications, it is convenient to calculate the molar exergy (mole basis) or specific exergy (mass basis) of a system by substituting either molar or specific thermodynamic quantities, respectively.

ILLUSTRATIVE EXAMPLE 19.2

One lbmol of inert nitrogen gas contained in a "weightless piston" and cylinder assembly is being utilized in a cyclic process. The gas conditions at its current state are 400 °F, and 3 atm (absolute). **(a)** Draw a schematic of the assembly, including all important data. **(b)** If ambient conditions are at 70 °F and 1 atm, calculate the exergy of the system in Btu at its current state. (Assume ideal gas, between the current state and equilibrium state: $\overline{C_P} = 6.986\,\text{Btu/lbmol} \cdot {}^\circ\text{R}$ and $\overline{C_V} = 5.000\,\text{Btu/lbmol} \cdot {}^\circ\text{R}$.) **(c)** If the piston has a cross sectional area of 0.5 ft^2 and is under a constant external load of 100 lb$_f$, calculate V_{eq}.

SOLUTION

(a) The system is shown below:

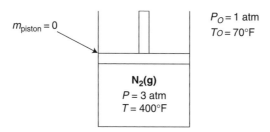

Figure 19.1 Piston-cylinder schematic.

(b) This is a closed system, therefore Equation (19.4) is employed. Since no chemical reactions occur while the closed system is brought to equilibrium with the reference environment, $n = n_{eq}$, reducing the equation to:

$$X = (U - U_{eq}) + P_O(V - V_{eq}) - T_O(S - S_{eq})$$

In lieu of tabulated thermodynamic property data, the values of $(V - V_{eq})$, $(U - U_{eq})$, and $(S - S_{eq})$ may be calculated directly by using ideal gas approximations:

$$V = \frac{nRT}{P} = \frac{(1 \text{ lbmol})\left(0.7302 \dfrac{\text{ft}^3 \cdot \text{atm}}{\text{lbmol} \cdot {}^\circ\text{R}}\right)(860^\circ\text{R})}{3 \text{ atm}} = 209 \text{ ft}^3$$

$$V_{eq} = \frac{n_{eq}RT_{eq}}{P_{eq}} = \frac{(1 \text{ lbmol})\left(0.7302 \dfrac{\text{ft}^3 \cdot \text{atm}}{\text{lbmol} \cdot {}^\circ\text{R}}\right)(530^\circ\text{R})}{1 \text{ atm}} = 387 \text{ ft}^3$$

$$V - V_{eq} = 207 \text{ ft}^3 - 387 \text{ ft}^3 = -180 \text{ ft}^3$$

$$U - U_{eq} = n \int_{T_{eq}}^{T} C_V \, dT \approx n \overline{C_V} (T - T_{eq})$$

$$U - U_{eq} = (1 \text{ lbmol})\left(5.00 \frac{\text{Btu}}{\text{lbmol} \cdot {}^\circ\text{R}}\right)(860^\circ\text{R} - 530^\circ\text{R}) = 1650 \text{ Btu}$$

$$S - S_{eq} = n \int_{T_{eq}}^{T} C_P \frac{dT}{T} + nR \ln\left(\frac{P_{eq}}{P}\right) \approx n \overline{C_P} \ln\left(\frac{T}{T_{eq}}\right) + nR \ln\left(\frac{P_{eq}}{P}\right)$$

$$S - S_{eq} = (1 \text{ lbmol})\left(6.986 \frac{\text{Btu}}{\text{lbmol} \cdot {}^\circ\text{R}}\right)\ln\left(\frac{860}{530}\right)$$
$$+ (1 \text{ lbmol})\left(1.986 \frac{\text{Btu}}{\text{lbmol} \cdot {}^\circ\text{R}}\right)\ln\left(\frac{1}{3}\right) = 1.20 \frac{\text{Btu}}{{}^\circ\text{R}}$$

Substituting the calculated values into the exergy equation,

$$X = 530 \text{ Btu}$$

This value of exergy represents the maximum amount of energy available within the steam to perform useful work at the given state *and* reference conditions.

(c) The equilibrium pressure of the system is now corrected for the piston load

$$P_{eq} = P_O + P_{load} = 14.7 \text{ psia} + \left(\frac{100 \text{ lb}_f}{0.5 \text{ ft}^2}\right)\left[\frac{\text{ft}^2}{144 \text{ in}^2}\right] = 16.1 \text{ psia}$$

$$V_{eq} = \frac{n_{eq}RT_{eq}}{P_{eq}} = \frac{(1 \text{ lbmol})\left(10.73 \dfrac{\text{ft}^3 \cdot \text{psia}}{\text{lbmol} \cdot {}^\circ\text{R}}\right)(530^\circ\text{R})}{16.1 \text{ psia}} = 353.2 \text{ ft}^3$$

ILLUSTRATIVE EXAMPLE 19.3

Steam enters a steam turbine irreversibly at 40,000 kg/hr under steady-state conditions. The inlet temperature and pressure are 500°C and 2500 kPa. As a result of poor insulation, the turbine loses heat to the surroundings at a rate of 0.25 MW. The outlet temperature and pressure of the steam are 175°C and 250 kPa, respectively. **(a)** Draw a schematic of the steam-turbine system, including all material and energy flows. **(b)** Determine the shaftwork produced by the turbine in kJ/(kg inlet steam). **(c)** If the reference environment is taken to be the surroundings at 25°C and 101.325 kPa, calculate the exergy of the inlet steam flow in kJ/kg. **(d)** Comment on the results of (b) and (c).

SOLUTION

(a)

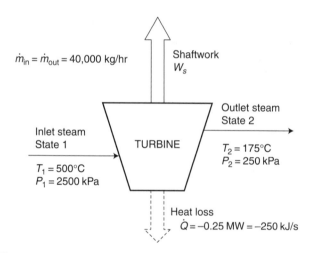

Figure 19.2   Steam turbine diagram.

(b) The general energy balance is:

$$\Delta E = \Delta KE + \Delta PE + \Delta H = Q + W_S$$

As is customary, potential and kinetic energy effects are ignored, and the resulting energy balance is solved for W_S,

$$W_S = (H_2 - H_1) - Q$$

Converting the mass flowrate to kg/s and Q to kJ/kg inlet steam,

$$\dot{m} = (40{,}000 \, \text{kg/hr}) \left[\frac{\text{hr}}{3600 \, \text{s}} \right] = 11.1 \, \text{kg/s}$$

$$Q = \frac{\dot{Q}}{\dot{m}} = \frac{-250 \text{ kJ/s}}{11.1 \text{ kg/s}} = -22.5 \text{ kJ/kg inlet steam}$$

Employing metric steam tables,

Inlet conditions (State 1: superheated vapor)	Outlet conditions (State 2: superheated vapor)
$H_1 = 3461.7 \text{ kJ/kg}$	$H_2 = 2816.7 \text{ kJ/kg}$

Solving for the shaftwork,

$$W_S = (2816.7 \text{ kJ/kg} - 3461.7 \text{ kJ/kg}) - (-22.5 \text{ kJ/kg})$$
$$W_S = -622.5 \text{ kJ/kg} = 622.5 \text{ kJ/kg}$$

This represents the actual shaftwork performed by the steam.

(c) In order to find the exergy of the steam at inlet conditions, employ (19.5), because the turbine is an open system. Since there are no chemical reactions, $n = n_{eq}$, and the equation for the exergy at State 1 may be expressed as:

$$X_1 = (H_1 - H_{eq}) - T_0(S_1 - S_{eq})$$

Utilize the metric steam tables again, noting that the steam may be approximated as saturated liquid at 25°C when in equilibrium (eq),

State 1 conditions (superheated vapor)	Equil. conditions (approx. sat. liquid)
$H_1 = 3461.7 \text{ kJ/kg}$	$H_{eq} \approx 104.8 \text{ kJ/kg}$
$S_1 = 7.324 \text{ kJ/kg} \cdot \text{K}$	$S_{eq} \approx 0.367 \text{ kJ/kg} \cdot \text{K}$

Applying these values to the exergy equation,

$$X_1 = (3461.7 \text{ kJ/kg} - 104.8 \text{ kJ/kg})$$
$$- 298\text{K}(7.324 \text{ kJ/kg} \cdot \text{K} - 0.367 \text{ kJ/kg} \cdot \text{K})$$
$$X_1 = 1283.7 \text{ kJ/kg}$$

This represents the exergy of the inlet steam, or the maximum amount of work possible for the steam to produce at the given state *and* reference conditions.

(d) When comparing the answers to parts (b) and (c), it is evident that approximately 48% of the inlet steam's maximum work potential is actually being utilized to perform shaftwork. It should be noted that the "exergy" of the inlet steam does *not* represent the reversible shaftwork for the process. This statement is made since the exergy of the inlet steam is a measure of how much work the steam could do if the process were modified to utilize *all* of the steam's quality energy (e.g., have the outlet conditions at equilibrium with the environment). In contrast, the reversible shaftwork must be determined by using the inlet and outlet conditions for the process (see Illustrative Example 19.4).   ∎

By simply changing the limits of integration in Equations (19.4) and (19.5) to that of any given process States 1 and 2, the change in exergy $(X_2 - X_1)$ of a system or process stream between conditions 1 and 2 may be calculated:

Closed System:

$$X_2 - X_1 = (U_2 - U_1) + P_O(V_2 - V_1) - T_O(S_2 - S_1)$$

$$- \sum_i \mu_{i,O}(n_{i,2} - n_{i,1}) \qquad (19.6)$$

Open System:

$$X_2 - X_1 = (H_2 - H_1) - T_O(S_2 - S_1) - \sum_i \mu_{i,O}(n_{i,2} - n_{i,1}) \qquad (19.7)$$

This difference in exergies is of great importance in determining the reversible work attainable between States 1 and 2. The total reversible work for a process is equivalent to the change in exergy between the outlet and inlet conditions as

$$W_{\text{rev}} = X_2 - X_1 \qquad (19.8)$$

Note that a negative value for reversible work according to Equation (19.8) means that the work is being performed *by* the system on the surroundings.

ILLUSTRATIVE EXAMPLE 19.4

Refer to Illustrative Example 19.3: **(a)** Determine the maximum shaftwork attainable by the steam turbine in kJ/kg. **(b)** Compare this result to that of the actual shaftwork produced, as well as the inlet steam's exergy content.

SOLUTION

(a) Equation (19.7) is employed which, when reduced to the appropriate form is,

$$X_2 - X_1 = (H_2 - H_1) - T_O(S_2 - S_1)$$

At the inlet and outlet conditions of the steam turbine,

Inlet conditions State 1 (superheated vapor)	Outlet conditions State 2 (superheated vapor)
$H_1 = 3461.7 \text{ kJ/kg}$	$H_2 = 2816.7 \text{ kJ/kg}$
$S_1 = 7.324 \text{ kJ/kg} \cdot \text{K}$	$S_2 = 7.289 \text{ kJ/kg} \cdot \text{K}$

Utilizing the relationship in Equation (19.8), it is noted that reversible work is the *maximum work* attainable for a *work producing system*, such as a turbine. The only work this system is capable of producing is shaftwork.

$$W_{S,rev} = \Delta X$$
$$= (2816.7 \, kJ/kg - 3461.7 \, kJ/kg)$$
$$- (298K)(7.289 \, kJ/kg \cdot K - 7.324 \, kJ/kg \cdot K)$$
$$W_{S,rev} = -634.6 \, kJ/kg = 634.6 \, kJ/kg$$

This represents the shaftwork that would be obtained for a reversible (ideal) process.

(b) Upon inspection, the relationship between actual work, reversible (ideal) work, and inlet exergy content becomes readily apparent. The actual work is determined from an energy balance, and is always smaller than the reversible work for a real process. In this case, the actual work is only slightly less than the reversible work, which illustrates a highly efficient process. The ratio $W_S/W_{S,rev} = 98\%$ shows that only 2% of the steam's potential to perform work at the given process conditions were lost due to irreversibilities. However, the inlet exergy value represents the maximum useful work contained in the steam. As previously mentioned, the exergy content could be utilized in its entirety if the process stream is brought into equilibrium with the reference environment, thus creating a perfect hypothetical process. ∎

ENVIRONMENTAL IMPACT

Exergy analysis is widely considered the best way to link the second law of thermodynamics with environmental impact by virtue of its definition. A major concern of engineers and environmentalists alike, is the efficiency with which an industrial scale process operates. While the engineer strives to maintain the highest efficiency in order to minimize costs, and abide by legislative regulation, those concerned with the environment realize that an inefficient process may result in ecological disturbance.

Consider an energy intensive process such as that of a nuclear power plant. Nuclear power plants are generally built near large bodies of water. This water is pumped into the plant, where it is superheated in a boiler by the nuclear reactor, and sent through a turbine which turns an electric generator. The exiting steam is normally cooled before it is re-introduced into the body of water. In the most energy efficient process, all of the excess energy would be drawn off of the exhausted water to be used elsewhere in the process. However, there are inevitable heat losses which will occur throughout the process.

If engineers choose not to cool this exiting water, large amounts of energy would be discarded into the environment. Moreover, the exiting stream would most likely be well in excess of acceptable ambient temperature conditions for the local body of water. Once this water is re-introduced into the lake or river, the water in the immediate area would undergo a sharp increase in temperature, resulting in thermal pollution. Note that it has been well documented that several forms of marine life cannot survive the sudden shock of such a temperature increase.

On the subject of exergy and environmental impact, Rosen and Dincer[6] have presented three correlations between exergy and environmental impact: order destruction, resource degradation, and waste exergy emissions. Each are briefly discussed below.

Order Destruction (Chaos Creation)

Entropy is generally understood as a measure of the disorder or "chaos" of a given system. Inversely, negentropy and exergy are measures of order. It therefore follows that as a system becomes more disordered, its entropy increases at the expense of exergy. In nature, there exist certain universal laws which dictate the tendencies of a natural system's thermodynamic behavior. Two of the main tendencies are that:

1 Nature favors lower potential energy states.

2 Nature moves towards randomness.

It is the second tendency in particular, which finds application with the second law of thermodynamics. The second law states that the entropy of the universe is constantly increasing. Lord Kelvin was among the first to loosely postulate that such movement towards randomness will one day result in what is now referred to as the "heat death" of the universe.[7] This "heat death" will result from the consumption of all conceivable exergy in the universe.

As a corollary for the planet Earth, order destruction results in environmental damage. It is known that for actual processes, the laws of nature dictate that the total entropy must increase to some extent, no matter how small. This assures that for every real process, the Earth will become somewhat more disordered, and therefore slightly lower in exergy. As civilizations continue to utilize more and more resources, many argue that it is of the utmost importance to reduce the order destruction caused by all man-made processes, so as to prolong the Earth's quality energy.

Resource Degradation

According to Kestin,[2,8] a resource is a material, found in nature or artificially produced, which is in a state of disequilibrium with the environment. This disequilibrium with the environment results in the resource's exergy. Perhaps the simplest example is that of petroleum-based fuel such as gasoline. Gasoline is considered a resource and has a certain exergy content because it is not in chemical equilibrium with the environment. When exposed to a spark, which provides the energy of activation, the gasoline spontaneously oxidizes until it achieves equilibrium. Once ignited in an engine cylinder, the system's internal energy, pressure, volume, entropy, and chemical composition change, and the resource may produce work by virtue of its chemical reactivity.

Resource degradation impacts the environment negatively, and should be reduced in a cost effective manner in order to help the conservationist effort. By increasing exergy efficiency, which essentially equates to operating the process as close to reversibly as possible, the net environmental damage would be reduced, since the same amount of work could be extracted from a smaller amount of fuel. This would result in a lower rate of combustion emission and slower resource degradation.

Waste Exergy Emissions

Most processes produce by-products, which may be released into the environment (e.g., stack gas off of an incinerator, the exhaust from a car engine, process cooling water in a power plant, etc.). Generally, these emissions are not in total equilibrium with the environment, and therefore contain exergy. The exergy, or ability to do work found in these by-products has the capacity to change the environment. In many cases, this change is damaging to the environment in some manner. For example, the oxides of sulfur and nitrogen often found in combustion stack gases rise high into the atmosphere, where they can react with oxygen and water vapor to form other acidic compounds such as sulfuric and nitric acids. These reactions reach chemical equilibrium with the environment and form acid rain. This acid affects alkaline or non-acidic substances in the environment such as limestone, soil, lakes, and streams.[9] These environmental impacts also affect life, both plant and animal, on land and in water.

The acid gases in the previous example are materials which are initially in a state of disequilibrium with the environment. Therefore, these are considered resources under the definition put forth by Kestin. However, they differ from the initial fuel material because the gases are *unconstrained* emissions of exergy.[2] The discussion of constrained vs unconstrained exergy may be explained in the following manner: many resources found in nature contain constrained exergy (e.g., coal), and generally may be put to good use, while unconstrained exergy emissions may cause environmental damage (e.g., SO_2). When unconstrained exergy is constrained, as when sulfurous compounds are removed from stack gases, two prospective advantages arise which are: (1) the pollutant has been removed and cannot damage the environment and (2) the constrained emission may become a source of exergy.

EXERGY EFFICIENCY

Engineers often make use of the term *efficiency* in order to describe how ideally a process operates. Energy based efficiencies are referred to as first law efficiencies, whereas exergy based efficiencies are second law efficiencies. From steady-state energy and exergy balances around a differential volume, fractional energy and exergy efficiencies may be defined, respectively, as

$$E_E = \frac{\text{energy out in product}}{\text{energy input}} = 1 - \frac{\text{energy loss}}{\text{energy input}} \tag{19.9}$$

$$E_X = \frac{\text{exergy out in product}}{\text{exergy input}} = 1 - \frac{\text{exergy loss} + \text{exergy destruction}}{\text{exergy input}} \tag{19.10}$$

The *exergy destruction* or *"quality energy destruction"* described in Equation (19.10) is due to the internal irreversibilities of the process. Exergy efficiency provides a more comprehensive analysis of performance, since it includes the "usefulness" or quality of the energy. Energy efficiencies, however, assign the same weight to energy whether it becomes shaft work or leaves in a stream of low temperature fluid.[1]

ILLUSTRATIVE EXAMPLE 19.5

Explain in qualitative terms the differences between using energy efficiencies alone, as well as why exergy efficiencies help to mediate these problems.

SOLUTION: Of course, the definition of efficiency may be modified from process to process. However, energy efficiency only illustrates quantities of energy, with no implication as to its quality. As a general rule: while energy efficiency implies quantity, exergy efficiency implies quality. For example, shaft work is high in quality energy whereas the aforementioned stream of low or near-ambient temperature liquid contains little quality energy. ■

During the second half of the 20th Century, the subjects of exergy and exergy efficiency analysis have taken a large role in the field of process efficiency evaluation. The applications of such analyses are not limited to the study of fundamental thermodynamic systems, or devices. Rather, exergy concepts may be applied to the analysis of a great many industrial processes, ranging from large-scale power plants to pulp and paper mills.

In summary, when properly applied, exergy analysis provides useful insights to how efficient a process may potentially become under ideal conditions. Calculations may be performed in order to observe how a process's energy utilization may be improved through exergy efficiency.[10]

REFERENCES

1. Y. A. Cengel and I. Dincer, "Energy, Entropy and Exergy Concepts and Their Roles in Thermal Engineering," *Entropy*, 116–149, 2001. <http://www.mdpi.org/entropy/papers/e3030116.pdf>
2. I. Dincer, "Thermodynamics, Exergy, and Environmental Impact," *Energy Sources*, 22, 723–732, 2000.
3. J. Smith, H. Van Ness, and M. Abbott, *Introduction to Chemical Engineering Thermodynamics*, 7th edition, McGraw-Hill, New York, 2005.
4. <http://www.holon.se/folke/kurs/Distans/Ekofys/fysbas/exergy/exergybasics.shtml>
5. G. Wall, "Exergy Flows in Industrial Processes," *Energy*, 13, 197–208, 1988. http://www.exergy.se/ftp/paper3.pdf
6. M. A. Rosen and I. Dincer, "On Exergy and Environmental Impact," *Int. J. Energy Res.*, 21, 643–654, 1997.
7. W. Thompson, "On the Dynamical Theory of Heat, with numerical results deduced from Mr. Joule's equivalent of a Thermal Unit & M. Regnault's Observations on Steam." *Transactions of the Royal Society of Edinburgh, March, 1851*; and *Philosophical Magazine IV*, 1852 [from *Mathematical and Physical Papers*, vol. i, art. XLVIII, pp. 174].

8. J. KESTIN, "Availability: The Concept and Associated Terminology." *Energy*, 5, 679–692, 1980.

9. <http://www.epa.gov/acidrain/education/site_students/whatcauses.html>

10. F. RICCI: personal notes, 2008.

NOTE: Additional problems for each chapter are available for all readers at www. These problems may be used for additional review or homework purposes.

Appendix

My job has never been to solve thermo problems for students. My job has always been to teach students how to solve thermo problems.

—L. Theodore, 2008

I. STEAM TABLES

The following excerpts from the literature[1–3] are divided into three parts: Steam Tables IA, IB, and IC. Table IA provides properties for saturated steam/water (32°F–700°F); Table IB contains properties for superheated steam (1.0 psia–600 psia). Table IC gives properties for saturated steam–ice [32°F–(−40°F)]. As indicated in Chapter 2, lowercase notation is employed for specific volume (v), enthalpy (h), and entropy (s) since the values are listed on a mass basis. English units are employed throughout the three tables. Those interested in using these values with other units should refer to Chapter 1 and Table III for the appropriate conversion constant(s). Linear interpolation should be employed where necessary. Additional data is available in the three cited references.

Thermodynamics for the Practicing Engineer. By L. Theodore, F. Ricci, and T. Van Vliet
Copyright © 2009 John Wiley & Sons, Inc.

Table IA Saturated Steam*

Temperature, °F, T	Absolute pressure, lb_f/in^2, P	Specific volume, ft³/lb			Enthalpy, Btu/lb			Entropy, Btu/lb·°R		
		Saturated liquid, v_l	Evaporation difference	Saturated vapor, v_g	Saturated liquid, h_l	Evaporation difference	Saturated vapor, h_g	Saturated liquid, s_l	Evaporation difference	Saturated vapor, s_g
32	0.08854	0.01602	3306	3306	0.00	1075.8	1075.8	0.0000	2.1877	2.1877
35	0.09995	0.01602	2947	2947	3.02	1074.1	1077.1	0.0061	2.1709	2.1770
40	0.12170	0.01602	2444	2444	8.05	1071.3	1079.3	0.0162	2.1435	2.1597
45	0.14752	0.01602	2036.4	2036.4	13.06	1068.4	1081.5	0.0262	2.1167	2.1429
50	0.17811	0.01603	1703.2	1703.2	18.07	1065.6	1083.7	0.0361	2.0903	2.1264
60	0.2563	0.01604	1206.6	1206.7	28.06	1059.9	1088.0	0.0555	2.0393	2.0948
70	0.3631	0.01606	867.8	867.9	38.04	1054.3	1092.3	0.0745	1.9902	2.0647
80	0.5069	0.01608	633.1	633.1	48.02	1048.6	1096.6	0.0932	1.9428	2.0360
90	0.6982	0.01610	468.0	468.0	57.99	1042.9	1100.9	0.1115	1.8972	2.0087
100	0.9492	0.01613	350.3	350.4	67.97	1037.2	1105.2	0.1295	1.8531	1.9826
110	1.2748	0.01617	265.3	265.4	77.94	1031.6	1109.5	0.1471	1.8106	1.9577
120	1.6924	0.01620	203.25	203.27	87.92	1025.8	1113.7	0.1645	1.7694	1.9339
130	2.2225	0.01625	157.32	157.34	97.90	1020.0	1117.9	0.1816	1.7296	1.9112
140	2.8886	0.01629	122.99	123.01	107.89	1014.1	1122.0	0.1984	1.6910	1.8894
150	3.718	0.01634	97.06	97.07	117.89	1008.2	1126.1	0.2149	1.6537	1.8685
160	4.741	0.01639	77.27	77.29	127.89	1002.3	1130.2	0.2311	1.6174	1.8485
170	5.992	0.01645	62.04	62.06	137.90	996.3	1134.2	0.2472	1.5822	1.8293
180	7.510	0.01651	50.21	50.23	147.92	990.2	1138.1	0.2630	1.5480	1.8109
190	9.339	0.01657	40.94	40.96	157.95	984.1	1142.0	0.2785	1.5147	1.7932
200	11.526	0.01663	33.62	33.64	167.99	977.9	1145.9	0.2938	1.4824	1.7762

(*Continued*)

Table IA *Continued*

Temperature, °F, T	Absolute pressure, lb_f/in^2, P	Specific volume, ft³/lb			Enthalpy, Btu/lb			Entropy, Btu/lb·°R		
		Saturated liquid, v_l	Evaporation difference	Saturated vapor, v_g	Saturated liquid, h_l	Evaporation difference	Saturated vapor, h_g	Saturated liquid, s_l	Evaporation difference	Saturated vapor, s_g
210	14.123	0.01670	27.80	27.82	178.05	971.6	1149.7	0.3090	1.4508	1.7598
212	14.696	0.01672	26.78	26.80	180.07	970.3	1150.4	0.3120	1.4446	1.7566
220	17.186	0.01677	23.13	23.15	188.13	965.2	1153.4	0.3239	1.4201	1.7440
230	20.780	0.01684	19.365	19.382	198.23	958.8	1157.0	0.3387	1.3901	1.7288
240	24.969	0.01692	16.306	16.323	208.34	952.2	1160.5	0.3531	1.3609	1.7140
250	29.825	0.01700	13.804	13.821	218.48	945.5	1164.0	0.3675	1.3323	1.6998
260	35.429	0.01709	11.746	11.763	228.64	938.7	1167.3	0.3817	1.3043	1.6860
270	41.858	0.01717	10.044	10.061	238.84	931.8	1170.6	0.3958	1.2769	1.6727
280	49.203	0.01726	8.628	8.645	249.06	924.7	1173.8	0.4096	1.2501	1.6597
290	57.556	0.01735	7.444	7.461	259.31	917.5	1176.8	0.4234	1.2238	1.6472
300	67.013	0.01745	6.449	6.466	269.59	910.1	1179.7	0.4369	1.1980	1.6350
310	77.68	0.01755	5.609	5.626	279.92	902.6	1182.5	0.4504	1.1727	1.6231
320	89.66	0.01765	4.896	4.914	290.28	894.9	1185.2	0.4637	1.1478	1.6115
330	103.06	0.01776	4.289	4.307	300.68	887.0	1187.7	0.4769	1.1233	1.6002
340	118.01	0.01787	3.770	3.788	311.13	879.0	1190.1	0.4900	1.0992	1.5891
350	134.63	0.01799	3.324	3.342	321.63	870.7	1192.3	0.5029	1.0754	1.5783
360	153.04	0.01811	2.939	2.957	332.18	862.2	1194.4	0.5158	1.0519	1.5677
370	173.37	0.01823	2.606	2.625	342.79	853.5	1196.3	0.5286	1.0287	1.5573
380	195.77	0.01836	2.317	2.335	353.45	844.6	1198.1	0.5413	1.0059	1.5471
390	220.37	0.01850	2.0651	2.0836	364.17	835.4	1199.6	0.5539	0.9832	1.5371
400	247.31	0.01864	1.8447	1.8633	374.97	826.0	1201.0	0.5664	0.9608	1.5272

(Continued)

Table 1A *Continued*

Temperature, °F, T	Absolute pressure, lb$_f$/in^2, P	Specific volume, ft^3/lb			Enthalpy, Btu/lb			Entropy, Btu/lb · °R		
		Saturated liquid, v_l	Evaporation difference	Saturated vapor, v_g	Saturated liquid, h_l	Evaporation difference	Saturated vapor, h_g	Saturated liquid, s_l	Evaporation difference	Saturated vapor, s_g
410	276.75	0.01878	1.6512	1.6700	385.83	816.3	1202.1	0.5788	0.9386	1.5174
420	308.83	0.01894	1.4811	1.5000	396.77	806.3	1203.1	0.5912	0.9166	1.5078
430	343.72	0.01910	1.3308	1.3499	407.79	796.0	1203.8	0.6035	0.8947	1.4982
440	381.59	0.01926	1.1979	1.2171	418.90	785.4	1204.3	0.6158	0.8730	1.4887
450	422.6	0.0194	1.0799	1.0993	430.1	774.5	1204.6	0.6280	0.8513	1.4793
460	466.9	0.0196	0.9748	0.9944	441.4	763.2	1204.6	0.6402	0.8298	1.4700
470	514.7	0.0198	0.8811	0.9009	452.8	751.5	1204.3	0.6523	0.8083	1.4606
480	566.1	0.0200	0.7972	0.8172	464.4	739.4	1203.7	0.6645	0.7868	1.4513
490	621.4	0.0202	0.7221	0.7423	476.0	726.8	1202.8	0.6766	0.7653	1.4419
500	680.8	0.0204	0.6545	0.6749	487.8	713.9	1201.7	0.6887	0.7438	1.4325
520	812.4	0.0209	0.5385	0.5594	511.9	686.4	1198.2	0.7130	0.7006	1.4136
540	962.5	0.0215	0.4434	0.4649	536.6	656.6	1193.2	0.7374	0.6568	1.3942
560	1133.1	0.0221	0.3647	0.3868	562.2	624.2	1186.4	0.7621	0.6121	1.3742
580	1325.8	0.0228	0.2989	0.3217	588.9	588.4	1177.3	0.7872	0.5659	1.3532
600	1542.9	0.0236	0.2432	0.2668	617.0	548.5	1165.5	0.8131	0.5176	1.3307
620	1786.6	0.0247	0.1955	0.2201	646.7	503.6	1150.3	0.8398	0.4664	1.3062
640	2059.7	0.0260	0.1538	0.1798	678.6	452.0	1130.5	0.8679	0.4110	1.2789
660	2365.4	0.0278	0.1165	0.1442	714.2	390.2	1104.4	0.8987	0.3485	1.2472
680	2708.1	0.0305	0.0810	0.1115	757.3	309.9	1067.2	0.9351	0.2719	1.2071
700	3093.7	0.0369	0.0392	0.0761	823.3	172.1	995.4	0.9905	0.1484	1.1389

*Abridged from *Thermodynamic Properties of Steam*, by Joseph H. Keenan and Frederick G. Keyes. Copyright 1936, by Joseph H. Keenan and Frederick G. Keyes. Published by John Wiley & Sons, Inc., New York.

Table IB Superheated Steam

Absolute pressure, lb_f/in² (Saturated temperature)		Temperature, °F												
		200	220	300	350	400	450	500	550	600	700	800	900	1000
1 (101.74)	v	392.6	404.5	452.3	482.2	512.0	541.8	571.6	601.4	631.2	690.8	750.4	809.9	869.5
	h	1150.4	1159.5	1195.8	1218.7	1241.7	1264.9	1288.3	1312.0	1335.7	1383.8	1432.8	1482.7	1533.5
	s	2.0512	2.0647	2.1153	2.1444	2.1720	2.1983	2.2233	2.2468	2.2702	2.3137	2.3542	2.3923	2.4283
5 (162.24)	v	78.16	80.59	90.25	96.26	102.26	108.24	114.22	120.19	126.16	138.10	150.03	161.95	173.87
	h	1148.8	1158.1	1195.0	1218.1	1241.2	1264.5	1288.0	1311.7	1335.4	1383.6	1432.7	1482.6	1533.4
	s	1.8718	1.8857	1.9370	1.9664	1.9942	2.0205	2.0456	2.0692	2.0927	2.1361	2.1767	2.2148	2.2509
10 (193.21)	v	38.85	40.09	45.00	48.03	51.04	54.05	57.05	60.04	63.03	69.01	74.98	80.95	86.92
	h	1146.6	1156.2	1193.9	1217.2	1240.6	1264.0	1287.5	1311.3	1335.1	1383.4	1432.5	1482.4	1533.2
	s	1.7927	1.8071	1.8595	1.8892	1.9172	1.9436	1.9689	1.9924	2.0160	2.0596	2.1002	2.1383	2.1744
14.696 (212.00)	v		27.15	30.53	32.62	34.68	36.73	38.78	40.82	42.86	46.94	51.00	55.07	59.13
	h		1154.4	1192.8	1216.4	1239.9	1263.5	1287.1	1310.9	1334.8	1383.2	1432.3	1482.3	1533.1
	s		1.7624	1.8160	1.8460	1.8743	1.9008	1.9261	1.9498	1.9734	2.0170	2.0576	2.0958	2.1319
20 (227.96)	v			22.36	23.91	25.43	26.95	28.46	29.97	31.47	34.47	37.46	40.45	43.44
	h			1191.6	1215.6	1239.2	1262.9	1286.6	1310.5	1334.4	1382.9	1432.1	1482.1	1533.0
	s			1.7808	1.8112	1.8396	1.8664	1.8918	1.9160	1.9392	1.9829	2.0235	2.0618	2.0978
40 (267.25)	v			11.040	11.843	12.628	13.401	14.168	14.93	15.688	17.198	18.702	20.20	21.70
	h			1186.8	1211.9	1236.5	1260.7	1284.8	1308.9	1333.1	1381.9	1431.3	1481.4	1532.4
	s			1.6994	1.7314	1.7608	1.7881	1.8140	1.8384	1.8619	1.9058	1.9467	1.9850	2.0214
	v			7.259	7.818	8.357	8.884	9.403	9.916	10.427	11.441	12.449	13.452	14.454

(Continued)

Table IB *Continued*

Absolute pressure, lbf/in² (Saturated temperature)		200	220	300	350	400	450	500	550	600	700	800	900	1000
								Temperature, °F						
60 (292.71)	h			1181.6	1208.2	1233.6	1258.5	1283.0	1307.4	1331.8	1380.9	1430.5	1480.8	1531.9
	s			1.6492	1.6830	1.7135	1.7416	1.7678	1.7926	1.8162	1.8605	1.9015	1.9400	1.9762
	v				5.803	6.220	6.624	7.020	7.410	7.797	8.562	9.322	10.077	10.830
80 (312.03)	h				1204.3	1230.7	1256.1	1281.1	1305.8	1330.5	1379.9	1429.7	1480.1	1531.3
	s				1.6475	1.6791	1.7078	1.7346	1.7598	1.7836	1.8281	1.8694	1.9079	1.9442
	v				4.592	4.937	5.268	5.589	5.905	6.218	6.835	7.446	8.052	8.656
100 (327.81)	h				1200.1	1227.6	1253.7	1279.1	1304.2	1329.1	1378.9	1428.9	1479.5	1530.8
	s				1.6188	1.6518	1.6813	1.7085	1.7339	1.7581	1.8029	1.8443	1.8829	1.9193
	v				3.783	4.081	4.363	4.636	4.902	5.165	5.683	6.195	6.702	7.207
120 (341.25)	h				1195.7	1224.4	1251.3	1277.2	1302.5	1327.7	1377.8	1428.1	1478.8	1530.2
	s				1.5944	1.6287	1.6591	1.6869	1.7127	1.7370	1.7822	1.8237	1.8625	1.8990
	v					3.468	3.715	3.954	4.186	4.413	4.861	5.301	5.738	6.172
140 (353.02)	h					1221.1	1248.7	1275.2	1300.9	1326.4	1376.8	1427.3	1478.2	1529.7
	s					1.6087	1.6399	1.6683	1.6945	1.7190	1.7645	1.8063	1.8451	1.8817
	v					3.008	3.230	3.443	3.648	3.849	4.244	4.631	5.015	5.396
160 (363.53)	h					1217.6	1246.1	1273.1	1299.3	1325.0	1375.7	1426.4	1477.5	1529.1
	s					1.5908	1.6230	1.6519	1.6785	1.7033	1.7491	1.7911	1.8301	1.8667
	v					2.649	2.852	3.044	3.229	3.411	3.764	4.110	4.452	4.792
180 (373.06)	h					1214.0	1243.5	1271.0	1297.6	1323.5	1374.7	1425.6	1476.8	1528.6
	s					1.5745	1.6077	1.6373	1.6642	1.6894	1.7355	1.7776	1.8167	1.8534
	v					2.361	2.549	2.726	2.895	3.060	3.380	3.693	4.002	4.309

(Continued)

Table IB *Continued*

Absolute pressure, lb_f/in² (Saturated temperature)		Temperature, °F												
		200	220	300	350	400	450	500	550	600	700	800	900	1000
200	h					1210.3	1240.7	1268.9	1295.8	1322.1	1373.6	1424.8	1476.2	1528.0
(381.79)	s					1.5594	1.5937	1.6240	1.6513	1.6767	1.7232	1.7655	1.8048	1.8415
	v					2.125	2.301	2.465	2.621	2.772	3.066	3.352	3.634	3.913
220	h					1206.5	1237.9	1266.7	1294.1	1320.7	1372.6	1424.0	1475.5	1527.5
(389.86)	s					1.5453	1.5808	1.6117	1.6395	1.6652	1.7120	1.7545	1.7939	1.8308
	v					1.9276	2.094	2.247	2.393	2.533	2.804	3.068	3.327	3.584
240	h					1202.5	1234.9	1264.5	1292.4	1319.2	1371.5	1423.2	1474.8	1526.9
(397.37)	s					1.5319	1.5686	1.6003	1.6286	1.6546	1.7017	1.7444	1.7839	1.8209
	v						1.9183	2.063	2.199	2.330	2.582	2.827	3.067	3.305
260	h						1232.0	1262.3	1290.5	1317.7	1370.4	1422.3	1474.2	1526.3
(404.42)	s						1.5573	1.5897	1.6184	1.6447	1.6922	1.7352	1.7748	1.8118
	v						1.7674	1.9047	2.033	2.156	2.392	2.621	2.845	3.066
280	h						1228.9	1260.0	1288.7	1316.2	1369.4	1421.5	1473.5	1525.8
(411.05)	s						1.5464	1.5796	1.6087	1.6354	1.6834	1.7265	1.7662	1.8033
	v						1.6364	1.7675	1.8891	2.005	2.227	2.442	2.652	2.859
300	h						1225.8	1257.6	1286.8	1314.7	1368.3	1420.6	1472.8	1525.2
(417.33)	s						1.5360	1.5701	1.5998	1.6268	1.6751	1.7184	1.7582	1.7954
	v						1.3734	1.4923	1.6010	1.7036	1.8980	2.084	2.266	2.445
350	h						1217.7	1251.5	1282.1	1310.9	1365.5	1418.5	1471.1	1523.8
(431.72)	s						1.5119	1.5481	1.5792	1.6070	1.6563	1.7002	1.7403	1.7777
	v						1.1744	1.2851	1.3843	1.4770	1.6508	1.8161	1.9767	2.134
400	h						1208.8	1245.1	1277.2	1306.9	1362.7	1416.4	1469.4	1522.4
(444.59)	s						1.4892	1.5281	1.5607	1.5894	1.6398	1.6842	1.7247	1.7623

Table IC Saturated Steam–Ice

Temperature, °F, T	Absolute pressure, lb$_f$/in^2, P	Specific volume, ft^3/lb		Enthalpy, Btu/lb			Entropy, Btu/lb · °R		
		Saturated ice, v_i	Saturated steam, $v_g \times 10^{-3}$	Saturated ice, h_i	Sublimation difference	Saturated steam, h_g	Saturated ice, s_i	Sublimation difference	Saturated steam, s_g
32	0.0885	0.01747	3.306	−143.35	1219.1	1075.8	−0.2916	2.4793	2.1877
30	0.0808	0.01747	3.609	−144.35	1219.3	1074.9	−0.2936	2.4897	2.1961
20	0.0505	0.01745	5.658	−149.31	1219.9	1070.6	−0.3038	2.5425	2.2387
10	0.0309	0.01744	9.05	−154.17	1220.4	1066.2	−0.3141	2.5977	2.2836
0	0.0185	0.01742	14.77	−158.93	1220.7	1061.8	−0.3241	2.6546	2.3305
−10	0.0108	0.01741	24.67	−163.59	1221.0	1057.4	−0.3346	2.7143	2.3797
−20	0.0062	0.01739	42.2	−168.16	1221.2	1053.0	−0.3448	2.7764	2.4316
−30	0.0035	0.01738	74.1	−172.63	1221.2	1048.6	−0.3551	2.8411	2.4860

II. SI UNITS

The need for a single worldwide coordinated measurement system was recognized over 300 years ago. In 1670, Gabriel Mouton, Vicar of St. Paul's church in Lyon, proposed a comprehensive decimal measurement system based on the length of one minute of arc of a great circle of the Earth. In 1671, Jean Picard, a French astronomer, proposed the length of a pendulum beating seconds as the unit of length. (Such a pendulum would have been fairly easy to reproduce, thus facilitating the wide-spread distribution of uniform standards.) Other proposals were made, but over a century elapsed before any action was taken.

In 1790, in the midst of the French Revolution, the National Assembly of France requested the French Academy of Sciences to "deduce an invariable standard for all the measures and weights." The Commission appointed by the Academy created a system that was, at once, simple and scientific. The unit of length was to be a portion of the Earth's circumference. Measures for capacity (volume) and mass (weight) were to be derived from the unit of length, thus relating the basic units of the system to each other and to nature. Furthermore, the larger and smaller versions of each unit were to be created by multiplying or dividing the basic units by 10 and its multiples. This feature provided a great convenience to users of the system, by eliminating the need for such calculating and dividing by 16 (to convert ounces to pounds) or by 12 (to convert inches to feet). Similar calculations in the metric system could be performed simply by shifting the decimal point. Thus, the metric system is a *base-10* or *decimal* system.

The Commission assigned the name *metre* (which we now spell *meter*) to the unit of length. This name was derived from the Greek word *metron* meaning "a measure." The physical standard representing the meter was to be constructed so that it would equal one ten-millionth of the distance from the north pole to the equator along the meridian of the Earth running near Dunkirk in France and Barcelona in Spain.

The metric unit of mass, called the *gram*, was defined as the mass of one cubic centimeter (a cube that is $1/100$ of a meter on each side) of water at its temperature of maximum density. The cubic decimeter (a cube $1/10$ of a meter on each side) was chosen as the unit of fluid capacity. This measure was given the name *liter*.

Although the metric system was not accepted with enthusiasm at first, adoption by other nations occurred steadily after France made its use compulsory in 1840. The standardized character and decimal features of the metric system made it well suited to scientific and engineering work. Consequently, it is not surprising that the rapid spread of the system coincided with an age of rapid technological development. In the United States, by Act of Congress in 1866, it was made "lawful throughout the United States of America to employ the weights and measures of the metric system in all contracts, dealings, or court proceedings."

By the late 1860s, even better metric standards were needed to keep pace with scientific advances. In 1875, an international treaty, the "Treaty of the Meter," set up well-defined metric standards for length and mass, and established permanent machinery to recommend and adopt further refinements in the metric system. This treaty, known as the *Metric Convention*, was signed by 17 countries, including the United States.

As a result of the treaty, metric standards were constructed and distributed to each nation that ratified the Convention. Since 1893, the internationally agreed to metric standards have served as the fundamental weights and measures standards of the United States.

By 1900, a total of 35 nations—including the major nations of continental Europe and most of South America—had officially accepted the metric system. Today, with the exception of the United States and a few small countries, the entire world is using predominantly the metric system or is committed to such use. In 1971, the Secretary of Commerce, in transmitting to Congress the results of a 3-year study authorized by the Metric Study Act of 1968, recommended that the U.S. change to predominant use of the metric system through a coordinated national program.

The International Bureau of Weights and Measures located at Sevres, France, serves as a permanent secretariat for the Metric Convention, coordinating the exchange of information about the use and refinement of the metric system. As measurement science develops more precise and easily reproducible ways of defining the measurement units, the General Conference of Weights and Measures—the diplomatic organization made up of adherents to the Convention—meets periodically to ratify improvements in the system and the standards.

The SI System

In 1960, the General Conference adopted an extensive revision and simplification of the system. The name *Le Systeme International d'Unites* (International System of Units), with the international abbreviation SI, was adopted for this modernized metric system. Further improvements in and additions to SI were made by the General Conference in 1964, 1968, and 1971.

The basic units in the SI system are the *kilogram* (mass), *meter* (length), *second* (time), *Kelvin* (temperature), *ampere* (electric current), *candela* (the unit of luminous intensity), and *radian* (angular measure). All are commonly used by the engineer. The Celsius scale of temperature ($0°C - 273.15K$) is commonly used with the absolute Kelvin scale. The important derived units are the *newton* (SI unit of force), the *joule* (SI unit of energy), the *watt* (SI unit of power), the *pascal* (SI unit of pressure), the *hertz* (unit of frequency). There are a number of electrical units: *coulomb* (charge), *farad* (capacitance), *henry* (inductance), *volt* (potential), and *weber* (magnetic flux). One of the major advantages of the metric system is that larger and smaller units are given in powers of ten. In the SI system, a further simplification is introduced by recommending only those units with multipliers of 10^3. Thus for lengths in engineering, the *micrometer* (previously *micron*), *millimeter*, and *kilometer* are recommended, and the *centimeter* is generally avoided. A further simplification is that the decimal point may be substituted by a comma (as in France, Germany, and South Africa), while the other number, before and after the comma, is separated by spaces between groups of three, that is, one million dollars will be $1 000 000,00. More details are provided below.

Seven Base Units

a Length—meter (m)

The meter (common international spelling, *metre*) is defined as 1 650 763.00 wavelengths in vacuum of the orange-red line of the spectrum of krypton-86. The SI unit of area is the *square meter* (m^2). The SI unit of volume is the *cubic meter* (m^3). The *liter* (0.001 cubic meter), although not an SI unit, is commonly used to measure fluid volume.

b Mass—kilogram (kg)

The standard for the unit of mass, the *kilogram*, is a cylinder of platinum–iridium alloy kept by the International Bureau of Weights and Measures at Paris. A duplicate in the custody of the National Bureau of Standards serves as the mass standard for the United States. This is the only base unit still defined by an artifact. The SI unit of force is the *newton* (N). One newton is the force which, when applied to a 1 kilogram mass, will give the kilogram mass an acceleration of 1 (meter per second) per second ($1\,N = 1\,kg \cdot m/s^2$). The SI unit for pressure is the *pascal* (Pa) ($1\,Pa = 1\,N/m^2$). The SI unit for work and energy of any kind is the *joule* (J) ($1\,J = 1\,N\text{-}m$). The SI unit for power of any kind is the *watt* (W) ($1\,W = 1\,J/s$).

c Time—second (s)

The *second* is defined as the duration of 9 192 632 770 cycles of the radiation associated with a specified transition of the cesium-133 atom. It is realized by tuning an oscillator to the resonance frequency of cesium-133 atoms as they pass through a system of magnets and a resonant cavity into a detector. The number of periods or cycles per second is call *frequency*. The SI unit for frequency is the *hertz* (Hz). One hertz equals one cycle per second. The SI unit for speed is the *meter per second* (m/s). The SI unit for acceleration is the *(meter per second) per second* (m/s^2).

d Electric current—ampere (A)

The *ampere* is defined as that current which, if maintained in each of two long parallel wires separated by one meter in free space, would produce a force between the two wires (due to their magnetic fields) of 2×10^{-7} newton for each meter of length. The SI unit of voltage is the *volt* (V) ($1\,V = 1\,W/A$). The SI unit of electrical resistance is the *ohm* (Ω) ($1\,\Omega = 1\,V/A$).

e Temperature—Kelvin (K)

The *Kelvin* is defined as the fraction $1/273.16$ of the thermodynamic temperature of the triple point of water. The temperature 0K is called *absolute zero*. On the commonly used Celsius temperature scale, water freezes at about 0°C and boils at about 100°C. The °C is defined as an interval of 1K, and the Celsius temperature 0°C is defined as 273.15K. 1.8 Fahrenheit scale degrees are equal to 1.0°C or 1.0K; the Fahrenheit scale uses 32°F as a temperature corresponding to 0°C.

f Amount of substance—mole (mol) (aka gram-mole)

The *mole* is the amount of substance of a system that contains as many elementary entities as there are atoms in 0.012 kilogram of carbon-12. When the mole is used, the elementary entities must be specified and may be atoms, molecules, ions, electrons, other particles, or specified groups of such particles. The SI unit of concentration (of amount of substance) is the *mole per cubic meter* (mol/m^3).

g Luminous intensity—candela (cd)

The *candela* is defined as the luminous intensity of 1/600 000 of a square meter of a blackbody at the temperature of freezing platinum (2045K). The SI unit of light flux is the *lumen* (lm). A source having an intensity of 1 candela in all directions radiates a light flux of 4π lumens.

There are also two supplementary units:

a Phase angle—radian (rad)

The *radian* is the plane angle with its vertex at the center of a circle that is subtended by an arc equal in length to the radius.

b Solid angle—steradian (sr)

The *steradian* is the solid angle with its vertex at the center of a sphere that is subtended by an area of the spherical surface equal to that of a square with sides equal in length to the radius.

III. CONVERSION CONSTANTS

To convert from	To	Multiply by
Length		
m	cm	100
m	mm	1000
m	microns (μm)	10^6
m	angstroms (Å)	10^{10}
m	in	39.37
m	ft	3.281
m	mi	6.214×10^{-4}
ft	in	12
ft	m	0.3048
ft	cm	30.48
ft	mi	1.894×10^{-4}
Mass		
kg	g	1000
kg	lb	2.205
kg	oz	35.24

kg	ton	2.268×10^{-4}
kg	grains	1.543×10^{4}
lb	oz	16
lb	ton	5×10^{-4}
lb	g	453.6
lb	kg	0.4536
lb	grains	7000

Time

s	min	0.01667
s	h	2.78×10^{-4}
s	day	1.157×10^{-7}
s	week	1.653×10^{-6}
s	yr	3.171×10^{-8}

Force

N	$kg \cdot m/s^2$	1
N	dynes	10^5
N	$g \cdot cm/s^2$	10^5
N	lb_f	0.2248
N	$lb \cdot ft/s^2$	7.233
lb_f	N	4.448
lb_f	dynes	4.448×10^5
lb_f	$g \cdot cm/s^2$	4.448×10^5
lb_f	$lb \cdot ft/s^2$	32.17

Pressure

atm	N/m^2 (Pa)	1.013×10^5
atm	kPa	101.3
atm	bars	1.013
atm	$dynes/cm^2$	1.013×10^6
atm	lb_f/in^2 (psi)	14.696
atm	mm Hg at 0°C (torr)	760
atm	in Hg at 0°C	29.92
atm	ft H_2O at 4°C	33.9
atm	in H_2O at 4°C	406.8
psi	atm	6.80×10^{-2}
psi	mm Hg at 0°C (torr)	51.71
psi	in H_2O at 4°C	27.70
in H_2O at 4°C	atm	2.458×10^{-3}
in H_2O at 4°C	psi	0.0361
in H_2O at 4°C	mm Hg at 0°C (torr)	1.868

Volume

m^3	L	1000
m^3	cm^3 (cc, mL)	10^6

m^3	ft^3	35.31
m^3	gal (U.S.)	264.2
m^3	qt	1057
ft^3	in^3	1728
ft^3	gal (U.S.)	7.48
ft^3	m^3	0.02832
ft^3	L	28.32

Energy

J	$N \cdot m$	1
J	erg	10^7
J	$dyne \cdot cm$	10^7
J	$kW \cdot h$	2.778×10^{-7}
J	cal	0.2390
J	$ft \cdot lb_f$	0.7376
J	Btu	9.486×10^{-4}
cal	J	4.186
cal	Btu	3.974×10^{-3}
cal	$ft \cdot lb_f$	3.088
Btu	$ft \cdot lb_f$	778
Btu	$hp \cdot h$	3.929×10^{-4}
Btu	cal	252
Btu	$kW \cdot h$	2.93×10^{-4}
$ft \cdot lb_f$	cal	0.3239
$ft \cdot lb_f$	J	1.356
$ft \cdot lb_f$	Btu	1.285×10^{-3}

Power

W	J/s	1
W	cal/s	0.2390
W	$ft \cdot lb/s$	0.7376
W	kW	10^{-3}
kW	Btu/s	0.949
kW	hp	1.341
hp	$ft \cdot lb/s$	550
hp	kW	0.7457
hp	cal/s	178.2
hp	Btu/s	0.707

Concentration

$\mu g/m^3$	lb/ft^3	6.243×10^{-11}
$\mu g/m^3$	lb/gal	8.346×10^{-12}
$\mu g/m^3$	gr/ft^3	4.370×10^{-7}
gr/ft^3	$\mu g/m^3$	2.288×10^6
gr/ft^3	g/m^3	2.288

lb/ft^3	μg/m^3	1.602×10^{10}
lb/ft^3	μg/L	1.602×10^{-8}
lb/ft^3	lb/gal	7.48

Viscosity

P (poise)	g/cm · s	1
P	cP (centipoise)	100
P	kg/m · h	360
P	lb/ft · s	6.72×10^{-2}
P	lb/ft · h	241.9
P	lb/m · s	5.6×10^{-3}
lb/ft · s	P	14.88
lb/ft · s	g/cm · s	14.88
lb/ft · s	kg/m · h	5.357×10^3
lb/ft · s	lb/ft · h	3600

Heat Capacity

cal/g · °C	Btu/lb · °F	1
cal/g · °C	kcal/kg · °C	1
cal/g · °C	cal/gmol · °C	Molecular weight
cal/gmol · °C	Btu/lbmol · °F	1
J/g · °C	Btu/lb · °F	0.2389
Btu/lb · °F	cal/g · °C	1
Btu/lb · °F	J/g · °C	4.186
Btu/lb · °F	Btu/lbmol · °F	Molecular weight

IV. SELECTED COMMON ABBREVIATIONS

Å, A	angstrom unit of length
abs	absolute
amb	ambient
app. MW, M	apparent molecular weight
atm	atmospheric
at. wt.	atomic weight
b.p.	boiling point
bbl	barrel
Btu	British thermal unit
cal	calorie
cg	centigram
cm	centimeter
cgs system	centimeter-gram-second system
conc	concentrated, concentration

cc, cm^3	cubic centimeter
cu ft, ft^3	cubic feet
cfh	cubic feet per hour
cfm	cubic feet per minute
cfs	cubic feet per second
m^3, M^3 (rarely)	cubic meter
°	degree
°C	degree Celsius, degree Centigrade
°F	degree Fahrenheit
°R	degree Reamur, degree Rankine
ft	foot
ft · lb	foot pound
fpm	feet per minute
fps	feet per second
fps system	foot-pound-second system
f.p.	freezing point
gr	grain
g, gm	gram
h	hour
in	inch
kcal	kilocalorie
kg	kilogram
km	kilometer
liq	liquid
L	liter
log	logarithm (common)
ln	logarithm (natural)
m.p.	melting point
m, M	meter
μm	micrometer (micron)
mks system	meter · kilogram · second system
mph	miles per hour
mg	milligram
ml	milliliter
mm	millimeter
mμ	millimicron
min	minute
mol wt, MW, M	molecular weight
oz	ounce
ppb	parts per billion
pphm	parts per hundred million

ppm	parts per million
lb	pound
psi	pounds per square inch
psia	pounds per square inch absolute
psig	pounds per square inch gage
rpm	revolutions per minute
s	second
sp gr	specific gravity
sp ht	specific heat
sp wt	specific weight
sq	square
scf	standard cubic foot
STP	standard temperature and pressure
t	time
T, temp.	temperature
wt	weight

REFERENCES

1. J. KEENAN and F. KEYES, "*Thermodynamic Properties of Steam*," John Wiley & Sons, Hoboken, NJ, 1936.
2. G. VAN WYLEN and R. SONNTAG, "*Fundamentals of Classical Thermodynamics*," John Wiley & Sons, Hoboken, NJ, 1965.
3. J. JONES and G. HAWKINS, "*Engineering Thermodynamics*," John Wiley & Sons, Hoboken, NJ, 1960.

Index

Thermodynamics for the Practicing Engineer. By L. Theodore, F. Ricci, and T. Van Vliet
Copyright © 2009 John Wiley & Sons, Inc.